Thermodynamic Equilibrium and Stability of Materials

Long-Qing Chen

Thermodynamic Equilibrium and Stability of Materials

 Springer

Long-Qing Chen
Department of Materials Science and Engineering
Department of Mathematics
Department of Engineering Science and Mechanics
Materials Research Institute
The Pennsylvania State University
University Park, PA, USA

ISBN 978-981-13-8693-0   ISBN 978-981-13-8691-6   (eBook)
https://doi.org/10.1007/978-981-13-8691-6

This Springer imprint is published by the registered company Springer Nature Singapore Pte Ltd.
The registered company address is: 152 Beach Road, #21-01/04 Gateway East, Singapore 189721, Singapore

Preface

This book on thermodynamics of materials is intended as a textbook for junior/senior undergraduate students and first-year graduate students in materials science and engineering and related fields as well as a reference for researchers who would like to refresh their understanding of basic thermodynamics and need to derive thermodynamic relations for their research. I made a conscious effort to explain the thermodynamic concepts and quantities using plain languages. The focus of this book is on the fundamentals of thermodynamics and their applications to generic thermodynamic systems and processes rather than specific materials systems. A reader after finishing the book is expected to achieve a high-level fundamental understanding of thermodynamics and acquire the analytical skills of applying thermodynamics to determining materials equilibria under different thermodynamic conditions and computing driving forces for materials processes.

One of the purposes for writing this book is to share my own understanding of thermodynamics through nearly 30 years of teaching thermodynamics of materials at the Pennsylvania State University. My understanding of thermodynamics has primarily been based on reading "Scientific Papers of Josiah Willard Gibbs" which laid the foundation of thermodynamics and illustrated the true mathematical rigor and beauty of thermodynamics.[1] I also benefited greatly reading a number of modern thermodynamics books, including *Thermodynamics and an Introduction to Thermostatistics* by Herbert B. Callen[2], *Thermodynamics and Its Applications* by Michael Modell and Robert C. Reid[3], *Modern Thermodynamics: From Heat Engines to Dissipative Structures* by Dilip Kondepudi and Ilya Prigogine[4], *The Dynamics of Heat* by Hans

---

[1] J. Willard Gibbs, "On the Equilibrium of Heterogeneous Substances," Transactions of the Connecticut Academy, III. pp. 108–248, October 1875–May 1876, and 343–524, May 1877–July 1878.

[2] Herbert B. Callen, "Thermodynamics and an Introduction to Thermostatistics," Second Edition, Wiley, 1985.

[3] Michael Modell and Robert C. Reid , "Thermodynamics and its Applications," Second Edition, Prentice Hall PTR, 1983.

[4] Dilip Kondepudi and Ilya Prigogine, "Modern Thermodynamics: From Heat Engines to Dissipative Structures," First Edition, John Wiley & Sons, 1998.

U. Fuchs[5], and *Phase Equilibria, Phase Diagrams and Phase Transitions* by Mats Hillert[6].

There are several existing excellent textbooks specifically devoted to applications of thermodynamics to materials science and engineering, e.g., *Introduction to Thermodynamics of Materials* by David R. Gaskell and David E. Laughlin[7], *Thermodynamics of Materials* by David V. Ragone[8], and *Thermodynamics in Materials Science* by Robert DeHoff[9]. The natural question would be "why another textbook on thermodynamics?" I try to answer this question by emphasizing several, which I believe, rather unique features of this book:

- Entropy S and chemical potential μ, considered by many as two of the most difficult and abstract concepts to fully comprehend in thermodynamics, are introduced very early in Chap. 1 as part of the definitions and terms, emphasizing the analog of chemical potential μ to temperature T and pressure p as a form of potential, and the analog of S to volume V and amount of chemical substance N as a type of matter. All seven basic thermodynamic variables (U, S, V, N, T, p, μ) associated with a simple thermodynamic system as defined by Gibbs are separated into three categories: energy (U), matter (S, V, N), and potential (T, p, μ), emphasizing not only the relations but also the differences among the three types of thermodynamic quantities: energy, matter, and potential.
- The discussion on the first law of thermodynamics is primarily focused on open systems rather than closed systems as it was done in essentially all modern textbooks on thermodynamics of materials. Open systems, by definition, allow exchanges of all three kinds of matter (entropy, volume, and chemical substance) and thus the exchanges of the three forms of energy (thermal, mechanical, and chemical) between a simple system and its surrounding. One of the benefits for discussing the first law of thermodynamics using open systems is the early introduction and discussion of chemical potential in terms of chemical energy or the Gibbs free energy per unit amount of matter or the derivatives of different thermodynamic potential energy functions with respect to the amount of substance under different thermodynamic conditions.
- The second law of thermodynamics, another concept often perceived to be difficult in thermodynamics, is quantified by directly connecting the amount of entropy produced to the amount of thermodynamic energy dissipated or driving force, and thus allowing one to quantitatively calculate the amount of entropy produced in any given irreversible thermodynamic process using thermodynamic quantities

---

[5] Hans U. Fuchs, "The Dynamics of Heat: A Unified Approach to Thermodynamics and Heat Transfer," Second Edition, Springer-Verlag New York, 2010.

[6] M. Hillert, "Phase Equilibria, Phase Diagrams and Phase Transformations," Second Edition, Cambridge University Press, Cambridge, 2008.

[7] David R. Gaskell and David E. Laughlin, "Introduction to the Thermodynamics of Materials," Sixth Edition, CRC Press LLC, 2018.

[8] David V. Ragone, "Thermodynamics of Materials," Volume I and II, John Wiley & Sons, Inc., 1995.

[9] Robert DeHoff, "Thermodynamics in Materials Science," CRC Press LLC, 2006.

associated with the system of interest. In the existing literature, the connection between entropy increase and energy dissipation is often discussed in irreversible thermodynamics[10].

- The book fully embraces, rather than shies away from, the mathematical beauty and rigor of Gibbs thermodynamics throughout the book by emphasizing the importance of fundamental equation of thermodynamics from which all thermodynamic properties of a material can be derived. However, a reader with basic first-year undergraduate calculus skills is expected to be able to get through the book without too much difficulty. It offers detailed descriptions of the step-by-step procedures for computing all the thermodynamic properties from the fundamental equation of thermodynamics or the construction of fundamental equations of thermodynamics in different alternative functional forms based on a set of common, experimentally measurable thermodynamic properties. The emphasis of the chapter on the brief introduction to statistical thermodynamics is also on the connection between the microscopic interactions in a system and the fundamental equation of thermodynamics associated with the system.

- By far, the most emphasized concept of this book is chemical potential and its applications to materials science and engineering. Not only the concept of chemical potential is introduced at the very beginning of the book together with temperature and pressure, but also essentially all the applications of thermodynamics are discussed based on the concept of chemical potentials, e.g., phase equilibria in single-component systems as well as binary or multicomponent solutions, chemical reaction equilibria, and lattice and electronic defects in crystals, and multiphysics problems including coupling among chemistry, mechanics, and electricity. One can even discuss chemical reaction equilibria without using the term of Gibbs free energy. The thermodynamic driving force for the reaction $2A + B = A_2B$ can be perfectly described by the difference between the chemical potential of the initial state (the reactants) with chemical potential $\mu_{A_2B}^r = 2\mu_A + \mu_B$ (in the state of mixture of A and B) and the chemical potential of the final state (the product) $\mu_{A_2B}^p = \mu_{A_2B}$ (in the form of a compound or molecule, A_2B). It is the chemical potential that determines the stability of chemical species, compounds, and phases and their tendency to chemically react to form new species, transform to new physical states, and migrate from one spatial location to another. Therefore, it is the chemical potential differences or gradients that drive essentially all materials processes of interest.

- The book avoids or at least minimizes the use of terms such as molar Gibbs free energy or partial molar Gibbs free energy because molar Gibbs free energy of a material or partial molar Gibbs free energy of a component is precisely the chemical potential of the corresponding material or the corresponding component regardless of whether the material is a single or multicomponent system. There is a common misconception in the literature that the statement, "molar Gibbs free energy of a system is equal to its chemical potential," or the equation, $G/N = \mu$,

---

[10] Sybren R. de Groot and Peter Mazur, "Non-Equilibrium Thermodynamics," Dover Publications, Inc., New. York, 1984.

where G is Gibbs free energy, N is number of moles of chemical substance, and μ is its chemical potential, is only true for single-component systems. Potentials are uniform in a homogeneous system at equilibrium, and they are also uniform within a heterogeneous system if the corresponding types of matter are allowed to freely flow or transport across the interfaces among the different homogeneous regions regardless of the external thermodynamic conditions. Molar Gibbs free energy of a system at a given overall composition is uniform in both homogeneous and heterogeneous systems in which all the chemical species are allowed to freely redistribute across the interfaces between the different homogeneous regions or phases and the whole system is in thermodynamic equilibrium. Therefore, the molar Gibbs free energy of a multicomponent system is a potential, the chemical potential. This point was also pointed out in the original paper by Gibbs as well as in the book by Hillert. According to Gibbs, for the purpose of defining chemical potential, "*any chemical element or combination of elements in given proportions may be considered a substance, whether capable or not of existing by itself as a homogeneous body.*" One can define the chemical potential of a component by "*choosing as one of the components the matter constituting the body itself,*" so that the molar Gibbs free energy of a system "*may always be considered as a potential.*" According to Hillert, "*Actually, one can define a component with the same composition as the whole system. The chemical potential of such a component is equal to G_m...*" where G_m is the molar Gibbs free energy.

- Another feature for this book is the rather extensive discussion on the thermodynamics of electron systems. Existing thermodynamics books in materials are largely focused on the thermodynamics of phase equilibria, and there is minimal coverage of thermodynamics of electronic systems. The concept of chemical potential or electrochemical potential of electronic defects is extremely important for understanding the thermodynamics of many processes involving electrons such as electronic transport in thermoelectrics and photovoltaics and electrochemical reactions in batteries, fuel cells, and corrosion, etc. This book discusses the contributions of electrons to thermodynamics by treating them an additional chemical, charged species.

It should be emphasized that a few topics that were emphasized in many of the existing thermodynamics textbooks are not treated in detail in this book. For example, although the concept of electrochemical potential is introduced, and its applications in energy conversions are briefly discussed, the discussions on thermodynamics of electrolyte solutions are not included. Furthermore, the concept of interfacial thermodynamics is touched upon in a number of sections in the book, but there is no separate chapter devoted to this topic. In addition,, the whole book has been focused on heterogeneous bulk systems consisting of several homogeneous regions, and the thermodynamics of diffuse interfaces is not discussed. Finally, the book does not include any tables of thermodynamic data since it is now much easier for a student to dig out data directly from Internet sources.

For adopting the book as a text for an undergraduate course in thermodynamics of materials, the following sections may be skipped: 2.4.6–2.4.10, 2.5.4–2.5.7; 3.2, 3.3,

3.5–3.6; Chap. 4; 5.3–5.6; 6.3.3–6.3.6; 7.8–7.9; 8.3–8.5; 9.6.3; 10.1.2, 10.2.1, 10.2.4, 10.3, 10.4.3–10.4.5; 11.9–11.10; 12.8, 12.10, 12.14, 12.24, 12.26, 12.28, 12.29.

I would like to acknowledge the help and contributions from many students in thermodynamics classes over the years, colleagues, current and former graduate students, and postdocs for proof-reading the different versions of book manuscript. In particular, I would like to thank Prof. Jiamian Hu (University of Wisconsin), Prof. Kasra Momeni (University of Alabama), Dr. Yanzhou Ji (Penn State), Prof. Yuhong Zhao (North University of China), Prof. Zijian Hong (Zhejiang University), Mr. Carter Dettor (Penn State), Mr. Erik Furton (Penn State), Dr. Chengchao Hu (Liaocheng University), Mr. Haowei Zhang (SDIC Unity Capital Co), Ms. Tina J. Chen (UC Berkeley), Prof. John Mauro (Penn State), Prof. Yi Wang (Penn State), Ms. Sandra Elder (Penn State), and many others.

The solutions to the exercise questions were primarily worked out by Dr. Yuhui Huang from Zhejiang University who visited Penn State for more than a year as a visiting scholar, and a separate solution manual will be published and available as a companion to this textbook.

Finally, I would like to express my sincere gratitude to my wife, Shuet-fun Mui, for her patience and understanding during the preparation of the book manuscript.

University Park, USA   Long-Qing Chen
                                                                       lqc3@psu.edu

Contents

About the author

Dr. Long-Qing Chen is Donald W. Hamer Professor of Materials Science and Engineering, Professor of Engineering Science and Mechanics, and Professor of Mathematics at the Pennsylvania State University, USA. He is currently Editor-in-Chief for npj Computational Materials by Nature Portfolio. He received his Ph.D. from the Massachusetts Institute of Technology, M.S. from Stony Brook University, and B.S. from Zhejiang University. He joined the faculty at Penn State in 1992 and has been teaching thermodynamics of materials for the past 29 years. He has published about 800 papers on the thermodynamics and kinetics of solid-state phase transformations as well as computational microstructure model development and applications for structural metallic alloys, functional oxide thin films, and energy materials. He is Clarivate Analytics Highly Cited Researcher since 2018. His research awards include Material Research Society (MRS) Materials Theory Award, Guggenheim Fellowship, Humboldt Research Award, TMS-FMD John Bardeen Award, TMS EMPMD Distinguished Scientist Award, American Ceramic Society (ACerS) Ross Coffin Purdy Award, Charles Hatchett Award, and ASM International Silver Medal. He is Fellow of TMS, MRS, American Physical Society (APS), American Association for the Advancement of Science (AAAS), ACerS, and ASM.

Chapter 1
Thermodynamic System and Its Quantification

1.1 Introduction

Thermodynamics is an important subject for almost every field of science and engineering ranging from chemistry, biology, physics, materials science, geoscience, meteorology, mechanics, health science, and environmental science to mechanical engineering, chemical engineering, civil engineering, electrical engineering, and agriculture.

Thermodynamics was born from the early studies of thermal energy of a system and its transfer as heat or its conversion into other forms of energy such as mechanical energy. The primary application of thermodynamics to materials science and engineering is to determine the chemical stability of a material with respect to transport of chemical species, chemical reactions, and transitions from one structural or physical state to another under different thermal and mechanical conditions.

Thermodynamics allows one to understand and predict the possible stable states, from which one can identify the stable phases and their compositions and volume fractions, under a given thermodynamic condition, e.g., overall chemical composition, temperature, and pressure. We may use thermodynamics to compute the thermodynamic driving forces for a given process, e.g., phase transitions such as solidification of liquids and precipitation in solids, and thus, thermodynamics is a key component for predicting the kinetics of a materials process. Thermodynamics can tell us why a solid is more stable than the corresponding liquid below its melting temperature by comparing their levels of chemical stability, quantified by a quantity called chemical potential. Using thermodynamics, we can understand why certain materials do not mix at room temperature while others can form complete solid solutions through the entire composition range at sufficiently high temperatures.

© Springer Nature Singapore Pte Ltd. 2022
L.-Q. Chen, *Thermodynamic Equilibrium and Stability of Materials*,
https://doi.org/10.1007/978-981-13-8691-6_1

One of the main contributions by Gibbs[1] was his formulation of the fundamental equation of thermodynamics by combining the first and second laws of thermodynamics. The fundamental equation of thermodynamics is a single mathematical expression that relates and connects all the basic thermodynamic variables or quantities of interest for a material. All the thermodynamic properties of a material can then be determined from the mathematical derivatives of thermodynamic energy or entropy representations of the fundamental equation of thermodynamics with respect to their natural thermodynamic variables. Based on the relations between properties derived from the fundamental equation of thermodynamics, one needs to simply measure or compute the properties that are easier or possible to measure or compute and relate them to those that are more difficult or not possible to measure or compute, which can save tremendous time and cost for carrying out thermodynamic property measurements. Therefore, a good understanding of the fundamental equation of thermodynamics is critical to the full comprehension and appreciation of thermodynamics and its applications to materials at equilibrium.

Another important concept introduced by Gibbs[1] is the chemical potential which measures the level of chemical stability of a chemical species, or a compound, a phase, or a combination of species, compounds, and phases. Therefore, one can use chemical potential to determine the stability and equilibrium of materials and analyze their tendency to transform from one physical state to another, e.g., from solid to liquid, and of chemical species to react to form new chemical species in chemical reactions or to migrate from one spatial location to another in mass transfer. Using the knowledge of chemical potentials of species and phases, we can quantitatively compute the thermodynamic driving forces for materials processes such as chemical mixing, phase transitions, formation of electronic and atomic defect species, and chemical reactions under different temperatures, pressures, and chemical compositions.

Before we discuss the fundamental equations of thermodynamics and chemical potentials and their applications to different materials processes, we will first introduce the definition of a thermodynamic system and its thermodynamic quantification.

1.2 Thermodynamic Systems

The first step in performing a thermodynamic analysis is to define a system. A thermodynamic system is generally defined as a collection of matter or space enclosed within a prescribed surface. For most of the materials applications, a system will usually be the material to be studied. The rest of the universe outside the system is called the surrounding (Fig. 1.1a).

---

[1] J. Willard Gibbs, "On the Equilibrium of Heterogeneous Substances," Transactions of the Connecticut Academy, III. pp. 108–248, October 1875–May 1876, and 343–524, May 1877–July 1878.

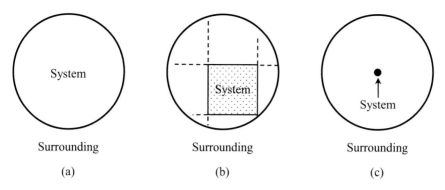

Fig. 1.1 Illustration of a thermodynamic system. **a** A system and its surrounding are separated by a physical wall schematically represented by the circle; **b** a system is separated from its surrounding by an imaginary wall represented by the square within a system; **c** a system is represented by a mathematically abstract point in the macroscopic continuum description of an inhomogeneous system

A system can also be defined using imaginary walls, e.g., a volume element used to discretize a three-dimensional (3D) volume of a substance in numerical simulations (Fig. 1.1b). In contemporary applications of thermodynamics to spatially inhomogeneous materials systems, a thermodynamic system can even be a point in space, which is assumed to possess its own well-defined thermodynamic properties measured in their densities (Fig. 1.1c).

It is important to remember that in thermodynamics, we are only concerned with the processes taking place in the chosen system of interest and how the system interacts with its surrounding. We do not consider any real or imaginary processes that are simultaneously happening in the surrounding. Very often, in materials science, the surrounding or the environment determines the thermodynamic conditions, e.g., temperature and pressure, of a material system. The choice of a system is based on the material of interest and a particular study to be performed. Examples include:

- A collection of valence electrons in a solid (Fig. 1.2a).

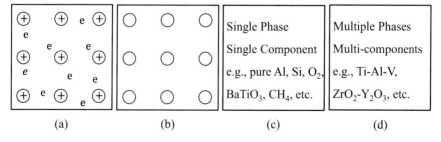

Fig. 1.2 Examples of a system: **a** a collection of valence electrons represented by e; **b** a collection of atoms; **c** single-component single-phase systems; and **d** multicomponent multiphase systems

- A collection of phonons or lattice vibrations in a solid.
- A collection of atoms in a solid (Fig. 1.2b).
- A homogeneous material with a single chemical species or elemental substance such as hydrogen gas (H_2) in a container, a piece of polycrystalline aluminum (Al) solid, or single-crystal silicon (Si) (Fig. 1.2c).
- A two-phase mixture of ice and water, etc.
- A multicomponent multiphase system (Fig. 1.2d).

The most common system in thermodynamics of materials is actually a very simple one: one mole of substance of interest which is usually the basis for most thermodynamic calculations.

1.3 Thermodynamic Variables

Once we decide on a thermodynamic system to study, we employ a set of basic thermodynamic variables to quantify the system. All relevant thermodynamic properties of interest can then be expressed in terms of this set of basic thermodynamic variables and their mutual mathematical derivatives.

We define three categories of basic thermodynamic variables to describe a system:

- Total energy of the system.
- The amounts of different types of matter inside the system.
- The levels of potentials representing the intensities of the different forms of energy and thus describing the corresponding thermodynamic stability of each type of matter in the system.

We employ simple systems to discuss the fundamentals of thermodynamics. A simple system is defined by Gibbs as a homogeneous system disregarding surface and gravitational effects and in the absence of stress, electric, or magnetic fields.

For a simple system, we introduce three basic variables to represent the three types of matter in the system: thermal, mechanical, and chemical:

- S—entropy representing the amount of thermal matter in the system.
- V—volume representing the amount of mechanical matter in the system.
- N—the number of moles representing the amount of chemical matter in the system.

For each type of matter, we define a corresponding potential measuring its thermodynamic stability, so we also have three types of potentials: thermal, mechanical, and chemical:

- T—Temperature representing the thermal potential and thus the thermodynamic stability of entropy in the system.
- p—Pressure representing the mechanical potential and thus the thermodynamic stability of volume.

- μ—Chemical potential representing the thermodynamic stability of chemical matter.

We can also define three forms of energy: thermal energy (U_T), mechanical energy (U_M), and chemical energy (U_C). The sum of the three forms of energy is the internal energy of the system U. Gibbs employed the internal energy U as the basic thermodynamic variable to represent the total energy of a system, and thus, he introduced a total of seven basic thermodynamic variables to quantify a simple system:

$$U, S, V, N, T, p, \text{ and } \mu.$$

These seven variables are the basic thermodynamic quantities of a simple thermodynamic system, from which the rest of the thermodynamic properties can be defined or derived; i.e., all other thermodynamic quantities of a simple system can be expressed in terms of these seven basic variables. It should be emphasized here without going to the details that the seven basic variables are related by the fundamental equation of thermodynamics and equations of state to be discussed in Chap. 3, and thus, they are not all independent.

For a simple system with n chemical components, there are $2n$ chemical variables with each component i described by its amount N_i and chemical potential μ_i. Therefore, for an n-component system, there are $2n + 5$ basic variables: internal energy (U), entropy (S), volume (V), the amount of each chemical component i (N_i, number of moles for each component i), temperature (T), pressure (p), chemical potential of each component (μ_i, with i from 1 to n).

1.3.1 Internal Energy

In thermodynamics, energy, in SI unit of joules (J), is a general term used to measure the transfer of different forms of energy such as thermal, mechanical, and chemical energies between a system and its surrounding or energy conversion from one form of energy to another, e.g., chemical energy to thermal energy during burning of a chemical fuel.

According to Einstein, mass and energy of a substance are equivalent, and for a mass at rest, they can be converted to each other through the relation,

$$U_o = mc^2 \tag{1.1}$$

where U_o is the rest mass energy of a body, m is mass, and c is the speed of light. A rough estimate for the rest energy of one mole of substance can easily show that the magnitude of rest mass energy is huge in unit of J/mol. For example, the molar mass of an elemental substance is on the order of 10^{-3}–10^{-1} kg/mol, and the speed of light is about 3×10^8 m/s, and hence, the rest mass energy is in the order of 10^{14}–10^{16} J/mol, which is many orders of magnitude larger than the typical changes in the energy of a substance associated with common processes such as changes in temperature, pressure, and atomic bonding in chemical reactions. Therefore, the applications of thermodynamics almost always involve the determination of the changes in energy associated with materials processes rather than measuring its absolute values. For all the applications discussed in this book, we will not include the rest mass energy. This is also the reason that when we write chemical reactions, we need to make sure that the mass of the reactants and products is balanced. For processes changing mass, e.g., isotope decay in nuclear reactions, the change in the rest mass energy is almost always dominant.

Macroscopically, the internal energy of a simple system U, excluding the rest mass energy U_o, may be considered as the total thermal, mechanical, and chemical energies required to create the system,

$$U = U_T + U_M + U_C \qquad (1.2)$$

where U_T is thermal energy, U_M is mechanical energy, and U_C is chemical energy. The internal energy U may increase or decrease through the exchanges of entropy, volume, and chemical substance, and thus the exchanges of thermal energy (or heat), mechanical energy (work), and chemical energy between the system and its surrounding.

In thermodynamics of materials, we usually do not include the kinetic energy associated with the translational or rotational motions of the entire macroscopic system in the internal energy U of the system although it can be incorporated in specific applications of thermodynamics. For describing the flow of fluid systems, the local kinetic energy has to be taken into account.

Microscopically, the internal energy of a material can be interpreted as the sum of two types of energies: (1) the interatomic and electronic interaction potential energy or chemical bonding energy E_p and (2) the kinetic energy E_k associated with the motion of electrons, motion or vibration of atoms, or motion and rotation of molecules within a material, i.e.,

$$U = E_p + E_k$$

Indeed, the macroscopic internal energy U of a system can be obtained as the average of the total microscopic kinetic and potential energies over many (or an infinite number of) systems which are thermodynamically and statistically the same, called an ensemble in statistical mechanics (see Chap. 4 for a brief introduction to statistical thermodynamics).

Throughout the whole book, Δ will be used to denote the finite change of a thermodynamic variable. As in calculus, d will be used to denote the infinitesimal change of a quantity; e.g., dU represents the infinitesimal change in the internal energy of a system.

1.3.2 Entropy

Macroscopically, entropy (S) is considered as the amount of thermal matter or thermal charge of a system. A system containing a large amount of entropy possesses a large amount of thermal energy. As a thermal matter, entropy can move around within a system and a system can exchange entropy with its surrounding. Adding entropy or thermal matter to a system increases the internal energy of the system. Since any chemical matter at finite temperature contains entropy, the transport through diffusion or flow of chemical matter is always accompanied by entropy transport. Entropy can also be created due to any of the materials kinetic processes taking place within a system such as phase transitions, chemical reactions, or entropy (or heat) transport, and mass or charge transport from one region to another within the system, which dissipates chemical or electric energy to become thermal energy.

Microscopically, a system of high entropy implies that the system can access a large number of energy states of different atomic/molecular/electronic configurations that the system may exhibit; i.e., the entropy of a system is a measure of the total number of accessible microscopic configurations (represented by, e.g., different vibrational modes of a crystalline lattice, spatial arrangements, etc.) of atoms/molecules by the system under a certain set of thermodynamic conditions. Adding entropy and thus thermal energy to a system increases the number of energy states, e.g., higher energy states, that the system can access. Based on this understanding, one mole of gas contains much more entropy than one mole of a solid, because one mole of gas atoms occupy a greater volume, and can move around the entire volume of the system rather than being confined to be around lattice sites in a solid, and thus can assume a much larger number of spatial arrangements than the same number of atoms in the solid. Similarly, one mole of atoms with two different types of species have more entropy than one mole of atoms with just a single type of species occupying the same lattice because the one mole of atoms with two types of species can be arranged into drastically more different atomic configurations on the lattice than the one mole of atoms with only one type of species.

The unit of entropy is J/K, which is the same unit as the Boltzmann constant, k_B, on a per particle basis and the same unit as the gas constant, R, on a per mole basis. The Boltzmann constant can be viewed as the elemental thermal charge just like e can be viewed as the elementary electric charge. One mole of elemental thermal charge is the gas constant R, an analog to the Faraday constant $\mathcal{F}$ for one mole of electric charge.

1.3.3 Volume

Volume (V) of a system, in SI unit of cubic meters (m^3), is the amount of space occupied by the system. It can be the entire volume of a piece of material or the volume of a volume element, (ΔV), in a numerical discretization of a computational domain,

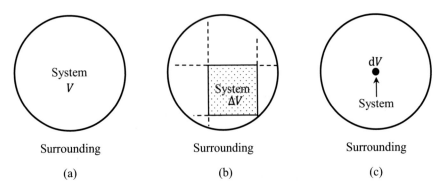

Fig. 1.3 Illustration of volume of a system. **a** The volume V of the entire system; **b** the volume ΔV of a discretized volume element; **c** the volume dV of a mathematically abstract point in the macroscopic continuum description of an inhomogeneous system

or dV of a spatial point within an inhomogeneous system (Fig. 1.3). ΔV is a finite volume element employed in numerical computations, whereas dV is an infinitesimal volume element in the language of calculus. As most thermodynamic calculations are performed on a per mole basis, volume generally appears as molar volume (v) in many applications of thermodynamics in materials. In physics, thermodynamic quantities are generally discussed on a per atom or per particle basis and the corresponding volume is the atomic volume or volume per particle.

1.3.4 Amount of Chemical Substance

Amounts ($N_1, N_2, \ldots N_i, \ldots, N_n$) of substances are generally measured in term of the number of moles, N_i, for each species i. Any substance or material consists of at least one chemical species that is either an atomic element or a molecular unit of more than one elements.

In a material with multiple chemical species involving chemical reactions or forming stoichiometric compounds, the amounts of chemical species may not be independent. Therefore, we need to specify another set of variables, the amounts of chemical components, which are independent variables by taking into account the relations among the amounts of chemical species due to chemical reactions or compound formation. For example, an Ag–Au alloy can be considered as a material with two atomic species or two components, while the CaO–BaO system contains three elemental atomic species, Ca, Ba, and O, but it is generally treated as a binary system consisting of two components, CaO and BaO. Therefore, in the discussions of fundamental equations of thermodynamics in the following chapters, the amounts of chemical substance $N_1, N_2, \ldots N_i, \ldots, N_n$ in a system are referred to the amounts of chemical components which are independent variables.

In the majority of practical applications, thermodynamic calculations are performed on per mole basis. In this case, we often use mole fractions for systems with more than one type of components, defined by $x_i = N_i/N$, where N is the total number of moles, to represent the chemical composition of a system.

1.3.5 Potentials Within a System

A potential represents the stability of its corresponding matter; the higher the potential level, the more unstable the corresponding matter. It is defined as the specific potential energy, or potential energy intensity, or the amount of potential energy per unit amount of matter of a homogeneous system.

$$\text{potential} = \frac{\text{energy}}{\text{matter}} \tag{1.3}$$

For example, the familiar electric potential, ϕ, is the amount of electrostatic potential energy, U_E, per unit amount of electric matter, i.e., the charge, q,

$$\text{electric potential } (\phi) = \frac{\text{electric energy } (U_E)}{\text{electric matter } (q)} \tag{1.4}$$

Electric potential can also be defined by the electrostatic energy U_E of a unit test charge ($q = 1\,C$ where C is Coulomb) in an electric potential field ϕ, or the infinitesimal amount of electrostatic energy dU_E when an infinitesimal amount of test charge dq is placed in a potential field ϕ, i.e.,

$$\phi = \frac{dU_E}{dq} \tag{1.5}$$

A positive charge is less stable at a high than lower potential, and thus, it will move from a high to a lower electrical potential region, and a negative charge will move from a low to a higher electrical potential region.

Similarly, the gravitational potential gz, where g is the gravitational acceleration (~ 9.8 m/s^2) and z is height of an object, is the amount of gravitational potential energy U_G per unit mass, m, of the object,

$$gz = \frac{U_G}{m} = \frac{mgz}{m} \tag{1.6}$$

i.e., the gravitational potential is the gravitational energy when one places a unit of mass in a gravitational potential gz. An object sitting at a high gravitational potential position is less stable than at a lower one position.

We can generalize the definition of potential in thermodynamics. For the system shown in Fig. 1.4a, we assume that the entire system is homogeneous with internal

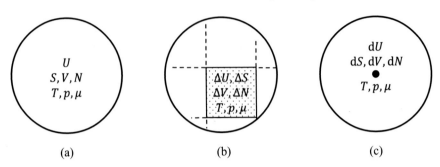

Fig. 1.4 Illustration of seven properties of a simple system: **a** the entire system; **b** a volume element within a system; and **c** a point inside a system

energy U, entropy S, volume V, and chemical substance N, respectively. According to Eq. (1.2), we can separate the total internal energy U into a sum of thermal energy U_T, mechanical energy U_M, and chemical energy U_C, i.e., $U = U_T + U_M + U_C$. Using Eq. (1.3), we can then define a thermal potential, temperature T, a mechanical potential, pressure p, and a chemical potential μ:

$$T = \frac{U_T}{S}, \quad p = -\frac{U_M}{V}, \quad \mu = \frac{U_C}{N} \tag{1.7}$$

The negative sign for pressure is due to the fact that if volume is added to a system, the mechanical energy and thus the internal energy of the system decrease. This differs from the case of temperature T for which the addition of entropy to a system leads to an increase in the thermal energy and thus the internal energy of the system, or the case of chemical potential for which the addition of chemical matter to a system increases the chemical energy of the system.

For the thermodynamic system specified in Fig. 1.4b, the system is a volume element with volume ΔV, amount of entropy ΔS, amount of chemical matter ΔN. The total internal energy ΔU within the system can be considered as a sum of thermal energy ΔU_T, mechanical energy ΔU_M, and chemical energy ΔU_C, i.e., $\Delta U = \Delta U_T + \Delta U_M + \Delta U_C$. The definitions for the corresponding temperature, pressure, and chemical potential for such a volume element are given by

$$T = \frac{\Delta U_T}{\Delta S}, \quad -p = \frac{\Delta U_M}{\Delta V}, \quad \mu = \frac{\Delta U_C}{\Delta N} \tag{1.8}$$

For an infinitesimal system, a point in an inhomogeneous system (Fig. 1.4c), it is a system with volume dV, amount of entropy dS, and amount of chemical matter dN. The total internal energy dU within the system can be considered as a sum of thermal energy dU_T, mechanical energy dU_M, and chemical energy dU_C, i.e., $dU = dU_T + dU_M + dU_C$, and thus, the differential definitions for temperature, pressure, and chemical potential are given by

$$T = \frac{dU_T}{dS}, \quad -p = \frac{dU_M}{dV}, \quad \mu = \frac{dU_C}{dN} \tag{1.9}$$

Since $dU = dU_T + dU_M + dU_C = TdS - pdV + \mu dN$ we can write the above definitions (Eq. 1.9) for T, p, and μ in terms of partial derivatives of internal energy U,

$$T = \left(\frac{\partial U}{\partial S}\right)_{V,N}, \quad -p = \left(\frac{\partial U}{\partial V}\right)_{S,N}, \quad \mu = \left(\frac{\partial U}{\partial N}\right)_{S,V} \tag{1.10}$$

i.e., temperature, T, can be considered as the rate of change in internal energy, U, with respect to entropy, S, at constant volume V and constant number of moles of matter, N, the negative pressure, $-p$, as the rate of change of U with respect to V at constant S and constant N, and chemical potential, μ, as the rate of change of U with respect to N at constant S and constant V.

The differential definition in Eq. (1.10) requires the conditions to be specified since the energy of a system also depends on all the thermodynamic variables: temperature or entropy, pressure or volume, chemical potential or the amount of other chemical species. Therefore, for understanding thermodynamics as a beginner, the integrated form, Eq. (1.7), is a simpler definition that allows a straightforward interpretation of potentials in terms of energy intensity or the amount of energy per unit amount of matter.

1.3.5.1 Temperature

According to Eq. (1.7), the temperature T of a system is defined as the amount of thermal potential energy U_T of the system per unit of entropy S. As we will learn in future chapters, the thermal energy U_T is actually a well-defined thermodynamic function which can be written as a combination of basic thermodynamic variables, $U_T = TS = U + pV - \mu N$.

Temperature has the unit of Kelvin (K). The Kelvin scale for temperature is the absolute scale and should be used in all thermodynamic calculations. It is related to another commonly used temperature scale, the Celsius scale, which is based on the behavior of water which freezes at 0 °C and boils at 100 °C at normal ambient pressure of 1 bar through

$$T\ (\text{K}) = T\ (°\text{C}) + 273.15 \tag{1.11}$$

Another commonly used temperature scale, primarily in the USA, is the Fahrenheit scale, given by

$$T\ (^{\circ}\mathrm{C}) = \frac{5}{9}\big[T\ (^{\circ}\mathrm{F}) - 32\big] \tag{1.12}$$

The lowest theoretical temperature is absolute 0 K, at which the atoms and molecules are static. At 0 K, the entropy of a homogeneous, stable material is zero (according to the third law of thermodynamics to be discussed in a future chapter). Therefore, at 0 K, there exists no thermal matter, i.e., no entropy, and no thermal energy in a material at equilibrium. Absolute zero on the Kelvin scale corresponds to $-273.15\ ^{\circ}\mathrm{C}$ on the Celsius scale and $-459.67\ ^{\circ}\mathrm{F}$ on the Fahrenheit scale.

Microscopically, temperature is a measure of the average kinetic energy arising from the thermal motion of the electrons, atoms, and molecules in the system.

We can draw an analogy between temperature and electric potential. Specifically, an electric potential difference or gradient, also referred as an electric field, results in electric conduction, or the transport of electric charges. Analogously, a difference in temperature or a temperature gradient leads to entropy transport, and thus thermal energy or heat transfer, from high temperature to low temperature regions.

1.3.5.2   Pressure

The mechanical potential or pressure p represents the amount of mechanical potential energy, U_{M}, stored per unit volume, V. It is a measure of mechanical energy intensity (Eq. 1.7). As we will learn later in the book that the mechanical potential energy U_{M} is the grand potential energy function which can be written as a combination of basic thermodynamic variables, $U_{\mathrm{M}} = -pV = U - TS - \mu N$.

In physics, pressure is defined as the normal force per unit area and has the unit of Pascal ($\mathrm{Pa} = \mathrm{N/m^2}$). Thus, it has the same unit as energy density, i.e., the amount of energy per unit volume, $\mathrm{J/m^3}$. Other commonly used units for pressure are

$$1\ \mathrm{bar} = 10^5\ \mathrm{Pa} \quad 1\ \mathrm{atmosphere\ (atm)} = 101.325\ \mathrm{kPa} \quad 1\ \mathrm{atm} = 760\ \mathrm{torr}$$

Pressure, like temperature, is a scalar that has no direction, and a pressure difference or gradient leads to the flow of volume of matter, e.g., a pressure difference or gradient leads to the flow of water in a pipe.

1.3.5.3   Chemical Potential

The chemical potential of a chemical species or component or a phase, μ, is the amount of chemical energy (U_{C}) per mole, or chemical energy intensity (Eq. 1.7), of a species or component or phase.

As we will learn later in the book, the chemical energy U_{C} is the Gibbs free energy G which can be expressed as a combination of basic thermodynamic variables, $U_{\mathrm{C}} = G = \mu N = U - TS + pV$. The magnitude of chemical potential is the same as the thermodynamic chemical energy or Gibbs free energy for one mole of a substance. Essentially all existing thermodynamic textbooks use the unit of J/mol for chemical potential. However, it has been suggested[2,3] to use the unit gibbs (G) for the unit of chemical potential where there is no confusion between the unit G and the variable G representing the Gibbs free energy. This suggestion is consistent with the units for other potentials, which are associated with the names of the scientists who invented them, e.g., volt (V) for electric potential named after Alessandro Volta, Kelvin (K) for temperature named after William Thomson, 1st Baron Kelvin, and Pascal (Pa) for pressure named after Blaise Pascal. We reserve J/mol as the unit for molar energy quantities such as molar internal energy or internal energy per mole. It should be pointed out that in physics, the most common unit used for chemical potential is energy per particle, e.g., eV/atom.

Chemical potential, introduced by Gibbs more than a century ago, is the central concept in thermodynamics of materials, chemistry, and physics. Chemical potential is a measure of the chemical stability of a component or a group of components with a fixed composition, e.g., a compound or a solution phase, in a material. The higher the chemical potential of a component, a phase, or a material is, the less stable the component, the phase or the material is. Therefore, chemical potential quantifies the thermodynamic tendency of a component or a group of components with a fixed composition to transform or react at a given spatial location or to transport from one location to another in a system. A species or a compound or a phase of high chemical potential will attempt to lower its chemical potential by reacting with other species, or going through phase transitions to phases with lower chemical potentials, or by moving to other locations with lower chemical potential.

One can define a chemical potential for essentially all types of particles including atoms, molecules, electrons, electron holes, phonons, photons, etc. For example, the Fermi level in semiconductor device physics is the chemical potential of electrons. A chemical potential can also be defined for a given combination of different atomic species.

Chemical potential is a central concept in thermodynamics of materials as almost all applications of thermodynamics to materials science can be formulated in terms of the changes in chemical potentials.

1.4 Densities

In applying thermodynamics to materials science, in particular to problems in inhomogeneous systems, we often deal with density quantities. There are mainly two types of density quantities: the amount of matter per mole, or molar quantities, and

---

[2] Hans U. Fuchs, "The Dynamics of Heat," Springer-Verlag, New York, 1996.

[3] G. Job and F. Herrmann, "Chemical potential—a quantity in search of recognition." Institute of Physics Publishing, Eur. J. Physics 27, 353 (2006).

the amount of matter per unit volume, simply called the density of a certain type of matter.

Let us use the lowercase letters to represent the molar quantities. For example, u, s, and v are the molar internal energy, molar entropy, and molar volume, respectively. For a simple system,

$$u = \frac{U}{N}, \quad s = \frac{S}{N}, \quad v = \frac{V}{N} \tag{1.13}$$

For an infinitesimal system with volume dV, e.g., a point in an inhomogeneous system,

$$u = \frac{dU}{dN}, \quad s = \frac{dS}{dN}, \quad v = \frac{dV}{dN} \tag{1.14}$$

Therefore, the molar quantities become derivatives of the amount of certain type of matter with respect to the number of moles in an inhomogeneous system. Later we will learn that U, S, and V are also functions of other thermodynamic quantities such as temperature T and pressure p. Therefore, in mathematics, they are typically expressed as partial derivatives. For example,

$$u = \left(\frac{\partial U}{\partial N}\right)_{T,p}, \quad s = \left(\frac{\partial S}{\partial N}\right)_{T,p}, \quad v = \left(\frac{\partial V}{\partial N}\right)_{T,p} \tag{1.15}$$

One can also define for each component of a multicomponent system,

$$u_i = \left(\frac{\partial U}{\partial N_i}\right)_{T,p,N_{j\neq i}}, \quad s_i = \left(\frac{\partial S}{\partial N_i}\right)_{T,p,N_{j\neq i}}, \quad v_i = \left(\frac{\partial V}{\partial N_i}\right)_{T,p,N_{j\neq i}} \tag{1.16}$$

where u_i, s_i, and v_i are molar energy, entropy, and volume of component i, respectively.

The densities representing the amount of matter per unit volume can be expressed similarly. We use lowercase letters with a subscript v to represent the volume densities. For example,

$$u_v = \frac{U}{V}, \quad s_v = \frac{S}{V}, \quad c = \frac{N}{V}, \quad c_i = \frac{N_i}{V} \tag{1.17}$$

where u_v, s_v, c, and c_i are the internal energy density or internal energy per unit volume, entropy density or entropy per unit volume, total number of moles per unit volume, and concentration of component i or the number of moles of i per unit volume, respectively. One can easily relate molar quantities and densities on a per unit volume basis. For example,

$$u_v = \frac{u}{v}, \quad s_v = \frac{s}{v}, \quad c = \frac{1}{v} \tag{1.18}$$

Sometimes, we define another density, called the mass density with the unit of kg/m^3 in SI unit, i.e., amount of mass per unit volume,

$$\rho = cm = \frac{m}{v} \tag{1.19}$$

where m is the mass per mole of chemical matter.

1.5 Extensive and Intensive Variables

A property or variable of a system is called extensive if it scales linearly with the size of the system. Examples of extensive properties and variables include the amounts of entropy S, volume V, chemical substances $N_1, N_2, \ldots N_n$, amount of electric charge, q, and the amount of energy U, etc. When the size of a system increases, the magnitudes of all the extensive properties increase proportionally.

A property or variable of a system is called intensive if it is independent of the size of the system. Examples of intensive properties and variables include all the potentials such as temperature T, pressure p, and chemical potentials, $\mu_1, \mu_2, \ldots \mu_n$ where the subscripts from 1 to n label the chemical components or a chemical potential of a phase, μ, as well as purely density quantities such as chemical concentration, molar volume, molar entropy, molar internal energy, etc.

It is important to realize the difference between a potential and a purely density quantity.[4] The value of a potential is uniform throughout a system if the corresponding matter is allowed to freely redistribute in the system, whereas a purely density quantity is usually different in different locations in a heterogeneous system consisting of different homogeneous regions. Therefore, if entropy, volume, and chemical species are allowed to redistribute within a homogeneous or heterogeneous system, at thermodynamic equilibrium, temperature T, pressure p, and chemical potentials, $\mu_1, \mu_2, \ldots \mu_n$ are all uniform, whereas molar internal energy, entropy, molar volume, concentrations of chemical components are generally not uniform within the system.

Intensive variables can be expressed as a ratio of two extensive variables. For example, molar volume is the ratio of volume to the amount of substance ($v = V/N$), and chemical potential (μ) is the ratio of chemical energy ($U_C = G$) to the amount of chemical matter (N).

Using calculus, an intensive property can be expressed as the derivative of one extensive variable with respect to another extensive variable. For example, T, p, and μ can be defined as

$$T = \left(\frac{\partial U}{\partial S}\right)_{V,N}, \quad -p = \left(\frac{\partial U}{\partial V}\right)_{S,N}, \quad \mu = \left(\frac{\partial U}{\partial N}\right)_{S,V} \tag{1.20}$$

---

[4] M. Hillert, "Phase Equilibria, Phase Diagrams and Phase Transformations," Second Edition, Cambridge University Press, Cambridge, 2008.

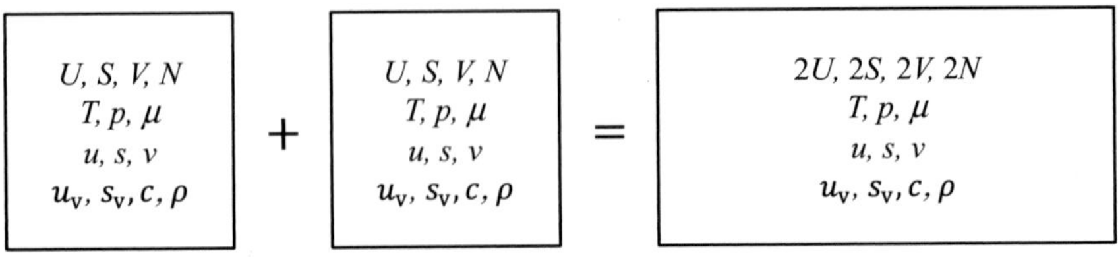

U, S, V, N		U, S, V, N		$2U, 2S, 2V, 2N$
T, p, μ	$+$	T, p, μ	$=$	T, p, μ
u, s, v		u, s, v		u, s, v
u_v, s_v, c, ρ		u_v, s_v, c, ρ		u_v, s_v, c, ρ

Fig. 1.5 Illustration of the changes in extensive and intensive properties when a system size is doubled

Figure 1.5 illustrates the behavior of extensive and intensive properties of a simple system as a system size doubles. All the values of the extensive properties (U, S, V, and N) double, whereas those of the intensive variables, T, p, μ, u, s, v, s_v, u_v, c, and ρ, remain the same.

1.6 Conjugate Variable Pairs

A pair of variables whose product is a quantity that represents a form of energy or energy density is called a conjugate variable pair. Conjugate variable pairs always appear together in the same term in thermodynamic energy functions or in the fundamental equations of thermodynamics to be discussed in Chap. 3. For a given fundamental equation of thermodynamics, only one of the two variables, never both, from each pair is chosen as an independent variable. Examples of thermodynamic variable conjugate pairs are given in the last rows of each table from Tables 1.1, 1.2, 1.3, 1.4, and 1.5.

Table 1.1 Basic conjugate variable pairs with their products having the unit of energy (J)

Thermal energy U_T	Mechanical energy U_M	Chemical energy U_C
$U_T = TS$	$U_M = -pV$	$U_C = \mu_i N_i$
$dU_T = T dS$	$dU_M = -p dV$	$dU_C = \mu_i dN_i$
(T, S)	$(-p, V)$	(μ_i, N_i)

Table 1.2 Basic conjugate variable pairs with their products having the unit of molar energy (J/mol)

Molar thermal energy u_T	Molar mechanical energy u_M	Molar chemical energy u_C
$u_T = Ts$	$u_M = -pv$	$u_C = \mu_i$
$du_T = T ds$	$du_M = -p dv$	N/A
(T, s)	$(-p, v)$	N/A

Table 1.3 Basic conjugate variable pairs with their products having the unit of energy density (J/m^3)

	Thermal energy density $u_{T,v}$	Mechanical energy density $u_{M,v}$	Chemical energy density $u_{C,v}$
	$u_{T,v} = T s_v$	$u_{M,v} = -p$	$u_{C,v} = \mu_i c_i$
	$du_{T,v} = T ds_v$	N/A	$du_{C,v} = \mu_i dc_i$
	(T, s_v)	N/A	(μ_i, c_i)

Table 1.4 Other conjugate variable pairs with their products having the unit of energy (J)

Electric energy U_E	Surface energy U_S	Gravitational energy U_G
$U_E = \phi q$	$U_S = \gamma A$	$U_G = gzm$
$dU_E = \phi dq$	$dU_S = \gamma dA$	$dU_C = gzdm$
(ϕ, q)	(γ, A)	(gz, m)
Electrical potential ϕ and charge q	Specific surface energy γ and surface area A	Gravitational potential gz and mass m

Table 1.5 Other conjugate variable pairs with their products having the unit of energy density (J/m^3)

Mechanical energy density	Dielectric energy density	Magnetic energy density
$\sigma_{ij}\varepsilon_{ij}$	$E_i D_i$	$H_i B_i$
$\sigma_{ij}d\varepsilon_{ij}$	$E_i dD_i$	$H_i dB_i$
$(\sigma_{ij}, \varepsilon_{ij})$	(E_i, D_i)	(H_i, B_i)
Mechanical stress σ_{ij} and mechanical strain ε_{ij} (i and j labeling directions)	Electrical field E_i and electric displacement D_i (i labeling direction)	Magnetic field H_i and magnetic induction B_i (i labeling direction)

1.7 Classical, Statistical, and Nonequilibrium Thermodynamics

This book is focused on the classical thermodynamics in which a system is described by the fundamental equation of thermodynamics that relates a small set of thermodynamic variables such as internal energy, entropy, volume, number of moles, temperature, pressure, chemical potentials, etc. It allows one to derive relationships among macroscopic properties of a system from the fundamental equations of thermodynamics formulated from the empirically verified laws of thermodynamics. Statistical thermodynamics relates the averages properties of microscopic thermodynamic systems containing a huge number of particles to the fundamental equation of thermodynamics of the corresponding macroscopic system specified by a small number of independent thermodynamic variables. In experiments, a measuring device does the statistical thermodynamic averaging.

In contrast to the equilibrium thermodynamics, nonequilibrium thermodynamics or irreversible thermodynamics relates the rate of entropy production or rate of energy dissipation to thermodynamic driving forces and rates of processes; i.e., nonequilibrium thermodynamics is more concerned with the kinetics of materials processes in systems out of equilibrium. Although our main interest here is the application of classical thermodynamics to determining the stability of materials and the tendency, direction, and driving forces of materials processes under a given set of thermodynamic conditions, we will attempt to make connections between equilibrium thermodynamics and irreversible thermodynamics through discussions on the entropy production and energy dissipation for irreversible processes.

1.8   Exercises

1. Identify whether the following thermodynamic quantities are extensive (E) or intensive (I) variables: T (temperature), p (pressure), V (volume), N (number of moles), U (internal energy), S (entropy), μ (chemical potential), charge (q), electric potential (ϕ).

2. Identify which of the following quantities can be considered "potentials" and which ones are "amounts of matter": S (entropy), T (temperature), p (pressure), V (volume), μ_i (chemical potential of species i), N_i (number of moles of specifies i), γ (specific surface energy), A (area), q (charge), ϕ (electric potential).

3. If one doubles the size of a system, the entropy of the system is also doubled: true or false?

4. If one doubles the size of a system, the chemical potential of the substance is also doubled: true or false?

5. Find a relationship among:

    (a) N_i (number of moles of species i where $i = 1, 2, \ldots n$), V (volume of the whole system), and c_i (volume concentration of species i).
    (b) c (total volume concentration), c_i (volume concentration of species i), and x_i (mole fraction of species i).

6. Translate the following texts into thermodynamic expressions using derivatives, differentials, or integrals:

    (a) The rate of change in volume with respect to the change in temperature while keeping pressure constant.
    (b) The rate of change of temperature with respect to the change in pressure while keeping volume constant.
    (c) The infinitesimal change in temperature due to an infinitesimal change in pressure while keeping volume constant.

(d) The infinitesimal change in temperature due to simultaneous infinitesimal changes in pressure and volume.

(e) The finite change in temperature due to a finite change in pressure while keeping volume constant.

(f) The finite change in temperature due to simultaneous finite changes in pressure and volume.

7. Translate the following thermodynamic expressions into text statements:

(a) $\left(\frac{\partial U}{\partial S}\right)_{V,N}$, $\left(\frac{\partial V}{\partial T}\right)_{p,N}$, $\left(\frac{\partial V}{\partial p}\right)_{T,N}$

(b) $dU = \left(\frac{\partial U}{\partial S}\right)_{V,N} dS$, $\quad dV = \left(\frac{\partial V}{\partial T}\right)_{p,N} dT + \left(\frac{\partial V}{\partial p}\right)_{T,N} dp$

(c) $\Delta U = \int_{S_o}^{S} \left(\frac{\partial U}{\partial S}\right)_{V,N} dS$, $\quad \Delta V = \int_{T_o,p_o}^{T,p} \left[\left(\frac{\partial V}{\partial T}\right)_{p,N} dT + \left(\frac{\partial V}{\partial p}\right)_{T,N} dp \right]$

8. Express the energy unit 1 bar $*$ L, where L represents unit of liter, in terms of J (joules).

9. The thermal expansion coefficient of a material describes how the material changes in volume with temperature. What is the unit of thermal expansion coefficient in the SI system?

10. For the relation $G = U - ST + pV$, calculate the value of G in joules (J) if U is 400 cal, S is 1.0 J/K, T is 500 °C, p is 1 bar, and V is 1 L.

11. Write down the SI units for the potentials: thermal, mechanical, gravitational, chemical, and electrical.

12. Write down the SI units for the amount of matter: entropy, volume, mass, amount of chemical matter, and electric charge.

13. Write down the total potential in units of Gibbs ($G = J/mol$), which is the sum of gravitational, chemical, and electrical potentials. Use quantities such as molar mass, and molar charge (Faraday constant) if necessary for unit conversions.

14. At 273 K and 1 bar($= 10^5$ Pa), the entropy difference between one mole of water and one mole of ice is $\Delta s = s^{water} - s^{ice} = 22.00$ J/K; the molar volumes of water and ice at 273 K and 1 bar are 18.02 cm^3/mol and 19.66 cm^3/mol, respectively; and the chemical potentials of ice and water are the same. Using the above information at 273 K and 1 bar($= 10^5$ Pa), please compute

(a) The difference in thermal energy between one mole of ice and one mole of water.

(b) The difference in mechanical energy between one mole of ice and one mole of water.

(c) The difference in chemical energy between one mole of ice and one mole of water.

(d) The difference in internal energy between one mole of ice and one mole of water.

15. For a two-phase mixture of ice and water at equilibrium, please indicate which of the following intensive thermodynamic quantities are uniform, and which are not, throughout the two-phase mixture: temperature (T), mass density $(\rho,$ kg per unit volume), concentration $(c,$ mole per unit volume), pressure (p), molar volume (v), chemical potential (μ) of H_2O, molar internal energy (u), molar entropy (s), internal energy density (u_v), and entropy density (s_v). For simplicity, please ignore the possible gravitational effects in your consideration for the problem.

Chapter 2
First and Second Laws of Thermodynamics

Early development in thermodynamics led to the laws of thermodynamics: the zeroth law, the first law, the second law, and the third law. The zeroth law states that if the temperature of system A is equal to the temperature of system C, and if the temperature of system B is equal to the temperature of system C, then the temperature of system A is equal to the temperature of system B, implying that temperature is measurable. The third law states that the entropy of a perfectly ordered, stable substance at 0 K is zero; i.e., entropy of a system has an absolute value just like the volume of a system. The discussions in this chapter will be entirely focused on the first and second laws of thermodynamics, based on which Gibbs formulated his fundamental equation of thermodynamics. Before we discuss the first and second laws of thermodynamics, we first introduce the concepts of thermodynamic states, state variables, and thermodynamic processes.

2.1 Thermodynamic States and State Variables

We follow Gibbs by quantifying a simple thermodynamic system using seven basic variables, internal energy U, entropy S, temperature T, volume V, pressure p, number of moles of chemical substance N, and chemical potential μ. Therefore, a given thermodynamic state (Fig. 2.1) is described by a specific set of values of U, S, T, V, p, N, and μ. It should be mentioned here that not all seven basic variables can be independently varied since they are mutually related by a fundamental equation and three equations of states as we will discuss them in Chap. 3.

The seven basic variables, U, S, T, V, p, N, and μ as well as their combinations and many of their mutual derivatives with respect to each other are all state variables. As a matter of fact, any variable which represents a property of a system is a state variable. For a given thermodynamic system, if we plot any of the state variables, e.g., U, as a function of another thermodynamic variable, e.g., entropy S or temperature T, or as a function of a set of variables, e.g., entropy S, volume V, and number of

© Springer Nature Singapore Pte Ltd. 2022
L.-Q. Chen, *Thermodynamic Equilibrium and Stability of Materials*,
https://doi.org/10.1007/978-981-13-8691-6_2

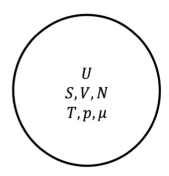

moles of chemical substance N, a state is represented by a point on such a plot. A given state is associated with a set of specific values for the seven basic variables, the tangent planes or lines, the local curvatures of the multidimensional surface at a given point. Figure 2.2 is an example of plotting internal U as a function of entropy S at fixed values of volume V and fixed number of moles N. The information about the temperature of the state is embedded in the slope $(T = \partial U/\partial S)_{V,N}$ while the information on the constant volume heat capacity of the system can be obtained from the curvature of internal energy U with respect to S at this state.

A state is considered uniform if all thermodynamic variables have uniform values throughout the system. A state consisting of regions of homogeneous states with different uniform thermodynamic properties is called a heterogeneous state.

A state in which thermodynamic variables depend on spatial positions is called a nonuniform state or inhomogeneous state. Let us assume that at a point r within a system, the values of basic thermodynamic variables are internal energy density $u_v(r)$, entropy density $s_v(r)$, chemical concentration, $c_i(r)$ or mass density $(\rho_i(r))$ of component i, temperature $T(r)$, pressure $p(r)$, and chemical potential $\mu_i(r)$ of component i. In such a continuous system, each point r can be considered as a subsystem with internal energy $u_v(r)dV$, entropy $s_v(r)dV$, volume dV, number of moles $c_i(r)dV$ of component i at temperature $T(r)$, pressure $p(r)$, and chemical potential $\mu_i(r)$. The subsystems constantly exchange chemical species, entropy, and volume among each other.

Fig. 2.2 Illustration of a
thermodynamic state which
is represented by a point in a
plot of internal U as a
function of entropy S while
keeping volume V and
number of moles N constant

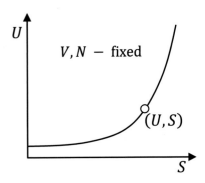

An equilibrium state is a stationary state at which the properties of a system or the state variables no longer change with time, and there are no fluxes of matter, e.g., chemical species, entropy, and electric charges, going through the system. Therefore, an equilibrium state is described by a unique set of specific values for the basic thermodynamic variables.

It should be noted that a steady state is a kinetic concept rather than a thermodynamic one. A steady-state represents a state of a system in which the spatial profiles of thermodynamic variables such as temperature, pressure, and chemical concentrations no longer vary with time under a constant applied field such as electric field, chemical potential gradient, and temperature gradient. However, in general, there is at least one flux (entropy or heat flux, chemical substance flux, electric current, etc.) that is nonzero. Those fluxes are constantly generating entropy for the system (entropy generation will be discussed below under irreversible processes). Therefore, a steady state is not an equilibrium state.

2.2 Thermodynamic Processes

A process in thermodynamics refers to the change of a system from one state described by one set of values for the thermodynamic variables to another state with another set of thermodynamic variables arising from the exchanges of entropy, volume, and chemical matter with its surrounding and/or from internal phase transitions, chemical reactions, diffusion, heat conduction, flow, fracture, plastic deformation, etc. For a given thermodynamic system, if we plot any of the state variables, e.g., U, as a function of another thermodynamic variable, e.g., entropy S or temperature T, or as a function of a set of variables, e.g., entropy S, volume V, and number of moles of chemical substance N, a process is described by a path connecting two states. The changes in state variables depend only on the values of state variables at the initial state and the final state and are independent of the process path that a system takes. Figure 2.3 illustrates a process from state 1 to state 2 leading to finite changes in internal energy ΔU and entropy ΔS at fixed volume and fixed number of moles. Figure 2.4 shows a corresponding infinitesimal process resulting in infinitesimal changes in thermodynamic variables dU, dS, etc.

Fig. 2.3 Schematic illustration of a finite process through which the state of the system is changed from state 1 to state 2

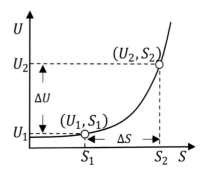

Fig. 2.4 Schematic
illustration of an
infinitesimal process
involving infinitesimal
changes in properties

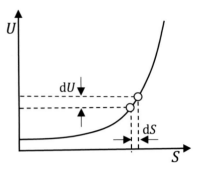

Processes taking place within a system (inside the prescribed boundaries) leading to a change of state are called internal processes. Some processes involve interactions between the system and its surrounding, which take place through the boundary, e.g., heat conduction and mass transfer across the boundary and volume change of the system through the mechanical displacement of the boundary. We may call such processes exchange processes. Distinguishing internal and exchange processes is useful in understanding the second law of thermodynamics and determining the changes in the entropy of a system during a process.

2.2.1 Spontaneous, or Natural, or Irreversible Processes

A process that involves the spontaneous movement of a system from an unstable equilibrium state to an equilibrium state is a spontaneous, or natural, or irreversible process. All real processes occurring in nature are irreversible processes. The driving force for an irreversible process, for example a chemical potential difference between initial state and transformed state in a phase transition or between reactants and products in a chemical reaction, is finite. During a spontaneous process, one form of potential energy, e.g., the chemical energy, is converted to thermal energy, and thus, it generates entropy.

2.2.2 Reversible Processes

A process that takes a system from one equilibrium state to another equilibrium without being out of equilibrium at any intermediate state during the process is a reversible process. A reversible process is described by a line or curve in the multidimensional space of state variables representing equilibrium states connecting the initial equilibrium and the final equilibrium states of the process. The states along which the system progresses during a reversible process are sometimes called quasi-static states. The driving force for a reversible process is infinitesimally small.

There are only entropy exchange and work interactions between the system and its surrounding, and there is no net entropy produced within the system, i.e., the total entropy, the combined entropy of the system and its surrounding, is conserved in a reversible process. Reversible processes set the theoretical limit for real processes. Reversible processes are idealized processes that can be designed and utilized to calculate thermodynamic quantities for irreversible processes based on the fact that the changes in state functions are independent of process path.

2.3 Thermodynamic Systems

A thermodynamic system is classified according to how it interacts with its surrounding. An isolated system has no interactions with its surrounding; i.e., there is no entropy exchange ($dS^e = 0$), no volume exchange ($dV^e = 0$) or other forms of work, and no mass exchange ($dN^e = 0$) across the walls between the system and its surrounding. A wall that does not permit entropy or heat transfer is called adiabatic, while a wall that does not allow mass transfer of chemical species is called impermeable. A wall that is not allowed to move is called a fixed wall. Therefore, an isolated system is considered to have adiabatic, fixed, and impermeable walls. The whole universe can be considered as an isolated system.

A system that can exchange entropy and volume with its surrounding ($dS^e \neq 0$, $dV^e \neq 0$) but not the amount of chemical substance ($dN^e = 0$) across the wall between the system and its surrounding is called a closed system. A wall that allows entropy or heat transfer is called a diathermal wall, and a wall that is allowed to move is called a movable wall.

A system that can exchange chemical substance, in addition to entropy and volume, with its surrounding ($dS^e \neq 0$, $dV^e \neq 0$, $dN^e \neq 0$) is an open system, and the corresponding wall that allows mass transfer is called a permeable wall (Fig. 2.5). In a materials system containing several phases or regions of different compositions and crystal structures, each phase within the multiple phase mixture can be considered

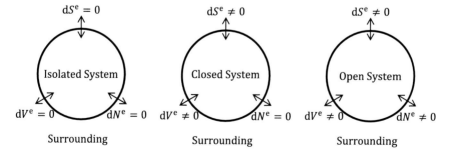

Fig. 2.5 Illustration of three types of thermodynamic systems

as an open system since the chemical species can move across the phase boundaries to reach chemical equilibrium.

2.4   First Law of Thermodynamics

The first law of thermodynamics is about the conservation of energy. In Chap. 1, we introduced the seven basic thermodynamic quantities of a simple thermodynamic system: $U, S, V, N, T, p,$ and μ. The main application of the first law of thermodynamics is to determine (either measure or calculate) the change in the internal energy of a system dU or ΔU for a given process. The first law also provides the governing principle of energy conservation for formulating kinetic theories and models for describing kinetic processes of materials, e.g., the heat conduction equation.

According to the first law of thermodynamics, the total energy of the whole universe (system + surrounding) (Fig. 2.6) is conserved,

$$U_{tot} = U + U_{sur} = \text{constant} \tag{2.1}$$

Therefore, the change in the total energy of the universe is zero for all processes,

$$dU_{tot} = dU + dU_{sur} = 0 \tag{2.2}$$

$$\Delta U_{tot} = \Delta U + \Delta U_{sur} = 0 \tag{2.3}$$

The first law of thermodynamics is valid regardless of whether a process is reversible or irreversible. In Eqs. (2.1)–(2.3), U_{tot} is the total internal energy of the universe (system plus the surrounding), U is the internal energy of the system, and U_{sur} is the internal energy of the surrounding. According to Eq. (2.2), when the energy of the system is increased by dU, there must be a corresponding decrease

Fig. 2.6 Illustration of a system and its surrounding

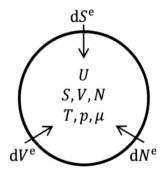

Surrounding

in energy in the surrounding by dU; i.e., energy can only be exchanged between the system and its surrounding or between two different systems, but it cannot be destroyed or created. Therefore, one can measure the change in the internal energy of a system by measuring the amount of energy exchanged, for example thermal and/or mechanical energy exchange, between a system and its surrounding. This can be especially useful when it is difficult or impossible to directly measure or calculate the change in the internal energy of a system.

A system exchanges energy with its surrounding through the transfer of various forms of matter, e.g., the transfer of entropy (thermal matter), volume (mechanical matter), substances (chemical matter), etc. For simplicity, let us assume that the system and its surrounding can only exchange the amount of entropy, volume, and substance, and ignore other forms of matter exchanges such as electric charges; i.e., we consider a simple system (Fig. 2.7). Let us discuss the first law of thermodynamics employing infinitesimal changes or exchanges while the temperature T, pressure p, and chemical potential μ of the system are fixed and uniform during the exchanges. The increase in the internal energy (dU) of a system is equal to the sum of the amount of thermal $(dU_T^e = T dS^e)$, mechanical $(dU_M^e = -p dV^e)$, and chemical $(dU_C^e = \mu dN^e)$ energy transfers from the surrounding to the system, i.e.,

$$dU = dU_T^e + dU_M^e + dU_C^e = T dS^e - p dV^e + \mu dN^e \qquad (2.4)$$

In most textbooks on thermodynamics of materials, the first law of thermodynamics is discussed almost exclusively for closed systems which do not allow chemical energy transfer $dU_C^e = 0$. For closed systems, the amount of thermal energy transfer (dU_T^e) is often called the amount of heat transfer (dQ), while the amount of mechanical energy transfer (dU_M) is called the amount of work (dW) done on the system by its surrounding. It should be emphasized that the amount of thermal energy or mechanical energy or chemical energy transfer to system is not necessarily equal to the respective amount of increase in the thermal energy or mechanical energy or chemical energy of the system.

Fig. 2.7 Illustration of changes in thermodynamic basic variables as a result of infinitesimal amounts of exchanges in entropy, volume, and chemical substance

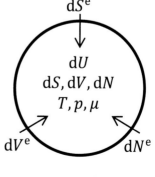

To understand the exchange processes for an open system, it is important to realize that a chemical substance possesses thermal energy and mechanical energy in addition to chemical energy. Therefore, for an open system, if we use dQ to represent an infinitesimal amount of thermal energy transfer purely due to heat transfer without any transfer of chemical substance, the total amount of thermal energy transfer dU_T^e to the system is

$$dU_T^e = dQ + TsdN^e \tag{2.5}$$

where T is the temperature of the system, s is the molar entropy of the chemical substance transferred, and dN^e is the amount of chemical substance transferred. The corresponding amount of entropy entering the system at temperature T from its surrounding is dS^e

$$dS^e = \frac{dU_T^e}{T} = dS^Q + sdN^e \tag{2.6}$$

where $dS^Q = dQ/T$ is the amount of entropy transfer from the surrounding to the system purely from heat transfer, and sdN^e is the amount of entropy transfer due to the transfer of the chemical substance.

Similarly, if we use dW to represent the amount of mechanical energy exchange purely due to the volume change arising from the mechanical deformation of the system under pressure p, the total amount of mechanical energy exchange between the system and its surrounding dU_M is

$$dU_M^e = dW - pvdN^e \tag{2.7}$$

where p is the pressure of the system, and v is the molar volume of the chemical substance transferred. Note that here dW represents the work done on the system by the surrounding.

The corresponding amount of volume exchange between the system and its surrounding dV^e is then

$$dV^e = -\frac{dU_M^e}{p} = -\frac{dW}{p} + vdN^e = dV^M + vdN^e \tag{2.8}$$

where dV^M is the amount of volume transfer from the surrounding to the system at pressure p purely from mechanical deformation, and vdN^e is the amount of volume exchange due to the transfer of chemical substance.

Therefore, the first law of thermodynamics for an open system can be written as

$$dU = TdS^e - pdV^e + \mu dN^e = T(dS^Q + sdN^e) - p(dV^M + vdN^e) + \mu dN^e \tag{2.9}$$

It is emphasized again here that for an open system, dS^e includes both the entropy transfer from the surrounding to the system purely due to heat transfer and the entropy of the chemical substance transferred from the surrounding to the system. Similarly, dV^e represents the volume transfer from both the volume exchange due to the mechanical deformation of the system and the volume of the chemical substance transferred from the surrounding to the system.

For the same initial state, a system can go through different process paths to reach the same final state. The associated changes in the thermodynamic properties of the system (or state functions) such as internal energy, entropy, temperature, volume, pressure, the number of moles and chemical potential are independent of the process path as long as the system starts from the same initial state and reaches the same final state for all the processes. However, the amounts of thermal, mechanical, and chemical energy added to (or subtracted from) the system as a result of finite amount of matter (entropy, volume, and substance) transfers depend on how the process is carried out. For example, if a process is carried out at constant and uniform potentials: constant and uniform temperature T, constant and uniform pressure p, and constant and uniform chemical potential, μ, for the system during the entire process, the internal energy change for the system due to the finite exchanges is given by the first law of thermodynamics,

$$\Delta U = T \Delta S^e - p \Delta V^e + \mu \Delta N^e \qquad (2.10)$$

where ΔS^e, ΔV^e and ΔN^e are the finite amounts of entropy, volume, and substance transferred from the surrounding to the system, respectively. $T \Delta S^e$, $-p \Delta V^e$ and $\mu \Delta N^e$ are the corresponding amounts of thermal, mechanical, and chemical energies transferred from the surrounding to the system.

Comparing Eq. (2.10) with the expressions for the first law of thermodynamics for closed systems in the existing literature, the first law of thermodynamics for a finite process for an open system is given by

$$\Delta U = (Q + Ts \Delta N^e) + (W - pv \Delta N^e) + \mu \Delta N^e \qquad (2.11)$$

It should be noted that the above dU or ΔU refers to the infinitesimal or final change in internal energy of the system, while the corresponding change in the energy of the surrounding is $dU_{sur} = -dU$, or $\Delta U_{sur} = -\Delta U$. The infinitesimal changes in the amounts of matter in the surrounding as a result of these infinitesimal exchanges are $-dS^e$, $-dV^e$, $-dN^e$, and the finite changes in the surrounding as a result of finite exchanges are $-\Delta S^e$, $-\Delta V^e$, $-\Delta N^e$. However, in most cases, we are not concerned with any processes taking place within the surrounding; i.e., we are only interested in changes for the system as a result of the processes.

2.4.1 First Law of Thermodynamics for Isolated Systems

For an isolated system, the first law of thermodynamics ($dS^e = 0$, $dV^e = 0$, $dN^e = 0$) is reduced to

$$dU = dU^e = -dU_{sur} = 0 \tag{2.12}$$

2.4.2 First Law of Thermodynamics for Closed Systems

For a closed system, the first law of thermodynamics ($dS^e \neq 0$, $dV^e \neq 0$, $dN^e = 0$) is

$$dU = T dS^e - p dV^e \tag{2.13}$$

where $T dS^e$ and $-p dV^e$ represent infinitesimal amounts of thermal energy (heat) exchange and mechanical energy (work) exchange, respectively.

2.4.2.1 Adiabatic Processes in a Closed System

For an adiabatic process, $dS^e = 0$. The amount of entropy produced inside a system during an adiabatic process depends on the pressure difference between the external pressure and internal pressure within the system during the volume exchange. If the pressure difference is zero, the process is reversible. In a reversible process, there is no entropy produced inside the system during the process, $dS^{ir} = 0$, and the entropy change for the system is entirely due to the entropy transfer from the surrounding to the system ($dS = dS^e$), and hence in a reversible adiabatic process

$$dS = dS^e = 0, \ dS^{ir} = 0 \tag{2.14}$$

For a reversible adiabatic expansion process in a closed system ($dS^e = 0$, $dV^e \neq 0$, $dN^e = 0$), the first law of thermodynamics is:

$$dU = -p^{ex} dV^e = -p dV \tag{2.15}$$

where p^{ex} is external pressure, and U, p and V are the internal energy, pressure, and volume of the system, respectively.

For an irreversible adiabatic expansion, $dS^e = 0$, $dS = dS^{ir}$, $p \neq p^{ex}$, $dV = dV^e$,

$$dU = T dS^e - p^{ex} dV^e = T dS - T dS^{ir} - p dV + (p - p^{ex}) dV \tag{2.16}$$

For the same entropy change dS and volume change dV, if both the initial state U, S, and V, and the final state $U + dU$, $S + dS$, and $V + dV$ are equilibrium states, the internal change dU is always equal to $T dS - p dV$ since U is a state function. The thermal energy $T dS^{ir}$ generated during the irreversible adiabatic expansion is then given by

$$T dS^{ir} = (p - p^{ex}) dV \qquad (2.17)$$

2.4.2.2 Constant Volume Processes in a Closed System

For constant volume ($dV^e = 0$) processes in a closed simple system ($dN^e = 0$):

$$dU = T dS^e = dQ \qquad (2.18)$$

For finite changes,

$$\Delta U = Q \qquad (2.19)$$

i.e., *the change in internal energy of a closed system at constant volume is equal to the thermal energy (or heat) absorbed or released.*

2.4.2.3 Constant Pressure Processes in a Closed System

For constant pressure processes in a closed system:

$$dU = T dS^e - p dV^e \qquad (2.20)$$

$$dU + p dV^e = T dS^e = dQ. \qquad (2.21)$$

If we assume $dV = dV^e$, i.e., there is no volume created or eliminated in the system other than the volume exchange with the surrounding during the process, we can rewrite Eq. (2.21) as

$$d(U + pV) = dH = dQ, \qquad (2.22)$$

where

$$H = U + pV \qquad (2.23)$$

is called the enthalpy of the system. H, a combination of state variables U, p, and V, is also a state function. Many thermodynamic processes of practical importance, e.g., chemical reactions and phase transitions, take place under a constant pressure.

For finite changes,

$$\Delta H = Q \tag{2.24}$$

Therefore, *the change in the enthalpy of a closed system is equal to the thermal energy (heat) absorbed or released at constant pressure.* The enthalpy of the system increases or the enthalpy change is positive if heat is absorbed by the system from its surrounding, and the enthalpy of the system decreases or enthalpy change is negative if heat is released from the system to the surrounding.

2.4.3   First Law of Thermodynamics for Reversible Processes in Open Systems

For reversible processes in open systems,

$$dN = dN^e \tag{2.25}$$

i.e., there are no internal chemical reactions taking place, and the change in the amount of substance of the system dN is entirely due to the exchange of chemical substance between the system and its surrounding dN^e. Here we also assume that the volume change of the system is due to both the reversible deformation of the system dV^M and the amount of chemical substance transfer from the surrounding to the system vdN, and there is no volume created or destroyed,

$$dV = dV^e = dV^M + vdN^e = dV^M + vdN \tag{2.26}$$

where v is the molar volume of the substance. For a reversible process,

$$dS = dS^e = dS^Q + sdN^e = dS^Q + sdN \tag{2.27}$$

where dS^Q is the amount of entropy transfer from the surrounding to the system entirely due to heat transfer, s is the molar entropy of the substance, and sdN is the amount of entropy transfer from the surrounding to the system due to the transfer of dN moles of chemical substance.

Therefore, for reversible processes in an open system, we can write down the first law of thermodynamics in two different ways,

$$dU = TdS^Q - pdV^M + udN, \tag{2.28}$$

where

$$u = Ts - pv + \mu \tag{2.29}$$

Equation (2.28) can also be written as

$$dU = TdS - pdV + \mu dN, \tag{2.30}$$

where

$$\mu = u - Ts + pv \tag{2.31}$$

In Eqs. (2.29) and (2.31), u, s, v, and μ are molar internal energy, molar entropy, molar volume, and chemical potential of the chemical substance, respectively. We can now write down the first law of thermodynamics for a number of special reversible processes in an open system.

2.4.3.1 Reversible Constant Entropy and Volume Processes in an Open System

From Eq. (2.30), it can easily be seen that for constant entropy ($dS = 0$) and constant volume ($dV = 0$), the first law of thermodynamics for a reversible process in an open system is then

$$dU = \mu dN \tag{2.32}$$

In this case, one can figure out the amounts of entropy and heat exchange between the system and its surrounding in order to keep the entropy constant ($dS = 0$) in the system during the exchange of dN amount of chemical substance between the system and its surrounding. The condition of constant entropy $dS = 0$ requires that

$$dS^Q = -sdN \tag{2.33}$$

The corresponding amount of heat exchange between the system and its surrounding is then

$$dQ = TdS^Q = -TsdN \tag{2.34}$$

which implies that dQ amount of heat must be removed from the system in order to keep its entropy constant.

The constant volume condition $dV = 0$ implies,

$$dV^M = -vdN \tag{2.35}$$

Therefore, the amount of work done on the system is then given by

$$dW = -pdV^M = pvdN \tag{2.36}$$

Therefore, $TsdN$ is the amount of heat flow from the system to the surrounding, and $pvdN$ is the amount of work done on the system by the surrounding during the reversible transfer of dN amount of substance from the surrounding to the system in order to maintain constant entropy and volume for the system.

2.4.3.2 Reversible Constant Entropy and Pressure Processes in an Open System

According to Eq. (2.30), if we keep both entropy and pressure constant, the first law of thermodynamics becomes

$$dU + pdV = dH = \mu dN$$

Therefore, the amount of increase in the enthalpy of the system dH at constant S and p is equal to the chemical energy transfer μdN from the surrounding to the system.

The requirement of constant entropy $dS = 0$ indicates that

$$dS^Q = -sdN \tag{2.37}$$

Hence, the corresponding amount of heat exchange between the system and its surrounding is then

$$dQ = TdS^Q = -TsdN \tag{2.38}$$

The volume change of the system is

$$dV = dV^M + vdN$$

The reversible work done on the system by the surrounding is

$$dW = -pdV^M = -p(dV - vdN) \tag{2.39}$$

2.4.3.3 Reversible Constant Temperature and Volume Processes in an Open System

For constant temperature and volume processes,

$$dU - TdS = dF = \mu dN \tag{2.40}$$

where

$$F = U - TS \tag{2.41}$$

is called the Helmholtz free energy.

Therefore, for constant temperature and volume, the first law of thermodynamics implies that the change in the Helmholtz free energy is simply equal to the amount of chemical energy transfer from the surrounding to the system,

$$dF = \mu dN \tag{2.42}$$

The corresponding amount of heat transfer to maintain constant temperature is given by

$$dQ = T dS - T s dN = T(dS - s dN) \tag{2.43}$$

The amount of work interaction in order to keep constant volume is given by

$$dW = -p dV^M = p v dN \tag{2.44}$$

2.4.3.4   Reversible Constant Temperature and Pressure Processes in an Open System

For a process at constant temperature and pressure, we have

$$dU - T dS + p dV = dG = \mu dN$$

where

$$G = U - TS + pV$$

is called the Gibbs free energy of the system.

Therefore, at constant temperature and pressure, the first law of thermodynamics can be stated in terms of the change in the Gibbs free energy of a system which is equal to the amount of chemical energy transfer from the surrounding to system at constant temperature and pressure,

$$dG = \mu dN$$

The amount of heat transfer involved is

$$dQ = T dS - T s dN = T(dS - s dN) \tag{2.45}$$

and the work done on the system by the surrounding is

$$dW = -pdV^M = -p(dV - vdN) \tag{2.46}$$

2.5 Second Law of Thermodynamics

While the first law of thermodynamics is about energy conservation, the second law of thermodynamics is about entropy conservation and creation. The second law of thermodynamics states that the total entropy of the universe (system + surrounding) is conserved during reversible processes and increases during irreversible processes.

If we assume that there is no volume or any new chemical species created or destroyed inside the system during a process, the corresponding infinitesimal changes for the system due to the exchanges are $dS = dS^e + dS^{ir}$, $dV = dV^e$, and $dN = dN^e$, where dS^{ir} is the entropy produced due to irreversible processes within the system during the exchange process. Thus, the entropy change for the system must take into account the possibility of not only the entropy transferred from the surrounding to the system due to the exchange between system and its surrounding but also the entropy produced due to any irreversible processes within the system. However, if all the processes taking place within the system are reversible during the exchange process, i.e., if the system never leaves equilibrium as the exchange takes place, $dS^{ir} = 0$, and the amount of entropy change for the system, dS, is the same as the amount of entropy transfer from the surrounding to the system, dS^e, so that $dS = dS^e$.

For any process, the entropy change dS can be separated into two contributions: (i) the entropy exchange dS^e between the system and its surrounding, and (ii) the entropy produced inside the system dS^{ir} due to internal irreversible processes (Fig. 2.8). Therefore, the entropy change of the system can be expressed as

$$dS = dS^e + dS^{ir} \tag{2.47}$$

Fig. 2.8 Illustration of entropy change of a system dS is equal to entropy exchange dS^e and creation dS^{ir}

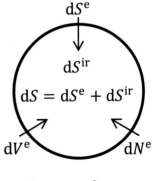

Surrounding

For an open system, dS^e includes the total entropy transfer due to heat transfer and the entropy carried by the chemical substance transferred to the system.

Since we are only concerned with the processes taking place within the system, the corresponding entropy change for the surrounding is simply

$$dS_{sur} = -dS^e \tag{2.48}$$

The total entropy change of the whole universe (system + surrounding) due to the processes taking place within the system is thus given by

$$dS_{tot} = dS + dS_{sur} = dS - dS^e = dS^{ir} \tag{2.49}$$

The amount of internal entropy production dS^{ir}, or total entropy change dS_{tot} for the system and surrounding, is always positive for irreversible processes and zero for reversible processes, i.e.,

$$dS_{tot} = dS^{ir} \geq 0. \tag{2.50}$$

Therefore, the total entropy is conserved for all reversible processes, and the total entropy change or the amount of entropy created within the system for all irreversible processes is positive. It should be noted that a process that leads to a negative total entropy change is an impossible process. A negative total entropy change is equivalent to having a system spontaneously evolve from a more stable state to a less stable state.

While the total entropy may never decrease, the entropy of a system may decrease as long as there is an increase in the entropy in the surrounding which must be at least as large as the decrease in the entropy of the system. One well-known example is a refrigerator, in which the entropy and thus temperature decreases inside the refrigerator by expending electric energy while there is a larger increase in the entropy of its outside surrounding.

If the processes within the system are reversible, the entropy change for the system is then equal to the entropy exchange from the surrounding,

$$dS = dS^e \tag{2.51}$$

It should be reminded that the entropy of a system is a state function; thus, the change in entropy of a system is independent of the path taken by the system between a set of specified initial and final states, regardless of whether the process is reversible or irreversible. This means that, for a given set of initial and final states of a system, one can always design a reversible path between the initial and final states to find the entropy difference of a system between any given two states. If we examine the entropy changes in the system and in the surrounding through a reversible versus an irreversible path for the same initial and final states, the entropy change for the system along an irreversible path will be the same as that along a reversible path. However, the entropy change along an irreversible path will be different for the surrounding as any entropy produced in an irreversible process within the system is transferred to the surrounding. More specific calculations of entropy and entropy changes will be discussed in more details in Chap. 8.

A system at equilibrium has higher entropy than the same system out of equilibrium under the same thermodynamic conditions. For example, one can imagine two boxes of a gas separated by an impermeable wall. Initially, one box contains one mole of oxygen gas, and the other box is a vacuum. Now we remove the wall between the two boxes to allow the oxygen gas molecules freely to move around between and within the two boxes. The single mole of oxygen gas is now out of equilibrium, and some of the oxygen molecules will move to the initially vacuum box. The process of oxygen gas molecule redistribution upon the removal of the wall between the two boxes is then an irreversible process which produces entropy, i.e., the entropy for the one mole of oxygen will keep increasing from the moment at which one removes the wall until a new equilibrium is established in the combined box. The new equilibrium state has the maximum entropy under the new constraint (the combined volume of the two boxes).

Any irreversible (real, spontaneous, natural) process creates entropy and thus converts some useful energy, e.g., mechanical or chemical, to thermal energy. The total entropy produced by a process is a representation of the degree of irreversibility of a process, which can also be interpreted as how much the initial system is out of equilibrium. The larger the amount of entropy produced, the more irreversible the process is, and the more out of equilibrium the initial system is.

2.5.1   Quantifying Second Law of Thermodynamics

To quantify the second law of thermodynamics or irreversibility of an internal process inside a system, we compute the total amount of entropy produced in the process. We rewrite the first law of thermodynamics by replacing all the exchange quantities in terms of quantities associated with the system, $dS^e (= dS - dS^{ir})$, $dV^e (= dV)$, and $dN^e (= dN)$ as

$$dU = T(dS - dS^{ir}) - pdV + \mu dN = TdS - pdV + \mu dN - TdS^{ir} \quad (2.52)$$

where dS^{ir} is the amount of entropy produced due to one or more internal processes within the system.

Following Kondepudi and Prigogine[1] and Hillert,[2] we define the last term of Equation (2.52) as

---

[1] Dilip Kondepudi and Ilya Prigogine, Modern Thermodynamics—From Heat Engines to Dissipative Structures, John Wiley and Sons 1998.

[2] Mats Hillert, Phase Equilibria, Phase Diagrams and Phase Transformations—Their Thermodynamic Basis, Second Edition, Cambridge University Press 2008.

$$T\,dS^{ir} = D\,d\xi \tag{2.53}$$

In the above equation, D measures the amount of internal energy dissipation for the entire process or called the driving force for the process,

$$D = -\left(\frac{\partial U}{\partial \xi}\right)_{S,V,N} \begin{bmatrix} = 0 \text{ for reversible processes} \\ > 0 \text{ for irreversible processes} \end{bmatrix} \tag{2.54}$$

and ξ measures the extent of the process,

$$\xi = \begin{bmatrix} 0 \text{ or } \xi_i \text{ at the start of the process} \\ 1 \text{ or } \xi_f \text{ at the end of the process} \end{bmatrix} \tag{2.55}$$

where ξ_i and ξ_f represent the extent of the process in the initial and final states, respectively. The internal process can be a phase transition, a chemical reaction, heat conduction, chemical diffusion, etc.

To make a connection between the thermodynamic driving force and the kinetics of a phase transition, ξ can also be regarded as an order parameter or a phase field in the phase-field method of microstructure evolution.[3] For example, in the phase-field method of phase transitions,

$$\begin{bmatrix} \xi = 0 & \text{represents the original, parent phase} \\ 0 < \xi < 1 & \text{represents a mixture of parent and new phases} \\ \xi = 1 & \text{represents the transformed, new phase} \end{bmatrix} \tag{2.56}$$

During a process, the amount of chemical energy $D\,d\xi$ is dissipated and transforms to $T\,dS^{ir}$ amount of thermal energy and produces dS^{ir} amount of entropy. The total entropy produced in a process is then

$$\Delta S^{ir} = \int_{\text{initial state}}^{\text{final state}} dS^{ir} = \int_{\xi_i}^{\xi_f} \frac{D}{T}\,d\xi \tag{2.57}$$

The relation among the amount of entropy produced, the amount of thermodynamic energy dissipated, and the magnitude of the thermodynamic driving force for a given process is further discussed below for different types of thermodynamic systems (isolated, closed, and open) and different types of thermodynamic processes, e.g., isothermal (constant temperature), isobaric (constant pressure), isoentropic (constant entropy), or isochoric (constant volume).

---

[3] L. Q. Chen, Phase-Field Models for Microstructure Evolution. Annual Review of Materials Research, 2002. **32**: P. 113–140.

2.5.2   Isolated Systems

The second law of thermodynamics for a process in an isolated system becomes

$$dS^{ir} = dS = \frac{Dd\xi}{T} \tag{2.58}$$

i.e., the amount of entropy produced can simply be determined by the change in the entropy of the system or the driving force and the extent of the evolution process from the initial nonequilibrium state to the final equilibrium state. During the infinitesimal process, $Dd\xi$ amount of useful energy is converted to TdS^{ir} amount of thermal energy.

For an isolated system, we can write the driving force for a process as

$$D = T\left(\frac{\partial S}{\partial \xi}\right)_{U,V,N} = T\left(\frac{\partial S^{ir}}{\partial \xi}\right)_{U,V,N} \begin{bmatrix} = 0 \text{ For an initially equilibrium state} \\ > 0 \text{ For an initially nonequilibrium state} \end{bmatrix} \tag{2.59}$$

2.5.3   Constant Entropy Processes in a Closed System

The amount of entropy produced during a constant entropy process in a closed system depends on the pressure difference between the external pressure and internal pressure within the system during the volume exchange. If the pressure difference is zero or infinitesimally small, the process is reversible, and the entropy produced is zero. Therefore, during a reversible constant entropy process, the internal energy change of the system is given by

$$dU = -pdV \tag{2.60}$$

Let us assume that the external pressure is p^{ex}, the internal energy change during an irreversible constant entropy process in a closed system can be written as

$$dU = -p^{ex}dV - TdS^{ir} \tag{2.61}$$

where $p^{ex}dV$ represents the work done by the system on the surrounding with pressure p^{ex}, and TdS^{ir} is the thermal energy generated due to the irreversible expansion of the system. Comparing Eqs. (2.60) and (2.61), we have

$$dS^{ir} = \frac{(p - p^{ex})dV}{T} \tag{2.62}$$

The amount of entropy produced in a finite process can then be expressed as

$$\Delta S^{ir} = \int dS^{ir} = \int \frac{(p - p^{ex})}{T} dV \begin{bmatrix} = 0 \text{ if } p = p^{ex} \\ > 0 \text{ if } p \neq p^{ex} \end{bmatrix} \quad (2.63)$$

while $(p - p^{ex})dV$ is the amount of internal energy converted into thermal energy within the system during the volume change dV.

2.5.4 Constant Entropy and Constant Volume Processes in a Closed System

For constant entropy and constant volume processes in a closed system, $dS = 0$, $dV = 0$, and $dN = 0$. Thus, the second law of thermodynamics becomes

$$dS^{ir} = -\frac{dU}{T} = \frac{Dd\xi}{T} \quad (2.64)$$

Equation (2.64) shows that *the amount of entropy produced dS^{ir} in a constant entropy and constant volume process within a closed system is equal to the amount of dissipated internal energy $Dd\xi$ or $- dU$ divided by temperature T.* In order to maintain constant entropy, the amount of entropy produced due to an irreversible process inside the system much be removed and transferred to the surrounding.

For a constant entropy and constant volume process in a closed system, we can write the driving force for a process as

$$D = -\left(\frac{\partial U}{\partial \xi}\right)_{S,V,N} = T\left(\frac{\partial S^{ir}}{\partial \xi}\right)_{S,V,N} \quad (2.65)$$

2.5.5 Constant Temperature and Constant Volume Processes in a Closed System

If both the volume and temperature of a closed system are fixed, the amount of entropy produced becomes

$$dS^{ir} = dS - \frac{dQ}{T} = dS - \frac{dU}{T} = -\frac{dU - TdS}{T} = -\frac{d(U - TS)}{T} = -\frac{dF}{T} \quad (2.66)$$

Or

$$\Delta S^{ir} = \frac{D\Delta\xi}{T} = -\frac{\Delta F}{T} \quad (2.67)$$

where

$$F = U - TS \tag{2.68}$$

is the Helmholtz free energy of the system. Equation (2.67) shows that *the amount of entropy produced ΔS^{ir} in a constant temperature and volume process within a closed system is given by the amount of dissipated Helmholtz free energy $-\Delta F$ or thermodynamic driving force $D\Delta\xi$ divided by temperature T.*

For a constant temperature and constant volume process in a closed system, we can write the driving force for a process as

$$D = -\left(\frac{\partial F}{\partial \xi}\right)_{T,V,N} = T\left(\frac{\partial S^{ir}}{\partial \xi}\right)_{T,V,N} \tag{2.69}$$

2.5.6 Constant Entropy and Pressure Processes in a Closed System

The second law of thermodynamics for a constant entropy and constant pressure process can be written as

$$dS^{ir} = -\frac{dQ}{T} = -\frac{dH}{T} = \frac{Dd\xi}{T} \tag{2.70}$$

Equation (2.70) demonstrates that *the amount of entropy produced in a constant entropy and constant pressure process within a closed system is equal to the amount of dissipated enthalpy $-dH$ or the thermodynamic driving force $Dd\xi$ divided by temperature T.*

For a constant entropy and constant pressure process in a closed system, we can write the driving force for a process as

$$D = -\left(\frac{\partial H}{\partial \xi}\right)_{S,p,N} = T\left(\frac{\partial S^{ir}}{\partial \xi}\right)_{S,p,N} \tag{2.71}$$

2.5.7 Constant Temperature and Pressure Processes in a Closed System

If both pressure and temperature are maintained constant, we have the second law of thermodynamics

$$dS^{ir} = dS - \frac{dQ}{T} = dS - \frac{dH}{T} = -\frac{d(H - TS)}{T} = -\frac{dG}{T} = \frac{Dd\xi}{T} \tag{2.72}$$

Or

$$\Delta S^{ir} = -\frac{\Delta G}{T} \qquad (2.73)$$

where

$$G = H - TS \qquad (2.74)$$

is the Gibbs free energy or free enthalpy of the system. Equation (2.73) indicates that *the amount of entropy produced in a constant temperature and pressure process within a closed system is given by the amount of dissipated Gibbs free energy or the thermodynamic driving force* $-\Delta G$ *divided by temperature* T.

For a constant temperature and constant pressure process in a closed system, we can write the driving force for a process as

$$D = -\left(\frac{\partial G}{\partial \xi}\right)_{T,p,N} = T\left(\frac{\partial S^{ir}}{\partial \xi}\right)_{T,p,N} \qquad (2.75)$$

2.5.8 Open Systems

We now discuss the applications of second laws of thermodynamics to open systems. For example, for an open system undergoing a constant temperature, constant volume, and constant chemical potential process, the first law of thermodynamics is given by

$$dU = TdS^e + \mu dN^e = TdS^e + \mu dN \qquad (2.76)$$

The second law of thermodynamics is given by

$$dS^{ir} = dS - dS^e \qquad (2.77)$$

Substituting Eq. (2.76) into Eq. (2.77), we have

$$dS^{ir} = dS - \frac{(dU - \mu dN)}{T} \qquad (2.78)$$

If the temperature and chemical potential of system are constant, we have

$$dS^{ir} = -\frac{d(U - TS - \mu N)}{T} = -\frac{d\Xi}{T} \qquad (2.79)$$

Or

$$\Delta S^{\text{ir}} = \frac{D d\xi}{T} = -\frac{\Delta \Xi}{T} \qquad (2.80)$$

where

$$\Xi = U - TS - \mu N \qquad (2.81)$$

is the grand potential energy of the system. Equation (2.80) shows that *the amount of entropy produced in a constant temperature, constant volume, and constant chemical potential process of an open system is given by the amount of dissipated grand potential energy or thermodynamic driving force, $-\Delta \Xi$, divided by temperature T.*

For a constant temperature, constant volume, and constant chemical potential process, we can write the driving force for a process as

$$D = -\left(\frac{\partial \Xi}{\partial \xi}\right)_{T,V,\mu} = T\left(\frac{\partial S^{\text{ir}}}{\partial \xi}\right)_{T,V,\mu} \qquad (2.82)$$

2.6 Summary on First and Second Laws of Thermodynamics

The following tables summarize the first and second laws of thermodynamics in closed systems. In the tables, the variables without a subscript represent the properties of a system; e.g., U, S, V, T, p, H, F, and G are the internal energy, entropy, volume, temperature, pressure, enthalpy, Helmholtz free energy, and Gibbs free energy of a system, respectively. dS^{ir} or ΔS^{ir} represents the entropy produced in the system, dQ is the infinitesimal amount of heat transfer, and dS^{e} represents the amount of entropy exchange between the system and its surrounding.

Table 2.1 First and second laws of thermodynamics for constant volume processes

	Constant V $dV = 0$	Constant V and S $dV = 0$, $dS = 0$	Constant V and T $dV = 0$, $dT = 0$
First law	$dU = dQ$	$dU = dQ$	$dU = dQ$
Second law	$dS^{\text{ir}} =$ $-\frac{dU - TdS}{T}$	$dS^{\text{ir}} = -\frac{dU}{T}$	$dS^{\text{ir}} =$ $-\frac{dF}{T}$, $\Delta S^{\text{ir}} =$ $-\frac{\Delta F}{T}$

Table 2.2 First and second laws of thermodynamics for constant pressure processes

	Constant p $\mathrm{d}p = 0$	Constant p and S $\mathrm{d}p = 0$, $\mathrm{d}S = 0$	Constant p and T $\mathrm{d}p = 0$, $\mathrm{d}T = 0$
First law	$\mathrm{d}H = \mathrm{d}Q$	$\mathrm{d}H = \mathrm{d}Q$	$\mathrm{d}H = \mathrm{d}Q$
Second law	$\mathrm{d}S^{\mathrm{ir}} = -\frac{\mathrm{d}H - T\mathrm{d}S}{T}$	$\mathrm{d}S^{\mathrm{ir}} = -\frac{\mathrm{d}H}{T}$	$\mathrm{d}S^{\mathrm{ir}} = -\frac{\mathrm{d}G}{T}$, $\Delta S^{\mathrm{ir}} = -\frac{\Delta G}{T}$

If a process is reversible, there will be no entropy produced. The following three tables summarize for first and second laws of thermodynamics for specific reversible processes of simple closed systems.

From Tables 2.1, 2.2, 2.3, 2.4, and 2.5, we can draw the following conclusions about first and second laws of thermodynamics for closed systems.

Table 2.3 First and second laws of thermodynamics for reversible adiabatic processes

	Adiabatic processes $\mathrm{d}S^{\mathrm{e}} = 0$	Isolated systems $\mathrm{d}S^{\mathrm{e}} = 0$, $\mathrm{d}V = 0$	Adiabatic constant p $\mathrm{d}S^{\mathrm{e}} = 0$, $\mathrm{d}p = 0$
First law	$\mathrm{d}U = -p\mathrm{d}V$	$\mathrm{d}U = 0$	$\mathrm{d}H = 0$
Second law	$\mathrm{d}S^{\mathrm{ir}} = \mathrm{d}S = 0$	$\mathrm{d}S^{\mathrm{ir}} = \mathrm{d}S = 0$	$\mathrm{d}S^{\mathrm{ir}} = \mathrm{d}S = 0$

Table 2.4 First and second laws of thermodynamics for reversible constant volume processes

	Constant V $\mathrm{d}V = 0$	Constant V and S $\mathrm{d}V = 0$, $\mathrm{d}S = 0$	Constant V and T $\mathrm{d}V = 0$, $\mathrm{d}T = 0$
First law	$\mathrm{d}U = \mathrm{d}Q = T\mathrm{d}S$	$\mathrm{d}U = 0$	$\mathrm{d}F = 0$
Second law	$\mathrm{d}S^{\mathrm{ir}} = 0$	$\mathrm{d}S^{\mathrm{ir}} = 0$	$\mathrm{d}S^{\mathrm{ir}} = 0$

Table 2.5 First and second laws of thermodynamics for reversible constant pressure processes

	Constant p $dp = 0$	Constant p and S $dp = 0,\ dS = 0$	Constant p and T $dp = 0,\ dT = 0$
First law	$dH = dQ = T\,dS$	$dH = 0$	$dG = 0$
Second law	$dS^{ir} = 0$	$dS^{ir} = 0$	$dS^{ir} = 0$

For the first law applied to simple closed systems,

$$dQ = T\,dS^e = \begin{bmatrix} 0 & \text{for all adiabatic processes} \\ dU & \text{for constant volume processes} \\ dH & \text{for constant pressure processes} \end{bmatrix}$$

$$dQ = T\,dS = \begin{bmatrix} 0 & \text{for all reversible adiabatic processes} \\ dU & \text{for reversible constant volume processes} \\ dH & \text{for reversible constant pressure processes} \end{bmatrix}$$

For the second law applied to simple closed systems,

$$dS^{ir} = \begin{bmatrix} 0 & \text{for all reversible processes} \\ dS & \text{for all adiabatic processes} \\ -dU/T & \text{for constant entropy and constant volume processes} \\ -dH/T & \text{for constant entropy and constant pressure processes} \\ -dF/T & \text{for constant temperature and constant volume processes} \\ -dG/T & \text{for constant temperature and constant pressure processes} \end{bmatrix}$$

For finite changes, the first law of thermodynamics can be written as

$$Q = \int T\,dS^e = \begin{bmatrix} 0 & \text{for all adiabatic processes} \\ \Delta U & \text{for constant volume processes} \\ \Delta H & \text{for constant pressure processes} \end{bmatrix}$$

$$Q = \int T\,dS = \begin{bmatrix} 0 & \text{for all reversible adiabatic processes} \\ \Delta U & \text{for reversible constant volume processes} \\ \Delta H & \text{for reversible constant pressure processes} \end{bmatrix}$$

We can also write the second law of thermodynamics for finite changes as

$$\Delta S^{ir} = \begin{cases} 0 & \text{for all reversible processes} \\ \Delta S & \text{for all adiabatic processes} \\ -\int (dU/T) & \text{for constant entropy and constant volume processes} \\ -\int (dH/T) & \text{for constant entropy and constant pressure processes} \\ -\Delta F/T & \text{for constant temperature and constant volume processes} \\ -\Delta G/T & \text{for constant temperature and constant pressure processes} \end{cases}$$

The second law for open systems at constant temperature, constant volume, and constant chemical potential can be written as

$$\Delta S^{ir} = -\frac{\Delta \Xi}{T}$$

where Ξ is the grand potential energy of the system.

Therefore, for isothermal processes, the irreversibility of a process can be simply quantified using either the amount of entropy produced or the magnitude of thermodynamic potential energy change.

Finally, the entropy production rate within a closed system undergoing a given process can also be quantified:

$$\frac{\partial S^{ir}}{\partial t} = \begin{cases} 0 & \text{for all reversible processes} \\ \partial S/\partial t & \text{for all adiabatic processes} \\ -(1/T)(\partial U/\partial t) & \text{for constant entropy and constant volume processes} \\ -(1/T)(\partial H/\partial t) & \text{for constant entropy and constant pressure processes} \\ -(1/T)(\partial F/\partial t) & \text{for constant temperature and constant volume processes} \\ -(1/T)(\partial G/\partial t) & \text{for constant temperature and constant pressure processes} \end{cases}$$

where t is time. Kinetic laws for different processes can be formulated based on either the entropy production rate $\partial S^{ir}/\partial t = (D/T)d\xi/dt$ or the potential energy dissipation rate given by $-(\partial U/\partial t) = Dd\xi/dt$, or $-(\partial H/\partial t) = Dd\xi/dt$, or $-(\partial F/\partial t) = Dd\xi/dt$, or $-(\partial G/\partial t) = Dd\xi/dt$.

2.7 Examples

Example 1 The amount of heat absorbed during melting of one mole of solid to become one mole of liquid at the equilibrium melting temperature and 1 bar is called the heat of melting or heat of fusion. The same amount of heat will be released when one mole of the corresponding liquid is solidified or crystallized to become one mole of solid. We use the convention that the amount of heat exchange between a system and its surrounding is positive if heat is added to the system and is negative if heat is released from the system to its surrounding.

The heat of melting of ice at equilibrium melting temperature 273 K and 1 bar is 6007 J/mol. The molar volumes of water and ice at 273 K and 1 bar are 18.02 cm³/mol and 19.66 cm³/mol, respectively.

For the process in which one mole of water is solidified to become one mole of ice at the fixed temperature of 273 K and fixed pressure of 1 bar,

(a)   What is the amount of heat released from the system to the surrounding?
(b)   What is the amount of entropy released from the system to the surrounding?
(c)   What is the amount of entropy change for the surrounding?
(d)   What is the amount of entropy change for the system?
(e)   What is the amount of entropy produced within the system?
(f)   What is the total entropy change of system plus the surrounding?
(g)   Is the process reversible or irreversible?
(h)   What is enthalpy difference, $h^{water} - h^{ice}$, between one mole of water and one mole of ice at 273 K and 1 bar?
(i)   What is the internal energy difference, $u^{water} - u^{ice}$, between one mole of water and one mole of ice at 273 K and 1 bar?
(j)   What is the entropy difference, $s^{water} - s^{ice}$, between one mole of water and one mole of ice at 273 K and 1 bar?

Let us consider a hypothetical process in which one mole of ice melts to become one mole of water at the fixed temperature of 298 K and fixed pressure of 1 bar, and the surrounding is also at 298 K and 1 bar. For simplicity, assume that the enthalpy and entropy differences between water and ice at 298 K and 1 bar are the same as those at 273 K and 1 bar.

(k)   What is the amount of heat absorbed by the system from the surrounding?
(l)   What is the entropy change for the system?
(m) What is the entropy change for the surrounding?
(n)   What is the entropy exchange between the system and the surrounding?
(o)   What is the total entropy change of the system plus surrounding?
(p)   What is the amount of entropy produced within the system?
(q)   What is the amount of useful energy converted to thermal energy?

Solution

(a)   The amount of heat exchange at the equilibrium melting temperature is

$$Q = -T_m \Delta S^e = -6007 \, J.$$

where T_m is the equilibrium melting temperature 273 K, and ΔS^e is the amount of entropy exchange. Since it is negative, the 6007 J of heat is released from the system to the surrounding during solidification of one mole of water to become ice at the equilibrium melting temperature of 273 K.

The 6007 J represents the amount of thermal energy stored in the liquid water that is transferred to the surrounding when the liquid water solidifies and becomes ice at the melting temperature.

(b) The amount of entropy exchange is given by

$$\Delta S^e = \frac{Q}{T_m} = \frac{-6007\,\text{J}}{273\,\text{K}} = -22.00\,\text{J/K}$$

Since the sign of entropy exchange is negative, the amount of entropy 22.00 J/K is released from the system to the surrounding.

(c) The amount of entropy change for the surrounding is

$$\Delta S_{sur} = -\Delta S^e = 22.00\,\text{J/K}$$

(d) The amount of entropy change for the system resulting from the solidification of one mole of water into one mole of ice at 273 K and 1 bar is

$$\Delta S = \Delta S^e = -22.00\,\text{J/K}$$

(e) The amount of entropy produced within the system is

$$\Delta S^{ir} = \Delta S - \Delta S^e = 0$$

(f) The total entropy change of system plus the surrounding is

$$\Delta S_{tot} = \Delta S + \Delta S_{sur} = \Delta S^{ir} = 0$$

(g) Since $\Delta S_{tot} = \Delta S^{ir} = 0$, the process is reversible.

(h) According to the first law of thermodynamic for a constant pressure process, the enthalpy change for the constant pressure solidification process is equal to the heat involved in the process,

$$\Delta h = h^{ice} - h^{water} = Q = -6007\,\text{J}$$

Therefore, the difference in enthalpy, $h^{water} - h^{ice}$, between one mole of water and one more of ice both at 273 K and 1 bar is

$$h^{water} - h^{ice} = 6007\,\text{J}$$

(i) To determine internal energy difference, $u^{water} - u^{ice}$, between one mole of water and one mole of ice both at 273 K and 1 bar, we use the definition of enthalpy,

$$h^{water} - h^{ice} = \left(u^{water} + pv^{water}\right) - \left(u^{ice} + pv^{ice}\right) = 6007\,\text{J}$$

$$u^{water} - u^{ice} = 6007\,\text{J} + p\left(v^{ice} - v^{water}\right)$$

$$u^{\text{water}} - u^{\text{ice}} = 6007 \, \text{J} + 1 \, \text{bar} \times (19.66 - 18.02) \, \text{cm}^3$$

$$u^{\text{water}} - u^{\text{ice}} = 6007 \, \text{J} + 1.0 \, \text{bar} \times 10^5 \, \text{Pa} \times (19.66 - 18.02) \times 10^{-6} \, \text{m}^3$$
$$= \sim 6007 \, \text{J}$$

Note that, due to the small difference in molar volume between the initial liquid state and the final solid state and the low pressure of 1 bar, the pv term contributes very little to the internal energy difference. Therefore, the internal energy difference, $u^{\text{water}} - u^{\text{ice}}$, between one mole of water and one mole of ice both at 273 K and 1 bar is approximately the same as the enthalpy difference,

$$u^{\text{water}} - u^{\text{ice}} \approx h^{\text{water}} - h^{\text{ice}} = 6007 \, \text{J}$$

(j)   The entropy difference, $s^{\text{water}} - s^{\text{ice}}$, between one mole of water and one mole of ice both at 273 K and 1 bar is

$$s^{\text{water}} - s^{\text{ice}} = 22.00 \, \text{J/K}$$

(k)   For the hypothetical process in which one mole of ice melts to become one mole of water at the fixed temperature of 298 K and fixed pressure of 1 bar, and the surrounding is also at 298 K and 1 bar, the amount of heat absorbed by the system from the surrounding is

$$Q = 6007 \, \text{J}$$

(l)   Since we assume that the enthalpy and entropy differences between water and ice at 298 K and 1 bar are the same as those at 273 K and 1 bar, the entropy change for melting at 298 K is the same as the entropy change at the equilibrium melting temperature,

$$\Delta S = 22.00 \, \text{J/K}$$

(m)   When one mole of ice melts to become one mole of water at the fixed temperature of 298 K and fixed pressure of 1 bar, the entropy change for the surrounding at 298 K is

$$\Delta S_{\text{sur}} = \frac{-Q}{T} = \frac{-6007 \, \text{J}}{298 \, \text{K}} = -20.16 \, \text{J/K}$$

(n)   When one mole of ice melts to become one mole of water at the fixed temperature of 298 K and fixed pressure of 1 bar with the surrounding also at 298 K, the entropy exchange between the system and surrounding is

$$\Delta S^{e} = 20.16 \, \text{J/K}$$

(o)   When one mole of ice melts to become one mole of water at the fixed temperature of 298 K and fixed pressure of 1 bar with the surrounding also at 298 K, the total entropy change of the system plus surrounding is

$$\Delta S_{tot} = \Delta S + \Delta S_{sur} = 22.00 - 20.16 = 1.84 \, \text{J/K}$$

(p)   When one mole of ice melts to become one mole of water at the fixed temperature of 298 K and fixed pressure of 1 bar with the surrounding also at 298 K, the amount of entropy produced within the system is

$$\Delta S^{ir} = 1.84 \, \text{J/K}$$

(q)   The Gibbs free energy change for the process of one mole of ice melting to become one mole of water at the fixed temperature of 298 K and fixed pressure of 1 bar with the surrounding also at 298 K is

$$\Delta G = -T \Delta S^{ir} = -298 \, \text{K} \times 1.84 \, \text{J/K} = -548.32 \, \text{J}$$

and thus, the amount of chemical energy converted to thermal energy at constant temperature and pressure is 548.32 J.

Example 2   One mole of monatomic ideal gas at room temperature 298 K and 1 bar ambient pressure is reversibly and isothermally compressed to half of its original volume. Calculate

(a)   The amount of work done on the system by the surrounding, i.e., the amount of mechanical energy transferred from the surrounding to the system.
(b)   The amount of entropy transfer from the system to the surrounding required for this process in order to maintain the same temperature as the initial temperature for the process.
(c)   The increase in the internal energy of the system during this process.

Solution

(a)   For one mole of ideal gas, the work done on the system by the surrounding through a reversible, isothermal compression of system volume from V_1 to V_2 is given by

$$W = -\int_{V_1}^{V_2} p \, dV = -\int_{V_1}^{V_2} \left(\frac{RT}{V} \right) dV = -RT \ln \frac{V_2}{V_1} = 298 R \ln 2 \approx 1717 \, \text{J}$$

(b)   For an ideal gas, by definition, atoms or molecules have no volume, and they do not interact with each other, so the internal energy is only a function of temperature. Since the system is maintained at a constant temperature, the internal energy change is zero, and the amount of thermal energy (heat) transferred from the system to the surrounding is the same as the amount of mechanical energy transferred from the surrounding to the system, i.e., the net result for the surrounding is the conversion of 1717.32 J of mechanical energy into thermal energy. Since the process is reversible, there is no entropy generated, and the amount of entropy change for the surrounding is the same as the amount of entropy exchange between the system and its surrounding. The amount of entropy exchange is given by

$$\Delta S^e = \frac{Q}{298} = \frac{-W}{298} = -R\ln 2 = -5.763 \text{ J/K}$$

Since the sign of ΔS^e negative, 5.763 J/K is the amount of entropy transferred from the system to the surrounding.

(c)   The increase in the internal energy of an ideal gas for this isothermal process,

$$\Delta U = T\Delta S^e + W = 298\Delta S^e + W = 0$$

2.8 Exercises

1.   Identify whether the following thermodynamic quantities are state functions: T (temperature), V (volume), U (internal energy), p (pressure), N (number of moles), S (entropy), μ (chemical potential), charge (q), electric potential (ϕ).

2.   Write down the equation for the first law of thermodynamics for a constant volume process in a closed system.

3.   Write down the equation for the first law of thermodynamics for a constant pressure process in a closed system.

4.   Write down the equation for the first law of thermodynamics for a reversible adiabatic process in a closed system.

5.   Write down the first law of thermodynamics for an isolated system.

6.   What is the difference between the internal energy and enthalpy for one mole of ideal gas at temperature T?

7.   When a closed system undergoes a reversible, isothermal expansion from volume V_1 to V_2 with $V_2 > V_1$, the total energy change of the system and its surrounding is

(A) > 0,   (B) < 0,   (C) $= 0$,   (D) Cannot be determined

8. When a closed system undergoes an irreversible, isothermal expansion from volume V_1 to V_2 with $V_2 > V_1$, the total energy change of the system and its surrounding is

   (A) > 0, (B) < 0, (C) = 0, (D) Cannot be determined

9. When a closed system undergoes a reversible, adiabatic expansion from volume V_1 to V_2 with $V_2 > V_1$, the internal energy change of the system is

   (A) > 0, (B) < 0, (C) = 0, (D) Cannot be determined

10. When a closed system undergoes a reversible, adiabatic expansion from volume V_1 to V_2 with $V_2 > V_1$, the work done by the system is

   (A) > 0, (B) < 0, (C) = 0, (D) Cannot be determined

11. When a closed system undergoes a reversible, adiabatic expansion from volume V_1 to V_2 with $V_2 > V_1$, the amount of heat absorbed by system is

   (A) > 0, (B) < 0, (C) = 0, (D) Cannot be determined

12. When a closed system undergoes an irreversible, adiabatic expansion against a vacuum from volume V_1 to V_2 with $V_2 > V_1$, the internal energy change of the system is

   (A) > 0, (B) < 0, (C) = 0, (D) Cannot be determined

13. For 1 mol of ideal gas initially at 298 K and 100 kPa in a compressible container, calculate the amount of mechanical energy transfer from the surrounding to the gas, the amount of thermal energy transfer from the gas to the surrounding, and the internal energy change ΔU of the gas in compressing the gas from 100 to 500 kPa reversibly while keeping the temperature constant (an isothermal process). Please note that the internal energy U of an ideal gas is only a function of temperature.

14. A Cu wire with a diameter of 2 mm and a length of 1 m is initially at room temperature (300 K). A constant voltage of 100 V is applied for 1000 s. The temperature of the wire rises due to Joule heating (heating due to the conversion of electric energy to thermal energy) during the application of the voltage, and so we wait for a sufficiently long time after the voltage is removed for the wire to cool back to the room temperature of 300 K. The conductivity of Cu is 6×10^7 S/m (Siemens per meter). Please answer the following questions, assuming the surrounding is at a room temperature of 300 K:

   a. What is the internal energy change for the wire from the very initial state to the final state after the entire process ends?
   b. What is the internal energy change for the surrounding for the entire process?

c.   What is the amount of thermal energy that was generated in the wire and transferred to the surrounding as heat during the entire process?
d.   What is the total internal energy change of the wire plus the surrounding during the entire process?
e.   What is the entropy change for the wire for the entire process?
f.   What is the entropy change for the surrounding for the entire process?
g.   What is the total entropy generated in the wire and transferred to the surrounding during the entire process?
h.   What is the total entropy change of the wire plus the surrounding during the entire process?
i.   Is this entire process reversible or irreversible?

15. Write down the second law of thermodynamics for a reversible adiabatic process in a closed system.

16. Write down the second law of thermodynamics for a reversible constant volume process in a closed system.

17. Write down the second law of thermodynamics for a reversible constant pressure process in a closed system.

18. Write down the second law of thermodynamics for an irreversible constant temperature and constant volume process in a closed system.

19. Write down the second law of thermodynamics for an irreversible constant temperature and constant pressure process in a closed system.

20. When a system undergoes a reversible, adiabatic expansion from volume V_1 to V_2 with $V_2 > V_1$, the entropy change of the system is

(A) > 0,   (B) < 0,   (C) $= 0$,   (D) Cannot be determined

21. When a system undergoes a reversible, isothermal expansion from volume V_1 to V_2 with $V_2 > V_1$, the total entropy change of the system and its surrounding is

(A) > 0,   (B) < 0,   (C) $= 0$,   (D) Cannot be determined

22. When a system undergoes an irreversible, isothermal contraction (volume decrease), the total entropy change of the system plus surrounding is

(A) > 0,   (B) < 0,   (C) $= 0$,   (D) Cannot be determined

23. Consider the entropy change for the isothermal solidification of Al at 1 bar. The equilibrium melting temperature of Al at 1 bar is $T_m = 932$ K. Since solid Al and liquid Al are at equilibrium at T_m, solidification of one mole of Al at T_m is a reversible process. However, solidification of one mole of Al below T_m is an irreversible process since it is a spontaneous, natural process. The amount of thermal energy or heat released from the system during solidification of one mole of Al at T_m, called the heat of solidification, is $-10,700$ J/mol. The corresponding entropy change for the solidification of one mole of Al is called

the entropy of solidification. We assume the heat and entropy of solidification are independent of temperature. We can separate the entropy change, ΔS, for the one mole of Al due to solidification into two contributions, ΔS^e (entropy exchange with its surrounding) and ΔS^{ir} (entropy generated in the system by the solidification). Please answer the following questions:

a. What are ΔS, ΔS^e, and ΔS^{ir} for solidifying one mole of Al at T_m in the system?

b. What are ΔS, ΔS^e, and ΔS^{ir} for solidifying one mole of Al at 800 K in the system?

c. What are the corresponding ΔS and T_m for the surrounding when one mole of Al is solidified at T_m in the system?

d. What are the corresponding ΔS and ΔS^{ir} for the surrounding when one mole of Al is solidified at 800 K in the system?

e. What is the total entropy change, ΔS_{tot}, for the one mole of Al in the system and the surrounding when the one mole of Al is solidified at T_m?

f. What is the enthalpy change of the system, ΔH, when the one mole of Al is solidified at T_m?

g. What is the enthalpy change of the system, ΔH, when the one mole of Al is solidified at 800 K?

h. What is the Gibbs free energy change of the system, ΔG, when the one mole of Al is solidified at T_m?

i. What is the Gibbs free energy change of the system, ΔG, when the one mole of Al is solidified at 800 K?

j. What is the amount of useful energy dissipated when one mole of liquid Al is solidified to become solid Al at T_m?

k. What is the amount of useful energy dissipated when one mole of liquid Al is solidified to become solid Al at 800 K?

Chapter 3
Fundamental Equation of Thermodynamics

3.1 Differential Form of Fundamental Equation of Thermodynamics

One of Gibbs' most important contributions to classical thermodynamics is the establishment of the fundamental equation of thermodynamics by combining the first and second laws of thermodynamics. A fundamental equation describes the relation among the basic thermodynamic variables of equilibrium states in a homogeneous system.

According to the first law of thermodynamics (Eq. 2.4), the internal energy change dU of a simple system due to the exchange of thermal, mechanical, and chemical energies between the system and its surrounding is

$$dU = T dS^e - p dV^e + \mu dN^e \tag{3.1}$$

where dS^e, dV^e, and dN^e are the infinitesimal amount of entropy, volume, and substance transferred from the surrounding to the system at uniform temperature T, pressure p and chemical potential μ for the system, respectively. $T dS^e$, $-p dV^e$, and μdN^e are the corresponding infinitesimal amounts of thermal, mechanical, and chemical energy transferred from the surrounding to the system.

Now let us replace the exchange quantities in Eq. (3.1) using the property changes of the system by utilizing the second law of thermodynamics (Eq. 2.47),

$$dS^e = dS - dS^{ir}, \ dV^e = dV, \ dN^e = dN, \tag{3.2}$$

we have,

$$dU = T\left(dS - dS^{ir}\right) - p dV + \mu dN \tag{3.3}$$

where dS, dV, and dN are the changes in entropy, volume, and the number of moles for the system due to a process, and dS^{ir} is the amount of entropy produced by

© Springer Nature Singapore Pte Ltd. 2022

L.-Q. Chen, *Thermodynamic Equilibrium and Stability of Materials*,

https://doi.org/10.1007/978-981-13-8691-6_3

an irreversible internal process or a set of irreversible internal processes within the system during the exchanges.

We can rewrite Eq. (3.3) as

$$dU = T\,dS - p\,dV + \mu\,dN - T\,dS^{ir} \tag{3.4}$$

or

$$dU = T\,dS - p\,dV + \mu\,dN - D\,d\xi \tag{3.5}$$

with

$$D = -\left(\frac{\partial U}{\partial \xi}\right)_{S,V,N} \tag{3.6}$$

where D is the driving force for an internal irreversible process, and ξ is the extent which can be considered as an order parameter describing the difference between the initial state and final state connected by the process.

If we are only concerned with equilibrium or quasi-static states of a system as the thermodynamic conditions (i.e., temperature, pressure, amount of matter, etc.) change, the system never deviates from equilibrium. A quasi-static process implies that the temperature, pressure, and chemical potential of the system are equal to the external temperature, pressure, and chemical potential, or the difference in these intensive potentials between the system and its surrounding are infinitesimally small, and thus, the equilibrium inside the system is maintained during the entire process. Therefore, the entropy change for the system is the same as the entropy exchange between the system and its surrounding because there is no internal entropy produced as the thermodynamic conditions change, i.e.,

$$T\,dS^{ir} = D\,d\xi = 0. \tag{3.7}$$

Therefore, the energy conservation from the first law of thermodynamics (Eq. 3.1) in the absence of internal entropy generation becomes

$$dU = T\,dS - p\,dV + \mu\,dN \tag{3.8}$$

where S, V, and N are the entropy, volume, and the amount of chemical matter of the system. It is important to realize that now all the quantities in Eq. (3.8) belong to the system; i.e., they represent the properties of the system.

Equation (3.8) is the differential form for the fundamental equation of thermodynamics that relates the seven basic variables U, S, V, N, T, p, and μ of a simple homogeneous system. As discussed in Chap. 1, Eq. (3.8) can also be viewed as the first law of thermodynamics applied to reversible processes of a system described by a point in space with volume dV, entropy dS, and amount of substance dN at temperature T, pressure p, and chemical potential μ. In some textbooks, Eq. (3.8) is also called the combined first and second law of thermodynamics. It should be remembered that Eq. (3.8) is only valid when there are no internal irreversible processes taking place inside the system.

3.2 Integrated Form of Fundamental Equation of Thermodynamics

If the changes in entropy dS, volume dV, and amount of matter dN are performed at fixed temperature T, pressure p, and chemical potential μ for the system, then we can integrate the differential form (given by Eq. 3.8) of the fundamental equation of thermodynamics to obtain the explicit, integrated form for the fundamental equation of thermodynamics,

$$U = TS - pV + \mu N \tag{3.9}$$

From Eq. (3.9), we can see clearly that the total internal energy U of a homogeneous system can be conceptually decoupled to a sum of different forms of energy contributions: thermal (TS), mechanical ($-pV$), and chemical (μN).

There are several ways to interpret the different energy contributions in Eq. (3.9). For example, to establish a system with internal energy U, entropy S, volume V, and number of moles of substance N with the corresponding temperature T, pressure p and chemical potential μ, one can think of S as the amount of entropy that has to be added to the system and TS as the amount of thermal energy required to maintain the temperature of the system at T. Similarly, pV is the amount of energy that the system has to spend to reversibly create a volume of V against external pressure p which is equal to the internal pressure. Another way to interpret the mechanical energy contribution is that V is the amount of spatial volume required to maintain the system at pressure p. Finally, if N is the amount of material, μN is the amount of chemical energy required to maintain the chemical potential of the system at μ.

It should be noted that adding entropy S at fixed volume V and fixed amount of substance N leads to an increase in temperature T while increasing the amount of substance N in the system at fixed volume V and fixed entropy S leads to an increase in chemical potential μ. However, expanding volume V of the system at fixed entropy S and amount of substance N decreases pressure p, thus resulting in a different sign (negative) for the pV term from the (positive) TS and μN terms in the integrated form of the fundamental equation of thermodynamics.

3.3   Equations of States

From the differential form of Eq. (3.8), one can deduce that S, V, and N are the natural variables of U; i.e., the internal energy U of a system is a function of its entropy S, volume V, and number of moles N,

$$dU(S, V, N) = T(S, V, N)dS - p(S, V, N)dV + \mu(S, V, N)dN \qquad (3.10)$$

The corresponding integrated form of the fundamental equation of thermodynamics expressing internal energy U as a function of S, V, and N is

$$U(S, V, N) = T(S, V, N)S - p(S, V, N)V + \mu(S, V, N)N \qquad (3.11)$$

All the thermodynamic properties including the equations of states can be derived from the fundamental equation of thermodynamics (Eq. 3.11) in combination with Eq. (3.10).

The total differential of the fundamental Eq. (3.11) can be written as

$$dU(S, V, N) = \left(\frac{\partial U}{\partial S}\right)_{V,N} dS + \left(\frac{\partial U}{\partial V}\right)_{S,N} dV + \left(\frac{\partial U}{\partial N}\right)_{S,V} dN \qquad (3.12)$$

By comparing Eqs. (3.10) and (3.12), we have

$$\left(\frac{\partial U}{\partial S}\right)_{V,N} = T(S, V, N) \qquad (3.13)$$

$$\left(\frac{\partial U}{\partial V}\right)_{S,N} = -p(S, V, N) \qquad (3.14)$$

$$\left(\frac{\partial U}{\partial N}\right)_{S,V} = \mu(S, V, N) \qquad (3.15)$$

These three equations represent the formal thermodynamic definitions of temperature (thermal potential) T, pressure (mechanical potential) p, and chemical potential μ. According to Eqs. (3.13)–(3.15), temperature is the rate of change in internal energy with respect to entropy at constant volume and constant number of moles, the negative pressure is the rate of change in internal energy with respect to volume at constant entropy and constant number of moles, and chemical potential is the rate of change in internal energy with respect to the number of moles at constant entropy and constant volume.

Equations (3.13)–(3.15) are three equations of state for a given material. Each equation describes a relation among a set of thermodynamic properties of a system at equilibrium states. However, each equation by itself contains less information about a thermodynamic system than the fundamental equation of thermodynamics. One can easily understand this by simply considering mathematics. For example, while one can directly obtain an equation of state from the fundamental equation, it is not possible to fully recover the fundamental equation from one of the equations of states by integration due to the underdetermined integration constant.

3.4 Independent Variables

For a simple system, we have four relations, i.e., Eqs. (3.11) and (3.13)–(3.15), that relate the seven variables U, S, V, N, T, p, and μ, and thus, there are three independent variables. If we fix three of the seven thermodynamic variables, the rest of the four variables can be obtained from the four relations. If we further consider a closed system, i.e., if N is also fixed, the number of independent variables is now reduced to two, and hence, we can choose a maximum of two quantities to independently vary for a closed system. Which two variables that we choose to represent a fundamental equation will depend on the thermodynamic conditions. For example, if we choose S and V as independent variables, $U(S, V)$ is a fundamental equation for a single-component system with a fixed amount of substance N. We can also choose T and p as independent variables and express U as a function of T and p. However, $U(T, p)$ is not a fundamental equation since T and p are not the natural variables of U.

3.5 Alternative Forms of Fundamental Equation of Thermodynamics

We can rewrite Eq. (3.9) in different forms by moving one or two terms from the right-hand side to the left-hand side. For example, if we move the pV term from right to left, we have

$$H = U + pV = TS + \mu N \qquad (3.16)$$

which is the enthalpy of a system. Based on Eq. (3.16), enthalpy H can be written either as $U + pV$ or $TS + \mu N$.

If we move the TS term from the right-hand side to the left-hand side of Eq. (3.9), we get

$$F = U - TS = -pV + \mu N \qquad (3.17)$$

which is the Helmholtz free energy of a system. According to Eq. (3.17), F can be written as either $U - TS$ or $-pV + \mu N$.

If we move both TS and μN terms from the right-hand side to the left-hand side of Eq. (3.9), we obtain

$$\Xi = U - TS - \mu N = -pV \tag{3.18}$$

which is the grand potential energy. According to Eq. (3.18), the grand potential energy is simply the mechanical energy $-pV$.

If we move both pV and TS terms from the right-hand side to the left-hand side of Eq. (3.9), we obtain

$$G = U - TS + pV = \mu N \tag{3.19}$$

which is called the Gibbs free energy or simply free enthalpy of a system. It is easy to see from Eq. (3.19) that Gibbs free energy or sometimes called free enthalpy G is essentially the chemical energy, μN.

We can also rewrite Eq. (3.19) in terms of molar quantities,

$$\mu = G/N = u - Ts + pv \tag{3.20}$$

where u, s, and v are the molar internal energy (U/N), molar entropy (S/N), and molar volume (V/N), respectively. Since Eqs. (3.16)–(3.20) result from the rearrangement of Eq. (3.9), all six equations, Eqs. (3.9) and (3.16)–(3.20), are alternative forms of the fundamental equation of thermodynamics.

3.6   Differential Forms of Alternative Fundamental Equations of Thermodynamics

We can easily write down the differential forms for H, F, Ξ, and G defined by Eqs. (3.16)–(3.19). For example, the differential form of enthalpy H is given by

$$dH = d(U + pV) = dU + pdV + Vdp = TdS - pdV + \mu dN + pdV + Vdp$$
$$dH = TdS + Vdp + \mu dN \tag{3.21}$$

The differential form for the Helmholtz free energy F can be written as

$$dF = d(U - TS) = dU - SdT - TdS = TdS - pdV + \mu dN - SdT - TdS$$
$$dF = -SdT - pdV + \mu dN \tag{3.22}$$

Similarly, the differential form for the grand potential energy Ξ is

$$d\Xi = d(U - TS - \mu N) = dU - SdT - TdS - \mu dN - Nd\mu$$
$$d\Xi = -SdT - pdV - Nd\mu \tag{3.23}$$

Finally, the differential form for the Gibbs free energy is given by

$$dG = d(U - TS + pV) = dU - SdT - TdS + pdV + Vdp$$
$$dG = -SdT + Vdp + \mu dN \qquad (3.24)$$

3.7 Interpretation of Fundamental Equation of Thermodynamics

From Eqs. (3.21)–(3.24), the general functional forms of H, F, Ξ, and G are:

$$H = H(S, p, N) \qquad (3.25)$$

$$F = F(T, V, N) \qquad (3.26)$$

$$\Xi = \Xi(T, V, \mu) \qquad (3.27)$$

$$G = G(T, p, N) \qquad (3.28)$$

Therefore, the natural variables for enthalpy H are S, p, N; the natural variables for Helmholtz free energy F are T, V, and N; the natural variables for the grand potential energy Ξ are T, V, and μ; and the natural variables for Gibbs free energy G are T, p, N. These equations are summarized in Table 3.1.

There are other ways that one can interpret the different forms of fundamental equations of thermodynamics. For example, a system under constant pressure can be thought of as a system being subject to a mechanical potential field $(-p)$. Then enthalpy, $U + pV = H$, can be understood as the sum of the internal energy (U) and

Table 3.1 Several common expressions for the fundamental equation of thermodynamics and their differential forms

Energy functions	Integrated form	Differential form
Internal energy (U)	$U = TS - pV + \mu N$	$dU =$ $TdS - pdV + \mu dN$
Enthalpy (H)	$H = U + pV =$ $TS + \mu N$	$dH =$ $TdS + Vdp + \mu dN$
Helmholtz free energy (F)	$F = U - TS =$ $-pV + \mu N$	$dF = -SdT -$ $pdV + \mu dN$
Grand potential energy (Ξ)	$\Xi = U - TS -$ $\mu N = -pV$	$d\Xi = -SdT -$ $pdV - Nd\mu$
Gibbs free energy (G)	$G =$ $U - TS + pV = \mu N$	$dG = -SdT +$ $Vdp + \mu dN$

mechanical interaction energy, pV, between the volume (V) of the system and the mechanical field $(-p)$ specified by the surrounding. The interaction energy is given by $-Xx$ where X is an internal extensive parameter, e.g., volume V, and x is an external potential, e.g., mechanical potential field, $-p$. Another way to understand enthalpy is to consider enthalpy as the available energy to do useful work under constant entropy and pressure; i.e., the available energy to do work under constant pressure is the internal energy $(U$, the total amount of energy under pressure $p)$ plus the volume mechanical energy (pV). Under a constant volume, the mechanical energy pV is not available to do useful work, but at constant pressure, there will be an additional energy pV available to perform useful work that should be added to the internal energy $(H = U + pV)$.

We next consider how to interpret the Helmholtz free energy F. It is the maximum available energy that can be converted to work under constant temperature and volume. If the walls between a system and its surrounding are allowed to conduct heat, and if the system is connected to a heat bath, the system is considered to be at constant temperature. We can then think of this system at constant temperature as being subject to a thermal potential (T). The Helmholtz energy or Helmholtz free energy of the system can be understood as the sum of the internal energy (U) and the thermal interaction energy between the entropy of a system and the temperature specified by the surrounding, $-TS$, i.e., $U-TS = F$. The thermal energy "TS" contribution to the total energy is not free or not available to do work, so the free or available energy is $F = U-TS$. The amount of thermal energy TS (and thus the amount of entropy S) is required to maintain the system at temperature T. Or one may think that if the temperature of a system is fixed at a constant value, the amount of thermal energy TS is not available to perform any useful work, and thus, it should be deducted from the internal energy to obtain the available energy. A final way to understand why F is an appropriate thermodynamic free energy for a system at constant temperature and constant volume is the fact that a system always wants to minimize its internal energy and maximize its entropy toward equilibrium. The combination of minimizing internal energy and maximizing entropy at constant temperature and constant volume is the Helmholtz free energy, $U-TS$.

The Gibbs free energy, $G = U-ST + pV$, of a system represents the available energy to do useful work at constant T and p; i.e., the available energy or free energy to do work is the total energy (U) minus the thermal energy (TS) required to maintain the system temperature at T and plus the available volume mechanical energy (pV) under the system pressure p. The thermal energy "TS" contribution to the total enthalpy is not free or available to do work, so the Gibbs free energy $G = H-TS$ is sometimes also called the free or available enthalpy to do useful work at constant temperature and pressure. The amount of thermal energy TS (and thus the amount of entropy S) is required to maintain the system at temperature T. Another way to understand why G is an appropriate thermodynamic free energy for a system at constant temperature and pressure is the fact that a system always wants to lower its enthalpy and raise its entropy toward equilibrium. The combination of minimizing enthalpy and maximizing entropy at constant temperature and constant pressure is the Gibbs free energy, $H-TS$.

Finally, it should be noted that the Gibbs free energy G is essentially the chemical energy μN of a system, and hence, the thermodynamics based on the chemical energy is sometimes also called chemical thermodynamics. Similarly, the grand potential energy Ξ is simply the mechanical energy term $-pV$. The value of Ξ can be understood as the available energy of the system if we reduce the volume of the system down to zero. Since Ξ is negative, $-pV$, it implies that it actually costs energy to make the volume of the system to go to zero.

Which form of the fundamental equation of thermodynamics to be employed for a given thermodynamic problem depends entirely on the thermodynamic condition. For example, internal energy, U, is the most useful thermodynamic potential energy function under constant entropy and volume, Helmholtz free energy is the most useful thermodynamic potential energy function under constant temperature and volume, enthalpy is the most useful thermodynamic potential energy function under constant entropy and pressure, and Gibbs free energy is the most useful thermodynamic potential energy function under constant temperature and pressure. The most common thermodynamic condition in practice is constant temperature and constant pressure as temperature and pressure are easier to control than their conjugate variables such as entropy and volume. Therefore, the Gibbs free energy, $G(T, p, N)$, is by far the most commonly used thermodynamic potential energy function as the fundamental equation of thermodynamics for a closed system since by minimizing the Gibbs free energy, one can obtain the equilibrium states. Under constant T and p, minimization of other thermodynamic potential energy functions such as internal energy U, Helmholtz free energy F, and enthalpy H will not lead to the equilibrium states.

In materials science, the majority of the thermodynamic calculations are performed on a one-mole basis. Therefore, since $G = \mu N$, and for one mole of materials, $G = \mu$, it is sufficient to employ chemical potentials to determine the equilibrium states of a material rather than Gibbs free energy. As a matter of fact, chemical potential μ is a more convenient quantity than the Gibbs free energy G to determine the stability of a component or a combination of components in materials. This is due to the fact that the Gibbs free energy G is proportional to the amount of substance N while chemical potential μ is independent of size. For example, if we have two pieces of the same materials with different sizes. If we ignore the surface energy contribution or the size effect, all the chemical species or components have the same thermodynamic stability in the two pieces of materials since their chemical potentials are equal. However, their Gibbs free energy in these two pieces of materials is different since their amounts are different in these two pieces of materials, and thus comparing their Gibbs free energies will not be able to determine their relative stability in these two pieces of materials.

3.8   Gibbs–Duhem Relation

If we move all the terms in the right-hand side of the fundamental equation of thermodynamics (Eq. 3.9) to the left-hand side, we have

$$U - TS + pV - \mu N = 0 \tag{3.29}$$

The differential form of Eq. (3.29) is then

$$-SdT + Vdp - Nd\mu = 0, \tag{3.30}$$

which is called the Gibbs–Duhem relation relating the three potentials T, p, and μ of a system.

If we divide both sides of Eq. (3.30) by N, we have

$$-sdT + vdp - d\mu = 0 \tag{3.31}$$

We can rewrite equation the above equation as

$$d\mu = -sdT + vdp, \tag{3.32}$$

Equation (3.32) is another way of writing the Gibbs–Duhem relation to express the dependence of chemical potential on temperature and pressure of a simple system. Therefore, only two of the three potentials, T, p, and μ, are independent for a simple system. Once we fix two of the potentials, e.g., temperature and pressure, we can obtain the value of the third potential, e.g., the chemical potential.

3.9   Entropic Representation of Fundamental Equation of Thermodynamics

The differential and integrated forms of the fundamental equation of thermodynamics can also be written in the entropic representation by rewriting Eqs. (3.8) and (3.9) expressing entropy S as a function of internal energy U, volume V, and the number of moles N,

$$dS = \left(\frac{1}{T}\right)dU + \left(\frac{p}{T}\right)dV - \left(\frac{\mu}{T}\right)dN \tag{3.33}$$

$$S = \left(\frac{1}{T}\right)U + \left(\frac{p}{T}\right)V - \left(\frac{\mu}{T}\right)N \tag{3.34}$$

Using the relation between internal energy U and enthalpy H, Eqs. (3.33) and (3.34) can also be written as

$$dS = \left(\frac{1}{T}\right)dH - \left(\frac{V}{T}\right)dp - \left(\frac{\mu}{T}\right)dN \tag{3.35}$$

$$S = \left(\frac{1}{T}\right)H - \left(\frac{\mu}{T}\right)N \tag{3.36}$$

As fundamental equations, Eqs. (3.34) and (3.36) also contain all the information about the thermodynamic properties of a system. According to Eqs. (3.33) and (3.35), the natural variables for entropy S for a closed system can be either U and V or H and p.

From Eq. (3.33), we can write down the corresponding equations of state in the entropic representation,

$$\left(\frac{\partial S}{\partial U}\right)_{V,N} = \frac{1}{T} \tag{3.37}$$

$$\left(\frac{\partial S}{\partial V}\right)_{U,N} = \frac{p}{T} \tag{3.38}$$

$$\left(\frac{\partial S}{\partial N}\right)_{U,V} = -\frac{\mu}{T} \tag{3.39}$$

Similarly, we can write down another set of equations of state from Eq. (3.35),

$$\left(\frac{\partial S}{\partial H}\right)_{p,N} = \frac{1}{T} \tag{3.40}$$

$$\left(\frac{\partial S}{\partial p}\right)_{H,N} = -\frac{V}{T} \tag{3.41}$$

$$\left(\frac{\partial S}{\partial N}\right)_{H,V} = -\frac{\mu}{T} \tag{3.42}$$

We can also define other forms of fundamental equations of thermodynamics by changing the natural variables. For example,

$$\frac{F}{T} = \frac{U - TS}{T} = \left(\frac{1}{T}\right)U - S \tag{3.43}$$

$$\frac{G}{T} = \frac{U + pV - TS}{T} = \left(\frac{1}{T}\right)U + \left(\frac{p}{T}\right)V - S = \left(\frac{1}{T}\right)H - S \tag{3.44}$$

The differential forms for F/T and G/T are

$$d\left(\frac{F}{T}\right) = Ud\left(\frac{1}{T}\right) - \left(\frac{p}{T}\right)dV + \left(\frac{\mu}{T}\right)dN \qquad (3.45)$$

$$d\left(\frac{G}{T}\right) = Hd\left(\frac{1}{T}\right) + \left(\frac{V}{T}\right)dp + \left(\frac{\mu}{T}\right)dN \qquad (3.46)$$

Therefore, the natural variables for F/T are $1/T$, V, and N, and the natural variables for G/T are $1/T$, p, and N. One can easily see from Eqs. (3.45) and (3.46) that

$$\left[\frac{\partial(F/T)}{\partial(1/T)}\right]_{V,N} = U \qquad (3.47)$$

and

$$\left[\frac{\partial(G/T)}{\partial(1/T)}\right]_{p,N} = H \qquad (3.48)$$

Equations (3.47) and (3.48) are useful for obtaining the internal energy as a function of temperature from the temperature dependence of Helmholtz free energy or for obtaining the enthalpy as a function of temperature from the temperature dependence of Gibbs free energy.

3.10 Fundamental Equations of Thermodynamics Including Irreversible Internal Processes

We can write down the corresponding different forms of the fundamental equation of thermodynamics including an irreversible internal process or a set of internal processes. They are summarized in Table 3.2.

All these equations can be readily extended to multicomponent and multiple internal processes.

Table 3.2 Differential forms of fundamental equation of thermodynamics including an irreversible internal process with the extent of the process described by an order parameter ξ

Energy functions	Differential form
$U = U(S, V, N, \xi)$	$dU = TdS - pdV + \mu dN - Dd\xi$
$H = H(S, p, N, \xi)$	$dH = TdS + Vdp + \mu dN - Dd\xi$
$F = F(T, V, N, \xi)$	$dF = -SdT - pdV + \mu dN - Dd\xi$
$\Xi = \Xi(T, V, \mu, \xi)$	$d\Xi = -SdT - pdV - Nd\mu - Dd\xi$
$G = G(T, p, N, \xi)$	$dG = -SdT + Vdp + \mu dN - Dd\xi$

3.11   General Forms of Fundamental Equations of Thermodynamics

For systems subject to other potential fields such as electric potential, electric field, magnetic field, gravitational field, and stress field, the total energy of a system should include all the interaction energies with its surrounding, and the corresponding energy function has been given different names such as electric free energy or enthalpy, and magnetic free energy.

A more general integrated form for the fundamental equation of thermodynamics is

$$U = TS + \mu N + \gamma A + q\phi + V\sigma_{ij}\varepsilon_{ij} + VE_iP_i + VH_iM_i + gzm \qquad (3.49)$$

where γ and A are the specific surface energy and surface area, σ_{ij} and ε_{ij} are ijth components of stress and strain, P_i and E_i are the ith components of electrical field and electric polarization, H_i and M_i are the ith components of magnetic field and magnetization, g in this expression is the gravitational acceleration constant, z is the height, and m is mass. Here we use the so-called Einstein summation convention, where repeating indices in the same term in Eq. (3.49) imply summation over those indices, e.g.,

$$E_iP_i = E_1P_1 + E_2P_2 + E_3P_3 \qquad (3.50)$$

The differential form for Eq. (3.49) is

$$dU = TdS + \mu dN + \gamma dA + \phi dq + V\sigma_{ij}d\varepsilon_{ij} + VE_idP_i + VH_idM_i + gzdm \qquad (3.51)$$

In general, a material can assume different homogenous states of a substance, e.g., solid, liquid, gas, or a solid with different crystal structures. Each homogeneous state of a substance is described by a separate fundamental equation of thermodynamics. It is possible to link the fundamental equations of thermodynamics for each possible state together to obtain an overall fundamental equation of thermodynamics for a multiphase heterogeneous state using order parameters that describe the relations among the different possible phases for a particular materials system.

One can also define a set of fundamental equations on the per unit volume basis. For example,

$$u_v = Ts_v + \mu c + \gamma A_v + \phi \rho_q + \sigma_{ij}\varepsilon_{ij} + E_iP_i + H_iM_i + gz\rho_m \qquad (3.52)$$

where u_v is internal energy density, s_v entropy density, c concentration (moles per unit volume), ρ_q is charge density, and ρ_m is mass density. The differential form for Eq. (3.49) is

$$du_v = T\,ds_v + \mu\,dc + \gamma\,dA_v + \phi\,d\rho_q + \sigma_{ij}\,d\varepsilon_{ij} + E_i\,dP_i + H_i\,dM_i + gz\,d\rho_m$$
$$(3.53)$$

3.12 Legendre Transforms

The alternative forms of the fundamental equation of thermodynamics can be beautifully connected using mathematics based on the Legendre transform.

For simplicity, let us consider a closed system in which the amount of substance is fixed. In this case, the number of independent thermodynamic variables of a simple system is two. The internal energy, $U = U(S, V)$, the Helmholtz free energy, $F = F(T, V)$, enthalpy, $H = H(S, p)$, and Gibbs free energy, $G = G(T, p)$, are all equivalent forms of the same fundamental equation of thermodynamics. They contain the same, complete thermodynamic information for a particular system. All thermodynamic properties including the equations of states can be derived from the fundamental equation of thermodynamics.

To explain the Legendre transforms, let us first consider a curve in two dimensions, $y(x)$ (Fig. 3.1). There are two equivalent ways to describe such a curve. The first is simply by the continuous points (x, y) described by $y(x)$. The second is the envelope formed by the slope (dy/dx) of the curve at (x, y), and the intercept of the tangent at (x, y) with the y-axis, $z = y-(dy/dx)x$. In the slope-intercept description, the function is the intercept and the variable is the slope; i.e., the intercept z is a function of slope dy/dx, or $z(dy/dx)$. Therefore, the following two functions,

$$y = y(x)$$

Or

$$z = y - \frac{dy}{dx}x = z\left(\frac{dy}{dx}\right) \qquad (3.54)$$

Fig. 3.1 Illustration of a curve described by $[x, y(x)]$ or $[dy/dx, y - (dy/dx)x]$

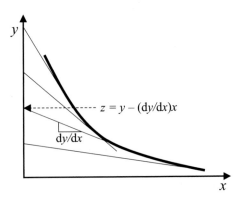

are two equivalent descriptions of the same curve. The transition of a function from $y(x)$ (y as a function of x) to $z(dy/dx)$ (z as a function of dy/dx) is called a Legendre transform in mathematics.

For example, the fundamental equation can be described by the internal energy $U = U(S, V)$ or by its Legendre transform, $F[(\partial U/\partial S)_V, V]$, the Helmholtz free energy. $F[(\partial U/\partial S)_V, V]$ can be considered as a Legendre transform of $U = U(S, V)$ by changing the variable from S to T,

$$F\left[\left(\frac{\partial U}{\partial S}\right)_V, V\right] = U(S, V) - \left(\frac{\partial U}{\partial S}\right)_V S \tag{3.55}$$

Substituting temperature T for $(\partial U/\partial S)_V$,

$$T = \left(\frac{\partial U}{\partial S}\right)_V, \tag{3.56}$$

we have

$$F(T, V) = U(S, V) - TS \tag{3.57}$$

where $F(T, V) = U - TS$ is the intercept of the tangent line to the curve of $U = U(S, V)$ intersected by a $V = $ constant plane, and T is the slope of the tangent line $T = (\partial U/\partial S)_V$.

To summarize the procedure of performing a Legendre transform from $U(S, V)$ to $F(T, V)$ by replacing the variable S with $T = (\partial U/\partial S)_V$, we perform the following steps:

Step 1: Perform the following Legendre transform,

$$F = F(T, V) = U(S, V) - \left(\frac{\partial U}{\partial S}\right)_V S = U(S, V) - TS \tag{3.58}$$

We then need to eliminate S from the above equation to obtain F as a function of its natural variables T and V. It is reminded that T in Eq. (3.58) is also a function of S and V.

Step 2: Solve the equation $T = (\partial U/\partial S)_V$ for S as a function of T and V, i.e.,

$$T = \left(\frac{\partial U}{\partial S}\right)_V = T(S, V) \Rightarrow S(T, V) \tag{3.59}$$

Step 3: Using the expression $S(T, V)$ to replace S in Eq. (3.58) to obtain $F(T, V)$,

$$F(T, V) = U(S(T, V), V) - TS(T, V) \tag{3.60}$$

Similarly, $H(S, V)$ can be obtained from $U(S, V)$ by replacing the variable V with $p = -(\partial U/\partial V)_S$ through the following Legendre transform,

$$H = H(S, p) = U(S, V) - \left(\frac{\partial U}{\partial V}\right)_S V = U + pV \tag{3.61}$$

Finally, the Gibbs free energy $G = G(T, p)$ can be obtained by a Legendre transformation of $U = U(S, V)$ by replacing the variable S with T and the variable V with p,

$$G = G(T, p) = U(S, V) - \left(\frac{\partial U}{\partial S}\right)_V S - \left(\frac{\partial U}{\partial V}\right)_S V = U - TS + pV \tag{3.62}$$

Therefore, mathematically, the different fundamental equations of thermodynamics, $U(S, V)$, $F(T, V)$, $H(S, p)$, and $G(T, p)$ describe the same relation among all the thermodynamic variables of an equilibrium state. Since they contain all the thermodynamic information, they are all called the fundamental equations of thermodynamics of a closed system at equilibrium.

As discussed in Sect. 3.7, different thermodynamic energy functions can also be understood in terms of interactions of a system with its surrounding. The thermodynamic equilibrium of a system that interacts with a potential or field is determined by the total energy U_{tot} which is the sum of internal energy U of the system and the interaction energy, $-\sum x_i X_i$, i.e.,

$$U_{tot} = U - \sum x_i X_i \tag{3.63}$$

where $-x_i X_i$ is an interaction energy, X_i is an extensive parameter such as entropy (S), volume (V), charge (q), electric polarization (P), and x_i is a potential or field such as temperature T (thermal potential), $-p$ (mechanical potential), ϕ (electric potential), E (electric field), etc. Based on the combined first and second law of thermodynamics, or the differential form of the fundamental equation of thermodynamics,

$$x_i = \left(\frac{\partial U}{\partial X_i}\right)_{X_{j \neq i}}, \tag{3.64}$$

so the Legendre-transformed new function $U'(x_i)$ is given by

$$U'(x_i) = U(X_i) - \sum x_i X_i \tag{3.65}$$

Each term in the second interaction term at the right-hand side of Eq. (3.65) represents a Legendre transform of function of U to U' and variable from X_i to x_i.

3.13 Examples

Example 1 The chemical potential μ (on a per atom basis rather than per mole) or chemical energy per atom u_C of a monatomic ideal gas as a function of temperature T and pressure p is given by

$$u_C(T, p) = \mu(T, p) = -k_B T \left\{ \ln \left[\frac{k_B T}{p} \left(\frac{2\pi m k_B T}{h^2} \right)^{3/2} \right] \right\}$$

where k_B is Boltzmann constant, m is atomic mass (mass per atom), and h is the Planck constant.

(a) Show that the atomic volume or volume per atom of the monatomic ideal gas is given by

$$v = \frac{k_B T}{p}$$

(b) Show that the entropy per atom of the monatomic gas is given by

$$s = k_B \left\{ \frac{5}{2} + \ln \left[\frac{k_B T}{p} \left(\frac{2\pi m k_B T}{h^2} \right)^{3/2} \right] \right\}$$

(c) Show that the enthalpy per atom of the monatomic gas is given by

$$h = \frac{5}{2} k_B T$$

(d) Show that the internal energy per atom of the monatomic gas is given by

$$u = \frac{3}{2} k_B T$$

(e) Show that the mechanical energy or grand potential energy per atom of the monatomic gas is given by

$$u_M = -pv = -k_B T$$

(f) Show that the thermal potential energy per atom of the monatomic gas is given by

$$u_T = k_B T \left\{ \frac{5}{2} + \ln \left[\frac{k_B T}{p} \left(\frac{2\pi m k_B T}{h^2} \right)^{3/2} \right] \right\}$$

(g)   Based on the above results, verify that $u = u_T + u_M + u_C$.

Solution

(a)   Based on the Gibbs–Duhem relation (Eq. 3.32), the atomic volume or volume per atom of the monatomic ideal gas can be obtained from the chemical potential as a function of temperature and pressure,

$$v = \left(\frac{\partial \mu}{\partial p}\right)_T = \frac{k_B T}{p}$$

(b)   From the Gibbs–Duhem relation (Eq. 3.32), the entropy per atom of the monatomic gas can be obtained from the chemical potential dependence on temperature and pressure,

$$s = -\left(\frac{\partial \mu}{\partial T}\right)_p = k_B \left\{ \frac{5}{2} + \ln\left[\frac{k_B T}{p}\left(\frac{2\pi m k_B T}{h^2}\right)^{3/2}\right]\right\}$$

(c)   The enthalpy per atom of the monatomic gas is

$$h = \mu + Ts = \frac{5}{2}k_B T$$

(d)   The internal energy per atom of the monatomic gas is

$$u = h - pv = \frac{3}{2}k_B T$$

(e)   The mechanical energy or grand potential energy per atom of the monatomic gas is

$$u_M = -pv = -k_B T$$

(f)   The thermal potential energy per atom of the monatomic gas is

$$u_T = Ts = k_B T \left\{ \frac{5}{2} + \ln\left[\frac{k_B T}{p}\left(\frac{2\pi m k_B T}{h^2}\right)^{3/2}\right]\right\}$$

(g)   Based on the above results, we have

$$u = u_T + u_M + u_C = \frac{3}{2}k_B T$$

Example 2 Given the following fundamental equation of thermodynamics expressing internal energy U of a monatomic ideal gas as a function of entropy

S, volume V, and number of atoms N,

$$U(S, V, N) = \frac{3h^2 N}{4\pi m} \left(\frac{N}{V}\right)^{2/3} e^{\left(\frac{2S}{3Nk_B} - \frac{5}{3}\right)}$$

where k_B Boltzmann constant, h the Planck constant, and m is atomic mass, please express:

(a) Enthalpy H as a function of entropy S, pressure p, and number of atoms N, $H(S, p, N)$.
(b) Helmholtz free energy F as a function of entropy T, pressure V, and number of atoms N, $F(T, V, N)$.
(c) Gibbs free energy G as a function of entropy T, pressure p, and number of atoms N, $G(T, p, N)$.

Solution

(a) To obtain $H(S, p, N)$, we perform the following Legendre transform,

$$H(S, p, N) = U(S, V, N) - \left(\frac{\partial U}{\partial V}\right)_{S,N} V = U(S, V, N) + pV$$

We need to eliminate V from the above equation by expressing V as a function of S, p, and N using the following relation

$$-p = \left(\frac{\partial U}{\partial V}\right)_{S,N} = -\frac{h^2}{2\pi m}\left(\frac{N}{V}\right)^{5/3} e^{\left(\frac{2S}{3Nk_B} - \frac{5}{3}\right)}$$

$$V = N\left(\frac{h^2}{2\pi mp}\right)^{3/5} e^{\left(\frac{2S}{5Nk_B} - 1\right)}$$

Substituting the above expression for V into the following equation

$$H(S, p, N) = \frac{3h^2 N}{4\pi m}\left(\frac{N}{V}\right)^{2/3} e^{\left(\frac{2S}{3Nk_B} - \frac{5}{3}\right)} + pV$$

we have

$$H(S, p, N) = \frac{5}{2}Np^{2/5}\left(\frac{h^2}{2\pi m}\right)^{3/5} e^{\left(\frac{2S}{5Nk_B} - 1\right)}$$

(b) To obtain $F(T, V, N)$, we perform the following Legendre transform,

$$F(T, V, N) = U(S, V, N) - \left(\frac{\partial U}{\partial S}\right)_{V,N} S = U(S, V, N) - TS$$

We need to eliminate S from the above equation by expressing S as a function of T, V, and N using the following relation

$$T = \left(\frac{\partial U}{\partial S}\right)_{V,N} = \frac{3h^2 N}{4\pi m}\left(\frac{N}{V}\right)^{2/3} e^{\left(\frac{2S}{3Nk_B} - \frac{5}{3}\right)} \times \frac{2}{3Nk_B}$$

$$S = Nk_B\left\{\frac{5}{2} + \ln\left[\frac{V}{N}\left(\frac{2\pi mk_B T}{h^2}\right)^{3/2}\right]\right\}$$

Substitute the above expression for S into the following equation,

$$F(T, V, N) = \frac{3h^2 N}{4\pi m}\left(\frac{N}{V}\right)^{2/3} e^{\left(\frac{2S}{3Nk_B} - \frac{5}{3}\right)} - TS$$

We have

$$F(T, V, N) = -Nk_B T\left\{1 + \ln\left[\frac{V}{N}\left(\frac{2\pi mk_B T}{h^2}\right)^{3/2}\right]\right\}$$

(c)   To obtain $G(T, p, N)$, we perform the following Legendre transform,

$$G(T, p, N) = F(T, V, N) - \left(\frac{\partial F}{\partial V}\right)_{T,N} V = F(T, V, N) + pV$$

We need to eliminate V from the above equation by expressing V as a function of T, p, and N using the following relation,

$$-p = \left(\frac{\partial F}{\partial V}\right)_{T,N} = -\frac{Nk_B T}{V}$$

$$V = \frac{Nk_B T}{p}$$

Substituting the above expression for V into the following equation,

$$G(T, p, N) = \frac{3Nk_B T}{2} - Nk_B T\left\{\frac{5}{2} + \ln\left[\frac{RT}{p}\left(\frac{2\pi mk_B T}{h^2}\right)^{3/2}\right]\right\} + pV$$

we have

$$G(T, p, N) = \frac{5Nk_B T}{2} - Nk_B T\left\{\frac{5}{2} + \ln\left[\frac{k_B T}{p}\left(\frac{2\pi mk_B T}{h^2}\right)^{3/2}\right]\right\}$$

or

$$G(T, p, N) = -Nk_B T \left\{ \ln \left[\frac{k_B T}{p} \left(\frac{2\pi m k_B T}{h^2} \right)^{3/2} \right] \right\}$$

3.14 Exercises

1. For a monatomic ideal gas with atomic mass m, the fundamental equation of thermodynamics in the entropic representation is given by the Sackur–Tetrode equation,

$$S = Nk_B \left\{ \frac{5}{2} + \ln \left[\frac{V}{N} \left(\frac{4\pi m U}{3h^2 N} \right)^{3/2} \right] \right\}$$

where S is entropy, N the number of atoms, k_B is Boltzmann constant, V volume, U internal energy, and h is the Planck constant.

(a) Using the Sackur–Tetrode equation, express the internal energy U as a function of temperature.

(b) Using the Sackur–Tetrode equation, derive the ideal gas law (an equation of state for an ideal gas).

(c) Using the Sackur–Tetrode equation, derive an expression for the chemical potential, μ, as a function of temperature and pressure.

(d) Using the Sackur–Tetrode equation, derive the internal energy, U, as a function of S, V, and N.

(e) Derive an expression for the enthalpy, H, of monatomic gas as a function of S, p, and N.

(f) Derive an expression for the enthalpy, H, of monatomic gas as a function of T, p, and N.

(g) Derive an expression for the Helmholtz free energy, F, of monatomic gas as a function of T, V, and N.

(h) Derive an expression for the Gibbs free energy or free enthalpy, G, of monatomic gas as a function of T, p, and N.

(i) Derive an expression for the grand potential energy, Ξ, of monatomic gas as a function of T, V, μ.

(j) Based on your results from part (c) and (h), write down the Gibbs free energy, G, as a function of chemical potential, μ, and the number of moles.

(k) Derive an expression for the entropy (S) of a monatomic ideal gas as a function of T, p, and N.

(l) Derive an expression for the Helmholtz free energy (F) of a monatomic ideal gas as a function of T, p, and N.

2. Given the vibrational Helmholtz free energy of a crystal,

$$F(T) = Nu_o - 3Nk_B T \ln \frac{\exp(-\theta_E/2T)}{1 - \exp(-\theta_E/T)}$$

where u_o and θ_E are constants independent of temperature, k_B is Boltzmann constant, N is the number of atoms, and T is temperature.

(a) Derive an expression for the entropy as a function of temperature T and number of atoms N.

(b) Derive an expression for the internal energy as a function of temperature T and N.

(c) Derive an expression for the chemical potential as a function of temperature T.

3. The fundamental equation of thermodynamics of a monatomic real gas with atomic mass m in terms of the Helmholtz free energy can be approximated as

$$F(T, V, N) = -\frac{aN^2}{V} - Nk_BT\left\{1 + \ln\left[\frac{(V - Nb)}{N}\left(\frac{2\pi mk_BT}{h^2}\right)^{3/2}\right]\right\}$$

where a and b are constants, N the number of atoms, T temperature, k_B the Boltzmann constant, V volume, and h the Planck constant.

(a) Derive an expression for entropy S as a function of T, V, and N.

(b) Derive an expression for pressure p as a function of T, V, and N.

(c) Derive an expression for internal energy U as a function of T, V, and N.

4. The Gibbs free energy of N electrons occupying the conduction band of a semiconductor can be approximated as

$$G(T, p, N) = NE_c - Nk_BT\left\{\ln\left[\frac{2k_BT}{p}\left(\frac{2\pi mk_BT}{h^2}\right)^{3/2}\right]\right\}$$

where E_c is the energy of conduction band bottom, k_B the Boltzmann constant, T is temperature, p pressure, m the effective mass of electrons, and h the Planck constant.

(a) Derive an expression for the entropy S of the conduction band electrons as a function of U, V, and N.

(b) Derive an expression for the Helmholtz free energy F of the conduction band electrons as a function of T, V, and N.

(c) Derive an expression for the chemical potential, often called the Fermi Level, of the conduction band electrons as a function of T and p.

(d) Derive an expression for the internal energy U of the conduction band electrons as a function of T and N.

(e) Derive an expression for the enthalpy H of the conduction band electrons as a function of T and N.

5. The Helmholtz free energy of a blackbody radiation contained in a volume V at temperature T is given by

$$F(T, V) = -\frac{4\sigma}{3c} V T^4$$

where σ is the Stefan–Boltzmann constant, and c is the speed of light in vacuum.

(a) Find entropy S as a function of T and V.
(b) Find internal energy U as a function of T and V.
(c) Find internal energy U as a function of S and V.
(d) Find entropy S as a function of U and V.
(e) Find enthalpy H as a function of T and V.
(f) Find enthalpy H as a function of S and p.
(g) Find grand potential energy Ξ as a function of T and V.
(h) Find pressure p as a function of T.
(i) Based on above answers, what is the Gibbs free energy of the blackbody radiation?
(j) Based on above answers, what is the chemical potential of the blackbody radiation?

6. Compare the molar quantities, internal energy (u), molar enthalpy (h), molar entropy (s), and chemical potential (μ) of ice and water at $0\ °C$ and 1 bar by indicating whether ice or water has a larger magnitude for a given quantity.

7. The equilibrium state of a system is determined by the competition between its internal energy (or enthalpy) minimization and entropy maximization. Since the thermal energy of a system is product of entropy and temperature, entropy maximization dominates over internal energy (or enthalpy minimization) at high temperatures while internal energy (or enthalpy) minimization dominates over entropy maximization at low temperatures. Using your knowledge about liquid and solid states of a material and intuitive reasoning, please explain why

(a) The Gibbs free energy of both solid and liquid decreases with temperature at a constant pressure.
(b) A liquid always exists at higher temperatures than the corresponding solid state at a given pressure.

Chapter 4
Introduction to Statistical Thermodynamics

All of our discussions on the first and second laws of thermodynamics and fundamental equations of thermodynamics so far are concerned with macroscopic systems. A macroscopic thermodynamic system contains a large number, on the order of Avogadro's number $\sim 10^{23}$, of particles such as atoms, molecules, and electrons. At any given moment of time, there are a huge number of possible spatial arrangements of such large number of particles within a system since atoms vibrate around their local equilibrium positions in a solid or move around in a fluid at finite temperatures. The experimentally measured macroscopic properties of a system under a given thermodynamic condition are the averages of the properties of all the instantaneous configurations over the time duration of the experimental measurement; i.e., the experimental device automatically does the statistical averaging of the properties over the large number of temporal microscopic configurations. The measured properties are dependent on the microscopic interactions among the particles within the system as well as the interactions between the system and its surrounding specified by the thermodynamic conditions in terms of temperature or entropy, pressure or volume, and chemical potential or the number of particles, etc.

As stated at the end of the introductory Chap. 1, it is the statistical thermodynamics that can be employed as the theoretical foundation and tool to connect the quantum mechanical energy states of a system or the classical microscopic interatomic interaction energies among the particles in the system and the corresponding fundamental equation of thermodynamics for the macroscopic system, from which all the macroscopic thermodynamic properties can be derived.

4.1 Macrostates and Microstates

In order to perform statistical averaging over microstates, let us first briefly discuss the definitions of microstates and macrostates. As an illustration, let us use a simple example of an isolated system (constant internal energy U) with N total number

© Springer Nature Singapore Pte Ltd. 2022
L.-Q. Chen, *Thermodynamic Equilibrium and Stability of Materials*,
https://doi.org/10.1007/978-981-13-8691-6_4

Fig. 4.1 Illustration of
energy levels ε_i and the
occupation of each level by
N_i number of particles

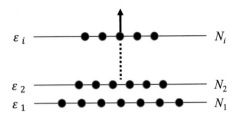

of independent particles in a given volume V. According to quantum mechanics,
the energy of particles in a system is quantized with discrete energy levels ε_i with
$i = 0, 1, 2$, etc. (Fig. 4.1). A distribution $\{N_1, N_2, \ldots, N_i, \ldots\}$ of the number of
particles, N_i, out of a total N number of particles, over the different energy levels ε_i
with the same total internal energy

$$U = \sum N_i \varepsilon_i \tag{4.1}$$

is called a macrostate consistent with the thermodynamic condition for the isolated
system, i.e., constant U, V, and N. The ratio, N_i/N, represents the fraction of
particles at the ith energy level.

There are typically many possible macrostates, or many possible different distributions of particles $\{N_1, N_2, \ldots, N_i, \ldots\}$ at different energy levels associated with
the N particles that can possess the same total energy U. Furthermore, for each
macrostate, i.e., for each specific particle distribution over the energy levels, there
are many different possible arrangements of N_i distinguishable particles at each
energy level, called microstates. In statistical thermodynamics, it is assumed that at
any given moment of time, all the accessible microstates of an isolated system with
the same energy are equally probable. This is often referred as the postulate of equal
a priori probability for all microstates of the system. Therefore, the probability of a
system in a given macrostate is determined by the number of microstates within the
macrostate.

In statistical thermodynamics, due to the huge number of particles in a system, on
the order of Avogadro's number, of particles in a system, it is also postulated that the
total number of microstates in a system is dominated by the most probable macrostate
with the greatest number of microstates, and therefore, the equilibrium state of the
system can be approximated by the most probable macrostate. This postulate dramat-
ically simplifies the statistical solution to the problem for connecting the microscopic interactions to macroscopic properties of a system because it reduces to the
problem of finding the most probable macrostate under a given set of thermodynamic
conditions.

4.2 Statistical Interpretation of Entropy and Boltzmann Equation

Microscopically, the entropy of a system is a measure of the number of microstates that are accessible to the system under a given thermodynamic condition. Therefore, an increase in entropy of a system implies an increase in the number of accessible microstates to the system. This can be understood by examining the change in the internal energy,

$$dU = \sum_i (N_i d\varepsilon_i + \varepsilon_i dN_i) = \sum_i N_i d\varepsilon_i + \sum_i \varepsilon_i dN_i \qquad (4.2)$$

For a given volume and number of particles, the energy levels, ε_i, are fixed. Therefore, the first term in Eq. (4.2) represents the changes in the energy levels due to the change in the volume of the system. The second term in Eq. (4.2) represents the change in the occupation of particles at different fixed energy levels at a given volume, and thus, the change in the entropy at temperature T is given by

$$dS = \frac{dU}{T} = \frac{\sum_i \varepsilon_i dN_i}{T} \qquad (4.3)$$

A shift in the distribution of N_i to higher energy levels, including energy levels which were originally inaccessible, leads to an increase in the number of accessible microstates and thus an increase in entropy.

According to thermodynamics, an isolated system evolves from macrostates of lower entropy to those of higher entropy until reaching its maximum. In statistical thermodynamics, an isolated system evolves from macrostates containing smaller numbers of microstates to those with larger numbers of microstates until the system assumes a macrostate with the maximum number of microstates. Therefore, many of the statistical mechanical solutions are obtained by either maximizing the entropy of an isolated system or finding the most probable macrostate.

The entropy of an isolated system is not directly proportional to the total number of accessible microstates, Z_{MC}, in the system. Entropy is an extensive property, so if a system size is doubled, the entropy is also doubled. However, doubling the system size will result in a total number of microstates of Z_{MC}^2. To make the connection between the amount of entropy and the total number of accessible microstates, Boltzmann postulated that the entropy, S, of a system is proportional to $\ln Z_{MC}$, i.e., entropy is a logarithmic measure of the total number of accessible microstates with significant probability of being occupied, and Max Planck wrote it down as

$$S(U, V, N) = k_B \ln Z_{MC}(U, V, N) \qquad (4.4)$$

where k_B is the Boltzmann constant equal to the ratio of gas constant R to the Avogadro's number N_A. The above equation is valid for an isolated system, i.e., the number of accessible microstates, and thus the entropy, is maximized at equilibrium at constant energy U, volume V, and the total number of particles N. This definition describes the entropy as being proportional to the natural logarithm of the number of possible microscopic configurations of the system that could give rise to the observed macroscopic state of the system. The constant of proportionality is the Boltzmann constant.

Equation (4.4) is a fundamental equation of thermodynamics in the entropic representation. Therefore, in principle, if we can derive the explicit expression $Z_{MC}(U, V, N)$ as a function of U, V, and N, we can compute all the thermodynamic properties of the system based on the differential form of entropy and thermodynamic definitions of properties.

4.3   Ensembles

As mentioned above, the macroscopic thermodynamic properties of a system can be obtained as the averages of properties over the temporal microstates of the system. However, we could also imagine that at a given moment of time we have a large number of identical systems specified by the same set of macroscopic thermodynamic variables for each system, e.g., number of particles or number of moles of particles N or chemical potential μ, temperature T or internal energy U, pressure p or volume V. This collection of large number of identical systems is called an ensemble by Gibbs.[1] Based on the ergodic hypothesis, the average properties of the systems in the ensemble are assumed to be identical to those obtained by averaging over the temporal microstates of a system. Depending on the specified thermodynamic conditions for the macrostates, there can be different types of ensembles. For example,

- *Microcanonical ensemble*: a collection of isolated simple systems with each system fixed at constant internal energy U, constant volume V, and constant number of particles N while the temperature T, pressure p, and chemical potential μ fluctuate in each system and from system to system in the ensemble.
- *Canonical ensemble*: a collection of closed simple systems connected to a thermal reservoir, so each system is fixed at temperature T, constant volume V, and constant number of particles N while the internal energy U, pressure p, and chemical potential μ fluctuate in each system and from system to system in the ensemble.
- *Grand canonical ensemble*: a collection of open simple systems connected simultaneously to a thermal reservoir and to a chemical reservoir, so each system is fixed at constant temperature T, constant volume V, and constant chemical potential μ while the internal energy U, pressure p, and the number of particles N fluctuate in each system and from system to system in the ensemble.

---

[1] J. W. Gibbs, Elementary Principles in Statistical Mechanics, New York: Charles Scribner's Sons, 1902.

- *Isothermal–isobaric ensemble*: a collection of simple systems simultaneously connected to a thermal reservoir and a mechanical reservoir, so each system is fixed at constant temperature T, constant pressure p, and constant number of particles N while the internal energy U, volume V, and the chemical potential μ fluctuate in each system and from system to system in the ensemble.

4.4 Partition Functions and Fundamental Equation of Thermodynamics

The key link between the microscopic interactions, and thus the quantum mechanical energy states, and the fundamental equation of thermodynamics of a system is accomplished by deriving the so-called partition functions in statistical thermodynamics. The partition function is in fact simply the total number of accessible microstates under a specified set of conditions. Therefore, the partition function is different for ensembles under different thermodynamic conditions, and thus, they are connected to different forms of the fundamental equation of thermodynamics. Detailed derivations can be found in many standard textbooks in statistical mechanics or statistical thermodynamics,[2] and thus in this brief introduction, here we will just outline the main results.

Microcanonical ensemble: The partition function Z_{MC} in the microcanonical ensemble is the total number of accessible microstates under constant U, V, and N for a system. In this approach, the partition function represents the system degeneracy, i.e., the number of systems having the same energy U, at constant V and N in the microcanonical ensemble. The probability of observing a specific microstate is then simply given by

$$P_i = \frac{1}{Z_{MC}} \tag{4.5}$$

The connection of the partition function $Z_{MC}(U, V, N)$ to the fundamental equation of thermodynamics in the entropic representation $S = S(U, V, N)$ is given by the same Boltzmann equation discussed in the last section using the total number of microstates in a given system.

$$S = k_B \ln[Z_{MC}(U, V, N)] \tag{4.6}$$

All other thermodynamic properties can be derived from Eq. (4.6) according to the differential form of the fundamental equation in the entropic representation,

$$dS = \frac{1}{T} dU + \frac{p}{T} dV - \frac{\mu}{T} dN \tag{4.7}$$

---

[2] Donald A. McQuarrie, Statistical Mechanics, Harper & Row, 1976.

or the equations of states. For example,

$$\frac{1}{T} = \left(\frac{\partial S}{\partial U}\right)_{V,N} = k_B \left(\frac{\partial \ln[Z_{MC}(U, V, N)]}{\partial U}\right)_{V,N} \tag{4.8}$$

$$\frac{p}{T} = \left(\frac{\partial S}{\partial V}\right)_{U,N} = k_B \left(\frac{\partial \ln[Z_{MC}(U, V, N)]}{\partial V}\right)_{U,N} \tag{4.9}$$

$$-\frac{\mu}{T} = \left(\frac{\partial S}{\partial N}\right)_{U,V} = k_B \left(\frac{\partial \ln[Z_{MC}(U, V, N)]}{\partial N}\right)_{U,V} \tag{4.10}$$

For the canonical ensemble: The natural variables for the partition function Z_C are T, V, and N, and

$$Z_C(T, V, N) = \sum_i \exp\left(-\frac{U_i}{k_B T}\right) \tag{4.11}$$

where U_i is the energy of a microstate i. The probability of a system at the microstate i is given by

$$P_i = \frac{\exp\left(-\frac{U_i}{k_B T}\right)}{Z_C(T, V, N)} \tag{4.12}$$

The macroscopic energy U, the average energy of the ensemble, can be obtained by

$$U = \frac{\sum_i U_i \exp\left(-\frac{U_i}{k_B T}\right)}{\sum_i \exp\left(-\frac{U_i}{k_B T}\right)} = -\left[\frac{\partial \ln[Z_C(T, V, N)]}{\partial(1/k_B T)}\right]_{V,N} \tag{4.13}$$

The internal energy can also be derived from the Helmholtz free energy through a Legendre transform as

$$U = F - T\left(\frac{\partial F}{\partial T}\right)_{V,N} = \left[\frac{\partial(F/k_B T)}{\partial(1/k_B T)}\right]_{V,N} \tag{4.14}$$

Comparing Eqs. (4.13) and (4.14), we have

$$F(T, V, N) = -k_B T \ln[Z_C(T, V, N)] \tag{4.15}$$

which connects the partition function $Z_C(T, V, N)$ to the fundamental equation of thermodynamics represented by the Helmholtz free energy $F(T, V, N)$ from which all the macroscopic thermodynamic properties can be derived based on the differential form for F,

$$dF = -SdT - pdV + \mu dN \tag{4.16}$$

For example,

$$S = -\left(\frac{\partial F}{\partial T}\right)_{V,N} = k_B \ln[Z_C(T, V, N)] + k_B T \left(\frac{\partial \ln[Z_C(T, V, N)]}{\partial T}\right)_{V,N} \tag{4.17}$$

$$p = -\left(\frac{\partial F}{\partial V}\right)_{T,N} = k_B T \left(\frac{\partial \ln[Z_C(T, V, N)]}{\partial V}\right)_{T,N} \tag{4.18}$$

$$\mu = \left(\frac{\partial F}{\partial N}\right)_{T,V} = -k_B T \left(\frac{\partial \ln[Z_C(T, V, N)]}{\partial N}\right)_{T,V} \tag{4.19}$$

The entropy of the system in terms of the partition function in the canonical ensemble can also be expressed as

$$S = \frac{U - F}{T} = \frac{U}{T} + k_B \ln[Z_C(T, V, N)] = k_B \ln\left[Z_C(T, V, N) \exp\left(\frac{U}{k_B T}\right)\right] \tag{4.20}$$

Therefore, the procedure to obtain a fundamental equation at constant T, V, and N starts with a list of energy associated with each microstate. We then obtain the partition function (Eq. 4.12) as a function of T as well as V and N which determine the energy levels of the microstates. Finally, we can use Eq. (4.15) to obtain the fundamental equation of thermodynamics in terms of Helmholtz free energy as a function of T, V, and N.

For the *grand canonical ensemble*: The natural variables for the partition function Z_G are T, V, and μ, and the partition function is given by

$$Z_G(T, V, \mu) = \sum_N \sum_i e^{-(U_{iN} - \mu N)/k_B T} \tag{4.21}$$

The connection of the partition function for this ensemble to the fundamental equation of thermodynamics in the energetic representation is given by grand potential energy,

$$\Xi = -pV = -k_B T \ln[Z_G(T, V, \mu)] \tag{4.22}$$

The probability that the system at microstate (i, N) is given by

$$P_{iN} = \frac{e^{-(U_{iN} - \mu N)/k_B T}}{Z_G} \tag{4.23}$$

All other thermodynamic properties can be derived from the differential form for the grand canonical ensemble,

$$d\Xi = -SdT - pdV - Nd\mu \tag{4.24}$$

For example,

$$S = -\left(\frac{\partial \Xi}{\partial T}\right)_{V,\mu} = k_B \ln[Z_G(T, V, \mu)] + k_B T\left(\frac{\partial \ln[Z_G(T, V, \mu)]}{\partial T}\right)_{V,\mu} \tag{4.25}$$

$$p = -\left(\frac{\partial \Xi}{\partial V}\right)_{T,\mu} = k_B T\left(\frac{\partial \ln[Z_G(T, V, \mu)]}{\partial V}\right)_{T,\mu} = k_B T\frac{\ln[Z_G(T, V, \mu)]}{V} \tag{4.26}$$

$$N = -\left(\frac{\partial \Xi}{\partial \mu}\right)_{T,V} = k_B T\left(\frac{\partial \ln[Z_G(T, V, \mu)]}{\partial \mu}\right)_{T,V} \tag{4.27}$$

The entropy for the system in the grand canonical ensemble can also be expressed as

$$S = \frac{U - N\mu + pV}{T} = \frac{U - N\mu}{T} + k_B \ln Z_G = k_B \ln\left[Z_G \exp\left(\frac{U - N\mu}{k_B T}\right)\right] \tag{4.28}$$

For the isothermal–isobaric ensemble: The natural variables for the partition function Z_{Tp} are T, p, and N, and the partition function is given by

$$Z_{Tp}(T, p, N) = \sum_V \sum_i e^{-(U_{iv}+pV)/k_B T} \tag{4.29}$$

The probability of the system at microstate (i, V) is

$$P_{iV} = \frac{e^{-(U_{iv}+pV)/k_B T}}{Z_{Tp}} \tag{4.30}$$

The connection of the partition function to the fundamental equation of thermodynamics in the energetic representation is given by

$$G = -k_B T \ln\left[Z_{Tp}(T, p, N)\right] \tag{4.31}$$

Other thermodynamic properties can be derived from the differential form of G, i.e.,

$$dG = -SdT + Vdp + \mu dN \tag{4.32}$$

For example,

$$S = -\left(\frac{\partial G}{\partial T}\right)_{p,N} = k_B \ln[Z_{Tp}(T, p, N)] + k_B T \left(\frac{\partial \ln[Z_{Tp}(T, p, N)]}{\partial T}\right)_{p,N}$$

$$(4.33)$$

$$V = \left(\frac{\partial G}{\partial p}\right)_{T,N} = -k_B T \left(\frac{\partial \ln[Z_{Tp}(T, p, N)]}{\partial p}\right)_{T,N} \qquad (4.34)$$

$$\mu = \left(\frac{\partial G}{\partial N}\right)_{T,p} = -k_B T \left(\frac{\partial \ln[Z_{Tp}(T, p, N)]}{\partial N}\right)_{T,p} \qquad (4.35)$$

The entropy in the isothermal–isobaric ensemble can also be expressed as

$$S = \frac{U + pV - G}{T} = \frac{H}{T} + k_B \ln Z_{Tp} = k_B \ln\left[Z_{Tp} \exp\left(\frac{H}{k_B T}\right)\right] \qquad (4.36)$$

where H is the enthalpy of the system. Therefore, if we are able to evaluate or derive the partition function of a system in a particular ensemble, we will have the fundamental equation of thermodynamics and thus can derive all the thermodynamic properties of the system. Theoretically, the most convenient ensemble in many cases is the canonic ensemble while for Bosons or Fermions, the grand canonical ensemble is more convenient. However, in general, to obtain an analytical expression for the partition function as a function of the macroscopic natural variables for a given ensemble is only possible for the very few simplest possible systems such as ideal gases, free electron systems, harmonic oscillators, photon gas, and van der Waals gases. For a realistic system, we have to rely on computer simulation techniques such as molecular dynamics and Monte Carlo techniques to generate a large number of temporal microscopic configurations using Newton's equations of motion or Metropolis types of algorithms and then relate the average properties over the computationally generated microscopic configurations to the macroscopic properties of a system.

4.5 Entropy and Microstate Probabilities

The entropy of a system can always be expressed in terms of the probability of appearance of microstates. If we use P_i to represent the probability of the appearance of a microstate i, the entropy is given by

$$S = -k_B \sum_i P_i \ln P_i \qquad (4.37)$$

For example, let us start with the example of microcanonical ensemble with fixed internal energy U, fixed N total number of independent particles in a fixed volume

V. For an isolated system, there is an equal probability for each of the accessible microstates, Z_{MC}, to appear, i.e.,

$$P_i = \frac{1}{Z_{MC}(U, V, N)} \tag{4.38}$$

Therefore,

$$S = -k_B \sum_i^{Z_{MC}} \frac{1}{Z_{MC}(U, V, N)} \ln \frac{1}{Z_{MC}(U, V, N)} = k_B \ln Z_{MC}(U, V, N) \tag{4.39}$$

For the canonical ensemble of closed systems at fixed temperature T, volume V, and number of particles N,

$$P_i = \frac{\exp\left(-\frac{U_i}{k_B T}\right)}{Z_C(T, V, N)} \tag{4.40}$$

Therefore, the entropy is

$$S = -k_B \sum_i P_i \ln P_i = -k_B \sum_i \frac{e^{-U_i/k_B T}}{Z_C} \ln \frac{e^{-U_i/k_B T}}{Z_C}$$

which can be rewritten as

$$S = -k_B \sum_i \frac{e^{-U_i/k_B T}}{Z_C} \left[-\frac{U_i}{k_B T} - \ln Z_C \right]$$

or

$$S = \frac{1}{T} \sum_i \frac{U_i e^{-E_i/k_B T}}{Z_C} + k_B \ln Z_C \sum_i \frac{e^{-U_i/k_B T}}{Z_C}$$

Using the definition of average internal energy U, we have

$$S = \frac{U}{T} + k_B \ln Z_C \tag{4.41}$$

which can be written as

$$S = k_B \ln\left[Z_C e^{U/(k_B T)} \right] \tag{4.42}$$

In terms of the partition function for the microcanonical ensemble Z_{MC}, we have

$$Z_{MC} = Z_C e^{U/(k_B T)} \tag{4.43}$$

or

$$Z_C = Z_{MC} e^{-U/(k_B T)} \tag{4.44}$$

For the case of a grand canonical ensemble,

$$P_{iN} = \frac{e^{-(U_{iN} - \mu N)/k_B T}}{\sum_N \sum_i e^{-(U_{iN} - \mu N)/k_B T}} = \frac{e^{-(U_{iN} - \mu N)/k_B T}}{Z_G} \tag{4.45}$$

Therefore, the entropy of the system

$$S = -k_B \sum_N \sum_i P_{iN} \ln P_{iN} = -k_B \sum_N \sum_i \frac{e^{-(U_{iN} - \mu N)/k_B T}}{Z_G} \ln \frac{e^{-(U_{iN} - \mu N)/k_B T}}{Z_G}$$

which can be rewritten as

$$S = \frac{(U - \mu N) + k_B T \ln \Xi}{T} = k_B \ln\left[Z_G e^{(U - \mu N)/(k_B T)} \right] \tag{4.46}$$

The relations among the partition functions for the grand canonical ensemble, canonical ensemble, and the microcanonical ensemble are

$$Z_{MC} = Z_C e^{U/(k_B T)} = Z_G e^{(U - \mu N)/(k_B T)} \tag{4.47}$$

or

$$Z_G = Z_C e^{(\mu N)/(k_B T)} = Z_{MC} e^{-(U - \mu N)/(k_B T)} \tag{4.48}$$

For the case of constant isothermal–isobaric ensemble,

$$P_{iV} = \frac{e^{-(U_{iV} + pV)/k_B T}}{\sum_V \sum_i e^{-(U_i + pV)/k_B T}} = \frac{e^{-(U_{iV} + pV)/k_B T}}{Z_{Tp}} \tag{4.49}$$

Therefore,

$$S = -k_B \sum_V \sum_i P_{iV} \ln P_{iV} = -k_B \sum_V \sum_i \frac{e^{-(U_{iV} + pV)/k_B T}}{Z_{Tp}} \ln \frac{e^{-(U_{iV} + pV)/k_B T}}{Z_{Tp}}$$

which can be written as

$$S = \frac{(U + pV) + k_B T \ln Z_{Tp}}{T} \tag{4.50}$$

or

$$S = k_B \ln\left[Z_{Tp} e^{\frac{(U+pV)}{k_B T}}\right] \tag{4.51}$$

The relations among the partition functions for the microcanonical ensemble, canonical ensemble, isothermal–isobaric ensemble are

$$Z_{MC} = Z_C e^{U/(k_B T)} = Z_{Tp} e^{\frac{(U+pV)}{k_B T}} \tag{4.52}$$

or

$$Z_{Tp} = Z_C e^{(-pV)/(k_B T)} = Z_{MC} e^{-(U+pV)/(k_B T)} \tag{4.53}$$

The above relations demonstrate that the entropy of a system depends solely on the probabilities of occurrence for the microstates P_i. For the same temperature, increasing entropy leads to the change in the probability distribution with increasingly higher probability to occupy higher energy states.

4.6 Examples

Example 1 A two-state system Consider a two-energy-level system with energy levels ε_0 and ε_1, volume V, and N total number of independent particles at temperature T, write down the Helmholtz free energy, entropy, internal energy, and chemical potential of the system in terms of the energy levels, the number of particles, and temperature.

Solution
Since each particle has two possible energy states, and all the particles are independent, the canonical partition function of the N particle two-level system at a constant temperature T and volume V is

$$Z_C = \sum_{\{\Omega\}} e^{-\frac{\sum_{i=1}^{N} \varepsilon_i}{k_B T}} = \left(e^{-\frac{\varepsilon_0}{k_B T}} + e^{-\frac{\varepsilon_1}{k_B T}}\right)^N$$

where Ω represents all possible configurations of the N independent particle system with each particle having two possible states.

Therefore, the Helmholtz free energy is given by

$$F = -k_B T \ln Z_C = -N k_B T \ln\left[e^{-\frac{\varepsilon_0}{k_B T}} + e^{-\frac{\varepsilon_1}{k_B T}}\right]$$

The entropy of the system is given by,

$$S(T, V, N) = -\left(\frac{\partial F}{\partial T}\right)_{V,N} = N k_B \ln\left[e^{-\frac{\varepsilon_0}{k_B T}} + e^{-\frac{\varepsilon_1}{k_B T}}\right]$$

$$+ Nk_\mathrm{B}T \frac{\frac{\varepsilon_0}{k_\mathrm{B}T^2}e^{-\frac{\varepsilon_0}{k_\mathrm{B}T}} + \frac{\varepsilon_1}{k_\mathrm{B}T^2}e^{-\frac{\varepsilon_1}{k_\mathrm{B}T}}}{\left[e^{-\frac{\varepsilon_0}{k_\mathrm{B}T}} + e^{-\frac{\varepsilon_1}{k_\mathrm{B}T}}\right]}$$

Simplifying the above equation, we have

$$S(T, V, N) = Nk_\mathrm{B} \ln\left[e^{-\frac{\varepsilon_0}{k_\mathrm{B}T}} + e^{-\frac{\varepsilon_1}{k_\mathrm{B}T}}\right] + \frac{N}{T}\frac{\varepsilon_0 e^{-\frac{\varepsilon_0}{k_\mathrm{B}T}} + \varepsilon_1 e^{-\frac{\varepsilon_1}{k_\mathrm{B}T}}}{\left[e^{-\frac{\varepsilon_0}{k_\mathrm{B}T}} + e^{-\frac{\varepsilon_1}{k_\mathrm{B}T}}\right]}$$

The internal energy of the system is

$$U(T, V, N) = -\left[\frac{\partial \ln Z_\mathrm{C}}{\partial(1/k_\mathrm{B}T)}\right]_{V,N} = N\frac{\varepsilon_0 e^{-\frac{\varepsilon_0}{k_\mathrm{B}T}} + \varepsilon_1 e^{-\frac{\varepsilon_1}{k_\mathrm{B}T}}}{\left[e^{-\frac{\varepsilon_0}{k_\mathrm{B}T}} + e^{-\frac{\varepsilon_1}{k_\mathrm{B}T}}\right]}$$

The chemical potential of the system is given by,

$$\mu(T, V, N) = \left(\frac{\partial F}{\partial N}\right)_{T,V} = -k_\mathrm{B}T \ln\left[e^{-\frac{\varepsilon_0}{k_\mathrm{B}T}} + e^{-\frac{\varepsilon_1}{k_\mathrm{B}T}}\right]$$

Example 2 Rubber statistics We consider a rubber at a certain temperature T as a polymer chain consisting of N rod-shaped monomers with length a. Assuming that each monomer can be pointing along one of the two directions: $\pm z$, and the two directions are degenerate in energy. Now we subject the rubber to a constant force along the $+z$-direction which lifts the degeneracy of the monomers with energy levels, $\mp af$ along the $\pm z$, directions, respectively.

(a) Write down the partition function for the system.
(b) Derive an expression for the Gibbs free energy for the system as a function of T, f, and N.
(c) Derive an expression for the length of the polymer chain as a function of T, f, and N.
(d) Derive an expression for the entropy of the polymer chain as a function of T, f, and N.

Solution

(a) Similar to Example 1, the partition function for the polymer chain in the ensemble of constant temperature T, constant force f, and constant number of monomers N with each monomer having two energy states is given by

$$Z_\mathrm{Tf}(T, f, N) = \left(e^{\frac{af}{k_\mathrm{B}T}} + e^{-\frac{af}{k_\mathrm{B}T}}\right)^N = \left[2\cosh\left(\frac{af}{k_\mathrm{B}T}\right)\right]^N$$

(b)   Since the thermodynamic condition is constant temperature T, constant force f, and constant number of monomers N, the Gibbs free energy G as a function of temperature T, force f, and number of monomers N is given by

$$G(T, f, N) = -k_B T \ln Z_{Tf} = -N k_B T \ln\left[2\cosh\left(\frac{af}{k_B T}\right)\right]$$

(c)   The length of the polymer chain as a function of temperature T, force f, and number of monomers N is given by

$$L = -\left(\frac{\partial G}{\partial f}\right)_{T,N} = Na\frac{2\sinh\left(\frac{af}{k_B T}\right)}{2\cosh\left(\frac{af}{k_B T}\right)} = N\tanh\left(\frac{af}{k_B T}\right)$$

(d)   The entropy of the polymer chain as a function of temperature T, force f, and number of monomers N is given by

$$S = -\left(\frac{\partial G}{\partial T}\right)_{f,N} = N k_B \ln\left[2\cosh\left(\frac{af}{k_B T}\right)\right] - \frac{Naf}{T}\tanh\left(\frac{af}{k_B T}\right)$$

Example 3 Ideal gas For a monatomic ideal gas in an isolated system with N number of atoms contained in a container with volume V and internal energy U, the partition function, or the number of accessible microstates, for the microcanonical ensemble is given by

$$Z_{MC}(U, V, N) = \frac{V^N}{N! h^{3N}}\frac{(2\pi m U)^{3N/2}}{(3N/2)!}$$

where h is the Planck constant, and m is the atomic mass.

(a)   Derive the fundamental equation of thermodynamics in terms of entropy S as a function of U, V, and N.
(b)   Derive an expression for the chemical potential μ as a function of U, V, and N.

Solution

(a)   The entropy of the system is simply connected to the number of accessible microstates or the partition function in the microcanonical ensemble through the Boltzmann equation,

$$S(U, V, N) = k_B \ln Z_{MC}(U, V, N) = k_B \ln\left[\frac{V^N}{N! h^{3N}}\frac{(2\pi m U)^{3N/2}}{(3N/2)!}\right]$$

where k_B is Boltzmann constant. Since N is a very large number, we can employ Stirling's approximation, $\ln N! = N \ln N - N$, and $\ln(3N/2)! = (3N/2) \ln(3N/2) - (3N/2)$. Therefore,

$$S(U, V, N) = Nk_B \left\{ \frac{5}{2} + \ln \left[\frac{V}{N} \left(\frac{4\pi mU}{3h^2N} \right)^{3/2} \right] \right\}$$

(b) The chemical potential can be obtained from entropy through one of the equations of state in the entropic representation, i.e.,

$$\left(\frac{\partial S}{\partial N} \right)_{U,V} = -\frac{\mu}{T} = k_B \left\{ \frac{5}{2} + \ln \left[\frac{V}{N} \left(\frac{4\pi mU}{3h^2N} \right)^{3/2} \right] \right\} - \frac{5}{2} k_B$$

Therefore,

$$\mu(U, V, N) = -k_B T \ln \left[\frac{V}{N} \left(\frac{4\pi mU}{3h^2N} \right)^{3/2} \right]$$

Example 4 Conduction band electrons in semiconductors By treating N conduction band electrons in a semiconductor of volume V at temperature T as an electron gas, the corresponding partition function in the canonical ensemble is given by

$$Z_C(T, V, N) = \frac{(2V)^N}{N!} \left(\frac{2\pi mk_B T}{h^2} \right)^{3N/2} e^{-\frac{NE_c}{k_B T}}$$

where m is the effective mass of an electron, h is Planck constant, k_B is Boltzmann constant, and E_c is the conduction band edge energy.

(a) Derive the fundamental equation of thermodynamics for the conduction band electrons in terms of the Helmholtz free energy as a function of temperature T, volume V, and the number of electrons N.
(b) Derive an expression for the chemical potential for the conduction band electrons as a function of temperature T, volume V, and the number of electrons N.

Solution

(a) The Helmholtz free energy is connected to the partition function in the canonical ensemble through the simple relation,

$$F(T, V, N) = -k_B T \ln Z_C(T, V, N)$$

$$= -k_B T \ln \left\{ \frac{(2V)^N}{N!} \left(\frac{2\pi mk_B T}{h^2} \right)^{3N/2} e^{-\frac{NE_c}{k_B T}} \right\}$$

Since N is a very large number, we can employ Stirling's approximation, $\ln N! = N \ln N - N$. Therefore,

$$F(T, V, N) = N E_c - N k_B T \left\{ 1 + \ln \left[\frac{2V}{N} \left(\frac{2\pi m k_B T}{h^2} \right)^{3/2} \right] \right\}$$

(b)   The chemical potential can be obtained from the Helmholtz free energy by one of the equations of state,

$$\mu(T, V, N) = \left(\frac{\partial F}{\partial N} \right)_{T,V} = E_c - k_B T \ln \left[\frac{2V}{N} \left(\frac{2\pi m k_B T}{h^2} \right)^{3/2} \right]$$

Example 5 A binary ideal gas mixture If the number of available quantum states far exceeds the number of particles in a gas, i.e., if the particles can be treated as independent, the partition function for the whole system, $Z_C(T, V, N)$, can be written in terms of the individual atomic partition functions, $q(T, V)$, i.e.,

$$Z_C(T, V, N) = \frac{[q(T, V)]^N}{N!}$$

For example, the canonical partition function for a binary mixture of monatomic ideal gases with N_A number of A atoms and N_B number of B atoms in a volume V at temperature T is given by

$$Z_C(T, V, N_A, N_B) = \frac{q_A^{N_A} q_B^{N_B}}{N_A! N_B!}$$

in which

$$q = \frac{V}{\Lambda^3}, \quad \Lambda = \left(\frac{h^2}{2\pi m k_B T} \right)^{1/2}$$

where m is the atomic mass, h is Planck constant, and k_B is Boltzmann constant.

(a)   Write down the Helmholtz free energy as a function of temperature T, volume V, and number of atoms N_A and N_B.

(b)   Write down the expression for the entropy of the binary system as a function of temperature T, volume V, and number of atoms N_A and N_B.

(c)   Write down the expression for the internal energy as a function of temperature T, volume V, and number of atoms N_A and N_B.

(d) Write down the chemical potentials of A atoms and B atoms as a function of temperature T, volume V, and number of atoms N_A and N_B.

Solution

(a) We write q_A and q_B as

$$q_A = \frac{V}{\Lambda_A^3} \text{ and } q_B = \frac{V}{\Lambda_B^3}$$

with

$$\Lambda_A = \left(\frac{h^2}{2\pi m_A k_B T}\right)^{1/2} \text{ and } \Lambda_B = \left(\frac{h^2}{2\pi m_B k_B T}\right)^{1/2}$$

Therefore, the partition function is

$$Z_C(T, V, N) = \frac{V^{N_A} V^{N_B}}{N_A! N_B! \Lambda_A^3 \Lambda_B^3}$$

The Helmholtz free energy is given by

$$F = -k_B T \ln Z_C(T, V, N_A, N_B) = -k_B T \ln\left[\frac{V^{N_A} V^{N_B}}{N_A! N_B! \Lambda_A^3 \Lambda_B^3}\right]$$

which can be simplified as

$$F(T, V, N_A, N_B) = -k_B T \left\{ (N_A + N_B) + N_A \ln \frac{V}{N_A \Lambda_A^3} + N_B \ln \frac{V}{N_B \Lambda_B^3} \right\}$$

(b) The entropy of the binary system as a function of temperature T, volume V, and number of atoms N_A and N_B is given by

$$S(T, V, N_A, N_B) = -\left(\frac{\partial F}{\partial T}\right)_{V, N_A, N_B}$$

$$= k_B \left\{ \frac{5}{2}(N_A + N_B) + N_A \ln \frac{V}{N_A \Lambda_A^3} + N_B \ln \frac{V}{N_B \Lambda_B^3} \right\}$$

(c) The internal energy of the binary system as a function of temperature T, volume V, and number of atoms N_A and N_B.

$$U = F + TS = \frac{3}{2} N k_B T = \frac{3}{2}(N_A + N_B) k_B T$$

(d)   The chemical potentials of A and B atoms in the binary system as a function of temperature T, volume V, and number of atoms N_A and N_B.

$$\mu_A(T, V, N_A, N_B) = \left(\frac{\partial F}{\partial N_A}\right)_{T,V,N_B} = -k_B T \ln \frac{V}{N_A \Lambda_A^3}$$

$$\mu_B(T, V, N_A, N_B) = \left(\frac{\partial F}{\partial N_B}\right)_{T,V,N_A} = -k_B T \ln \frac{V}{N_B \Lambda_B^3}$$

The chemical potential of the binary mixture is

$$\mu(T, V, N_A, N_B) = \frac{N_A}{N_A + N_B} \mu_A(T, V, N_A, N_B)$$
$$+ \frac{N_B}{N_A + N_B} \mu_B(T, V, N_A, N_B)$$

$$\mu(T, V, N_A, N_B) = -k_B T \left[\frac{N_A}{N_A + N_B} \ln \frac{V}{N_A \Lambda_A^3} + \frac{N_B}{N_A + N_B} \ln \frac{V}{N_B \Lambda_B^3} \right]$$

Example 6 Langmuir adsorption isotherm Consider a solid–vapor system in which a vapor phase is in equilibrium with the corresponding solid at temperature T. We model the solid surface as a simple two-dimensional lattice of N sites and assume that the gas can be approximated as a monatomic gas. Each site of the surface will be either occupied by an absorbed atom or vacant. Let us assume that the absorbed atoms do not interact with each other. The adsorption energy, the energy difference between an atom on the surface and in the gas phase, is assumed to be ε. Let the number of adsorbed atoms on the surface is n. The chemical potential of atoms in the vapor phase is given by

$$\mu^g = -k_B T \ln \left[\frac{(2\pi m)^{3/2} (k_B T)^{5/2}}{p h^3} \right]$$

where p is vapor pressure, T is temperature, m is the atomic mass, k_B is Boltzmann constant, and h is Planck constant.

Derive the fraction of the surface sites occupied, n/N, as a function of the vapor pressure p.

Solution
For n adsorbed atoms on a surface with N lattice sites, the total number of possible configurations of arrangement for adsorbed atoms and vacant surface sites is given by

$$\frac{N!}{(N-n)! n!}$$

The energy of each configuration is given by $n\varepsilon$. Therefore, the canonical partition function for the solid surface is

$$Z_c = \frac{N!}{(N-n)!n!} e^{-\frac{n\varepsilon}{k_B T}}$$

The Helmholtz free energy is then given by

$$F = -k_B T \ln Z_C = -k_B T \ln\left[\frac{N!}{(N-n)!n!} e^{-\frac{n\varepsilon}{k_B T}}\right]$$

Applying Stirling's approximation, we have

$$F = n\varepsilon - k_B T\left[-\left(\frac{N-n}{N}\right)\ln\left(\frac{N-n}{N}\right) - \frac{n}{N}\ln\frac{n}{N}\right]$$

Therefore, the chemical potential of atoms on the surface is given by

$$\mu^s = \left(\frac{\partial F}{\partial n}\right)_{T,N} = \varepsilon - k_B T[\ln(N-n) - \ln n]$$

The corresponding chemical potential of atoms in the vapor phase is given by

$$\mu^g = -k_B T \ln\left[\frac{(2\pi m)^{3/2}(k_B T)^{5/2}}{p h^3}\right]$$

At thermodynamic equilibrium, $\mu^s = \mu^g$, and hence,

$$\varepsilon - k_B T[\ln(N-n) - \ln n] = -k_B T \ln\left[\frac{(2\pi m)^{3/2}(k_B T)^{5/2}}{p h^3}\right]$$

Therefore,

$$\ln\frac{n}{(N-n)} = \ln\frac{\theta}{(1-\theta)} = \ln\frac{p h^3}{(2\pi m)^{3/2}(k_B T)^{5/2}} e^{-\frac{\varepsilon}{k_B T}}$$

where $\theta = n/N$. Solving the above equation for the surface coverage θ, we get

$$\theta = \frac{Kp}{Kp+1}$$

in which

$$K = \frac{h^3}{(2\pi m)^{3/2}(k_B T)^{5/2}} e^{-\frac{\varepsilon}{k_B T}}$$

In order to have significant adsorption, we need $\varepsilon < 0$ since the atoms in the vapor phase have a lot more entropy than those on the solid surface, i.e., energetically the atoms should be more favorable to be on the surface while entropically they prefer to be in the vapor phase.

4.7 Exercises

1. Consider the Einstein model for a solid, in which a solid of N atoms is assumed to exhibit $3N$ independent vibrational modes with the same frequency ω. The energy levels are quantized and are given by

$$\varepsilon_i = \left(i + \frac{1}{2}\right)h\omega$$

Calculate the internal energy U and entropy S as a function of temperature T in terms of the energy levels ε_i.

2. Consider a monatomic ideal gas of N atoms contained in a box of volume V at temperature T. The canonical partition function is given by

$$Z_C = \frac{V^N}{N!}\left(\frac{2\pi m k_B T}{h^2}\right)^{3N/2}$$

where m is mass, k_B is Boltzmann constant, and h is the Planck constant.

(a) Write down the Helmholtz free energy F of the system as a function of T, V, and N.
(b) Write down the entropy S of the system as a function of T, V, and N.
(c) Write down internal energy U of the system as a function of T, V, and N.
(d) Write down chemical potential μ of the system as a function of T, V, and N.

3. Consider a monatomic ideal gas of N atoms at constant temperature T and pressure p. The isothermal–isobaric partition function for the monatomic gas is

$$Z_{T,p}(T, p, N) = \left(\frac{k_B T}{p}\frac{(2\pi m k_B T)^{3/2}}{h^3}\right)^N$$

where m is mass, k_B is Boltzmann constant, and h is the Planck constant.

(a) Write down the Gibbs free energy G of the system as a function of T, p, and N.
(b) Write down the entropy S of the system as a function of T, p, and N.

(c) Write down internal energy U of the system as a function of T, p, and N.

(d) Write down chemical potential μ of the system as a function of T, p, and N.

4. Consider a monatomic ideal gas of N atoms contained in a volume V at constant temperature T and chemical potential μ. The grand canonical partition function for the monatomic gas is

$$Z_G(T, V, \mu) = e^{\left(\frac{2\pi m k_B T}{h^2}\right)^{3/2} V e^{\mu/k_B T}}$$

where m is mass, k_B is Boltzmann constant, and h is the Planck constant.

(a) Write down the grand potential energy Ξ of the system as a function of T, V, and μ.

(b) Write down the entropy S of the system as a function of T, V, and μ.

(c) Write down the average number of atoms N of the system as a function of T, V, and μ.

(d) Express the chemical potential μ as a function of T, V, and N.

5. Consider a Van der Waals gas of N atoms contained in a box of volume V at temperature T. The partition function is given by

$$Z_C = \frac{(V - Nb)^N}{N!} \left(\frac{2\pi m k_B T}{h^2}\right)^{3N/2} \exp\left(\frac{aN^2}{V k_B T}\right)$$

where m is mass, k_B is Boltzmann constant, h is the Planck constant, and a and b are constants.

(a) Write down the Helmholtz free energy F of the system as a function of T, V, and N.

(b) Write down the entropy S of the system as a function of T, V, and N.

(c) Write down the internal energy U of the system as a function of T, V, and N.

6. By treating the electrons in the conduction band of a semiconductor as an ideal electron gas, the microcanonical partition function can be approximated as

$$Z_{MC}(N, V, U) = \frac{(2V)^N}{N! h^{3N}} \frac{[2\pi m (U - N E_c)]^{3N/2}}{(3N/2)!}$$

where N is the number of electrons, V is the volume of the crystal, U is the internal energy of the conduction band electrons, m is the effective electron mass, h is Planck constant, and E_c is the conduction band edge energy.

(a)   Write down the fundamental equation of thermodynamics for the conduction band electrons in terms of entropy S as a function of U, V, and N.

(b)   Derive the chemical potential μ of the conduction band electrons as a function of U, V, and N.

7.   By treating the electrons in the conduction band of a semiconductor as an ideal electron gas, the constant temperature and constant pressure Gibbs partition function can be approximated as

$$Z_{\mathrm{Tp}}(T, p, N) = \left[\frac{k_{\mathrm{B}}T}{p} 2 \left(\frac{2\pi m k_{\mathrm{B}} T}{h^2} \right)^{3/2} \right]^N e^{-\frac{N E_{\mathrm{c}}}{k_{\mathrm{B}} T}}$$

where N is the number of electrons, p is pressure, T is temperature, m is the effective electron mass, k_{B} is Boltzmann constant, h is Planck constant, and E_{c} is the conduction band edge energy.

(a)   Write down the fundamental equation of thermodynamics for the conduction band electrons in terms of Gibbs free energy G as a function of T, p, and N.

(b)   Derive the chemical potential μ of the conduction band electrons as a function of T and p.

8.   By treating the electrons in the conduction band of a semiconductor as an ideal electron gas, the grand canonical partition function can be written as

$$Z_{\mathrm{G}}(T, V, \mu) = e^{\left[2V \left(\frac{2\pi m k_{\mathrm{B}} T}{h^2} \right)^{3/2} \right] e^{\frac{\mu - E_{\mathrm{c}}}{k_{\mathrm{B}} T}}}$$

where μ is chemical potential or the Fermi level of the conduction band electrons, V is the volume of the crystal, m is the effective electron mass, k_{B} is Boltzmann constant, h is Planck constant, and E_{c} is the conduction band edge energy.

(a)   Write down the fundamental equation of thermodynamics for the conduction band electrons in terms of grand potential energy Ξ as a function of T, V, and μ.

(b)   Write down an expression for the electron gas pressure p as a function of T and μ.

(c)   Write down an expression for the electron gas pressure p as a function of T, V, and N.

(d)   Write down an expression for the electron concentration (N/V) as a function of T and μ.

(e)   Write down an expression for the electron chemical potential μ as a function of T and electron concentration N/V.

(f) Write down an expression for the thermal energy (TS) of the conduction band electrons as a function of T, p, and N.

(g) Write down the internal energy per electron u of the conduction band electrons as a function of T.

9. Consider the formation of n_v vacancies in an elemental crystalline solid with N total number of lattice sites. The formation energy or free energy (excluding the configurational entropy contribution) is assumed to be Δg_v.

(a) Write down the canonical ensemble partition function for the system.

(b) Derive an expression for the chemical potential μ_v of vacancies as a function of temperature T, n_v, and N.

(c) It is known that at equilibrium, the chemical potential μ_v of vacancies is zero. Find the equilibrium vacancy fraction (n_v/N) as a function of temperature T.

10. Consider a binary system with N_A number of A atoms and N_B of B atoms distributed over a total of $N(= N_A + N_B)$ lattice sites at temperature T. The volume of system is assumed to be fixed by the number of lattice sites, i.e., $V = Nv$, where v is the volume of a lattice unit cell. The partition function is given by

$$Z_C(T, N_A, N_B) = \frac{N!}{N_A! N_B!} \exp\left[-\frac{z}{2k_B T}\left(N_A E_{AA} + N_B E_{BB} - \frac{N_A N_B \varepsilon_{AB}}{N}\right)\right]$$

$$\varepsilon_{AB} = E_{AA} + E_{BB} - 2E_{AB}$$

where E_{AA}, E_{BB}, and E_{AB} are AA, BB, and AB nearest neighbor bond energies, z is the number of nearest neighbors, k_B is Boltzmann constant, and ε_{AB} is called the effective exchange energy.

(a) Write down the Helmholtz free energy F of the system as a function of T, N_A, and N_B.

(b) Write down the entropy S of the system as a function of T, N_A, and N_B.

(c) Write down chemical potential μ_A of A atoms as a function of T, x_A, and x_B where $x_A = N_A/N$ and $x_B = N_B/N$.

(d) Write down chemical potential μ_B of B atoms as a function of T, x_A, and x_B.

(e) Write down internal energy U of the system as a function of T, N_A, and N_B.

Chapter 5
From Fundamental Equations to Thermodynamic Properties

The fundamental equations of thermodynamics contain all the information about the equilibrium states of a material, from which all the thermodynamic properties of the material can be derived. For example, all the familiar thermodynamic properties of a system such as heat capacity, mechanical compressibility, and thermal expansion coefficient can be obtained through the second derivatives of thermodynamic energy functions with respect to a certain thermodynamic variable or to a certain pair of variables.

5.1 First Derivatives of Thermodynamic Energy Functions

To facilitate the discussion, let us rewrite the more familiar differential forms of thermodynamic energy functions,

$$dU = TdS - pdV + \mu dN \qquad (5.1)$$

$$dF = -SdT - pdV + \mu dN \qquad (5.2)$$

$$dH = TdS + Vdp + \mu dN \qquad (5.3)$$

$$dG = -SdT + Vdp + \mu dN \qquad (5.4)$$

where U, F, H, and G are the internal energy, Helmholtz free energy, enthalpy, Gibbs free energy of a system. From the above differentials, it is easy to see that the thermodynamic variables, temperature T, pressure p, entropy S, volume V, and chemical potential μ, can be obtained from the first derivatives of a thermodynamic energy function with respect to their natural thermodynamic variables; i.e., the first

© Springer Nature Singapore Pte Ltd. 2022
L.-Q. Chen, *Thermodynamic Equilibrium and Stability of Materials*,
https://doi.org/10.1007/978-981-13-8691-6_5

derivative of a thermodynamic energy function with respect to a thermodynamic variable is another thermodynamic variable within the same thermodynamic variable conjugate pair. For example, according to Eq. (5.1), temperature T within the thermodynamic conjugate pair (T, S) can be obtained from the first derivative of internal energy U with respect to entropy S at fixed volume V and the fixed number of moles N,

$$T = \left(\frac{\partial U}{\partial S} \right)_{V,N} \tag{5.5}$$

Temperature can also be obtained from enthalpy according to Eq. (5.3),

$$T = \left(\frac{\partial H}{\partial S} \right)_{p,N} \tag{5.6}$$

Similarly,

$$p = -\left(\frac{\partial U}{\partial V} \right)_{S,N} = -\left(\frac{\partial F}{\partial V} \right)_{T,N} \tag{5.7}$$

$$S = -\left(\frac{\partial F}{\partial T} \right)_{V,N} = -\left(\frac{\partial G}{\partial T} \right)_{p,N} \tag{5.8}$$

$$V = \left(\frac{\partial H}{\partial p} \right)_{S,N} = \left(\frac{\partial G}{\partial p} \right)_{T,N} \tag{5.9}$$

$$\mu = \left(\frac{\partial U}{\partial N} \right)_{S,V} = \left(\frac{\partial H}{\partial N} \right)_{S,p} = \left(\frac{\partial F}{\partial N} \right)_{T,V} = \left(\frac{\partial G}{\partial N} \right)_{T,p} \tag{5.10}$$

All the above equations can be considered as equations of state, i.e., equations relating the thermodynamic variables of an equilibrium state of a system, which are derived by taking the first derivatives of the fundamental equations with respect to one of their natural variables.

The fundamental equation in terms of G can be alternatively represented by the chemical potential as a function of temperature and pressure. The differential form of chemical potential can simply be obtained by setting $N = 1$ mol in Eq. (5.4). For one mole of material, the magnitude of Gibbs free energy G is equal to that of chemical potential μ, entropy is molar entropy s, volume is molar volume v, and dN is zero since it is fixed at 1 mol. Therefore,

$$d\mu = -sdT + vdp \tag{5.11}$$

which is also the Gibbs–Duhem relation for a single-component system. Therefore, the molar entropy s and molar volume v can be obtained from chemical potentials,

$$s = -\left(\frac{\partial \mu}{\partial T}\right)_p \tag{5.12}$$

$$v = \left(\frac{\partial \mu}{\partial p}\right)_T \tag{5.13}$$

5.2 Second Derivatives of Energy Functions

A second derivative of a thermodynamic energy function with respect to a thermo-dynamic variable measures the curvature of an energy function surface and thus describes the capacitance (capacity) of a material to store matter or the susceptibility or stiffness of a material to changes in any of the thermodynamic variables. Essentially all familiar thermodynamic properties are related to the second derivatives of a certain energy function. For example, the capacitance or capacity of a material describes the general ability of a system to store matter and is defined as the increase in the amount of matter per unit of increase in the corresponding potential, i.e.,

$$\text{Capacity} \propto \frac{d(\text{amount of matter})}{d(\text{potential})} \tag{5.14}$$

where d represents differential. Since either the amount of matter or the potential in the above equation can be expressed as a first derivative of an energy function with respect to the potential or the amount of matter, respectively, the capacity and capacitance can be expressed as a second derivative of the same energy function with respect to the corresponding thermodynamic potential variable such as temperature, pressure or chemical potential for a simple system. Examples include thermal capacitance, i.e., heat capacity, and electric capacitance.

Similarly, a materials susceptibility represents the magnitude of a material's response to an external potential change or to an applied field. It is defined as the relative change in the amount of matter per unit of increase in the corresponding potential,

$$\text{Susceptibility} \propto \frac{d(\text{amount of matter})}{(\text{amount of matter}) \times d(\text{potential})} \tag{5.15}$$

or the change in the density of matter per unit change in the corresponding field,

$$\text{Susceptibility} \propto \frac{d(\text{densty of matter})}{d(\text{field})} \tag{5.16}$$

Therefore, they can also be expressed as second derivatives of a thermodynamic potential energy function with respect to one of the thermodynamic potential or field variables. Familiar examples include mechanical compressibility, elastic compliance, dielectric permittivity, and magnetic permeability.

The stiffness or modulus of a material is simply the inverse to the materials susceptibility. The best-known example is mechanical bulk modulus or elastic stiffness. Other types of moduli can also be similarly defined, such as dielectric stiffness or magnetic stiffness.

5.2.1   Heat Capacity or Thermal Capacitance

Following the definition of capacity using Eq. (5.14), the thermal energy capacity should be defined as the amount of thermal matter, i.e., the entropy, that a material can store or release per unit change of temperature (the thermal potential), i.e.,

$$C_T = \frac{dS}{dT} \tag{5.17}$$

which can be called entropy capacity. However, experimentally, the thermal capacity represents the amount of heat or thermal energy $dQ(= T\,dS)$ that a material can store or release per unit temperature change dT, which is called the heat capacity,

$$C = \frac{dQ}{dT} = T\frac{dS}{dT} \tag{5.18}$$

The heat capacity of a material is different under constant volume or constant pressure conditions. For example, according to the first law of thermodynamics, at constant volume, the amount of heat absorbed by a material is equal to the internal energy increase of the material; i.e., the heat capacity at constant volume of a material is defined as

$$C_V = \left(\frac{dQ}{dT}\right)_{V,N} = \left(\frac{\partial U}{\partial T}\right)_{V,N} = T\left(\frac{\partial S}{\partial T}\right)_{V,N} \tag{5.19}$$

Using Eq. (5.8), C_V can also be expressed as the second derivative of the Helmholtz free energy F with respect to temperature,

$$C_V = T\left(\frac{\partial S}{\partial T}\right)_{V,N} = -T\left(\frac{\partial^2 F}{\partial T^2}\right)_{V,N} \tag{5.20}$$

i.e., the constant volume heat capacity can be directly obtained by taking the second derivative of the Helmholtz free energy with respect to temperature at constant volume V and the number of moles N.

On the other hand, at constant pressure, the amount of heat absorbed by a material is equal to the enthalpy increase of the material, and thus, the constant pressure heat capacity of a material is defined as

$$C_p = \left(\frac{dQ}{dT}\right)_{p,N} = \left(\frac{\partial H}{\partial T}\right)_{p,N} = T\left(\frac{\partial S}{\partial T}\right)_{p,N} \tag{5.21}$$

Using Eq. (5.8), C_p can also be expressed as the second derivative of the Gibbs free energy G with respect to temperature,

$$C_p = T\left(\frac{\partial S}{\partial T}\right)_{p,N} = -T\left(\frac{\partial^2 G}{\partial T^2}\right)_{p,N} \tag{5.22}$$

i.e., the constant pressure heat capacity can be directly obtained by taking the second derivative of the Gibbs free energy with respect to temperature at constant pressure p and the number of moles N.

The molar heat capacities are defined as the heat capacities for one mole of material, and they can be obtained from

$$c_v = \left(\frac{\partial u}{\partial T}\right)_v = T\left(\frac{\partial s}{\partial T}\right)_v = -T\left(\frac{\partial^2 f}{\partial T^2}\right)_v \tag{5.23}$$

$$c_p = \left(\frac{\partial h}{\partial T}\right)_p = T\left(\frac{\partial s}{\partial T}\right)_p = -T\left(\frac{\partial^2 \mu}{\partial T^2}\right)_p \tag{5.24}$$

where u, s, f, h, and μ are the molar internal energy, molar entropy, molar Helmholtz free energy, molar enthalpy, and chemical potential, respectively. In Eq. (5.24), we use chemical potential μ rather than molar Gibbs free energy g.

The total heat capacity and the molar heat capacity are simply related by the number of moles as

$$C_p = N c_p \tag{5.25}$$

It should be noted that in the literature, the specific heat capacity is also often expressed in terms of per kilogram of matter or per gram rather than per mole.

5.2.2   Compressibility or Mechanical Susceptibility and Modulus

Compressibility describes the relative volume change per unit of change in pressure, which can be interpreted as the mechanical susceptibility. The compressibility of a material also depends on the thermodynamic conditions. For example, the isothermal compressibility β_T is the compressibility at constant temperature and defined as

$$\beta_T = -\frac{1}{V}\left(\frac{\partial V}{\partial p}\right)_{T,N} \tag{5.26}$$

Since the volume of a stable material decreases as the pressure increases, the negative sign is introduced in the definition (Eq. 5.26) for the compressibility to have a positive value. Using the relation (Eq. 5.9) between volume V and Gibbs free energy G, we can express the isothermal compressibility in terms of the second derivative of the Gibbs free energy with respect to pressure at constant temperature and number of moles,

$$\beta_T = -\frac{1}{V}\left(\frac{\partial^2 G}{\partial p^2}\right)_{T,N} \tag{5.27}$$

The adiabatic compressibility β_S is the compressibility of a material measured at constant entropy (i.e., under a reversible adiabatic condition with no entropy exchange between the system and its surrounding) and can be obtained from the second derivative of enthalpy of the material with respect to pressure at constant entropy and number of moles,

$$\beta_S = -\frac{1}{V}\left(\frac{\partial V}{\partial p}\right)_{S,N} = -\frac{1}{V}\left(\frac{\partial^2 H}{\partial p^2}\right)_{S,N} \tag{5.28}$$

The isothermal bulk modulus B_T and the adiabatic bulk modulus B_S of a material are simply the inverses to the corresponding compressibility i.e.,

$$B_T = \frac{1}{\beta_T} = -V\left(\frac{\partial p}{\partial V}\right)_{T,N} = V\left(\frac{\partial^2 F}{\partial V^2}\right)_{T,N} = -\frac{V}{\left(\frac{\partial^2 G}{\partial p^2}\right)_{T,N}} \tag{5.29}$$

and

$$B_S = \frac{1}{\beta_S} = -V\left(\frac{\partial p}{\partial V}\right)_{S,N} = V\left(\frac{\partial^2 U}{\partial V^2}\right)_{S,N} = -\frac{V}{\left(\frac{\partial^2 H}{\partial p^2}\right)_{S,N}} \tag{5.30}$$

The mechanical compressibility and isothermal bulk modulus can also be expressed in terms of second derives of chemical potential, molar Helmholtz free energy, molar enthalpy, and molar internal energy,

$$\beta_T = -\frac{1}{v}\left(\frac{\partial^2 \mu}{\partial p^2}\right)_T \tag{5.31}$$

$$\beta_s = -\frac{1}{v}\left(\frac{\partial^2 h}{\partial p^2}\right)_s \tag{5.32}$$

$$B_T = v\left(\frac{\partial^2 f}{\partial v^2}\right)_T = -\frac{v}{\left(\frac{\partial^2 \mu}{\partial p^2}\right)_T} \tag{5.33}$$

$$B_s = v \left(\frac{\partial^2 u}{\partial v^2} \right)_s = -\frac{v}{\left(\frac{\partial^2 h}{\partial p^2} \right)_s} \tag{5.34}$$

5.2.3 Chemical Capacitance[1]

Similar to heat capacity and compressibility, one may define a chemical capacitance,

$$C_C = \frac{\partial N}{\partial \mu} = \frac{1}{\frac{\partial \mu}{\partial N}} \tag{5.35}$$

The chemical capacitance also depends on the thermodynamic conditions. For example, for a single-component system, the chemical potential is only a function of temperature and pressure and is independent of the number of moles N,

$$\left(\frac{\partial \mu}{\partial N} \right)_{T,p} = 0$$

Hence,

$$C_C = \left(\frac{\partial N}{\partial \mu} \right)_{T,p} = \frac{1}{\left(\frac{\partial \mu}{\partial N} \right)_{T,p}} = \frac{1}{\left(\frac{\partial^2 G}{\partial N^2} \right)_{T,p}} = \infty \text{(undefined)} \tag{5.36}$$

Therefore, the chemical capacitance for a single-component system at constant temperature and pressure is not defined because of the Gibbs–Duhem relation among the potentials. However, the chemical capacitance of a single-component system for other thermodynamic conditions such as constant temperature (T) and volume (V), or constant entropy (S) and constant pressure (p), or constant entropy (S) and constant volume (V) or chemical capacitance of binary or multicomponent systems at constant temperature (T) and pressure (p) is well-defined. For example, the chemical capacitance of a simple-component system at constant temperature and volume can be defined as

$$C_C = \left(\frac{\partial N}{\partial \mu} \right)_{T,V} = -\left(\frac{\partial^2 \Xi}{\partial \mu^2} \right)_{T,V} = \frac{1}{\left(\frac{\partial \mu}{\partial N} \right)_{T,V}} = \frac{1}{\left(\frac{\partial^2 F}{\partial N^2} \right)_{T,V}} \tag{5.37}$$

where Ξ is the grand potential energy, and F is the Helmholtz free energy.

It should be emphasized that similar to the different magnitudes for heat capacity under the constant pressure and constant volume conditions or for compressibility

---

[1] Joachim Maier, "Chemical resistance and chemical capacitance," Zeitschrift für Naturforschung, 75(1–2)b: 15–22, (2020).

under the isothermal and adiabatic conditions, the chemical capacitance under different conditions, e.g., the adiabatic and constant volume condition, or the adiabatic and constant pressure condition, will be different from the chemical capacitance under the isothermal constant volume condition.

5.2.4   Electric Capacitance

The familiar electric capacitance, C_E, is equal to the amount of charge (the electric matter) dq that a dielectric capacitor can store or release per unit of electric potential change $d\phi$,

$$C_E = \frac{dq}{d\phi} \tag{5.38}$$

We can also express the electric capacitance in terms of second derivative of an energy function with respect to charge q,

$$C_E = \frac{dq}{d\phi} = -\left(\frac{\partial^2 G'}{\partial \phi^2}\right)_{T,p,N} \tag{5.39}$$

where G' is the electric potential energy function with electric potential ϕ as one of the natural variables.

5.3   Volume Thermal Expansion Coefficient

Thermal expansion coefficient describes the thermomechanical coupling effect of a material, i.e., the amount of relative volume change per degree of temperature change,

$$\alpha = \frac{1}{V}\left(\frac{\partial V}{\partial T}\right)_{p,N} \tag{5.40}$$

Replacing V in Eq. (5.40) using the first derivative of G with respect to pressure, we have the expression of thermal expansion coefficient in terms of the mixed second derivative of Gibbs free energy with respect to temperature and pressure,

$$\alpha = \frac{1}{V}\left(\frac{\partial^2 G}{\partial T \partial p}\right)_N \tag{5.41}$$

The thermal expansion coefficient can also be expressed in terms of the second derivative of chemical potential with respect to temperature and pressure,

$$\alpha = \frac{1}{v} \frac{\partial^2 \mu}{\partial T \partial p} \tag{5.42}$$

Typically, the thermal expansion coefficient of a material is experimentally measured using a dilatometer that measures the linear thermal expansion coefficient. Since the magnitude of thermal expansion coefficient of a solid is small, typically on the order of $10^{-6}/K$, one can obtain the volume thermal expansion coefficient α from the linear thermal expansion coefficient α_l using the following approximation,

$$\alpha = \frac{1}{v} \left(\frac{\partial v}{\partial T} \right)_p \approx 3\alpha_l = 3\frac{1}{L} \left(\frac{\partial L}{\partial T} \right)_{p,N} \tag{5.43}$$

where L is the length of a bar used to measure the linear thermal expansion coefficient.

5.4 Thermal, Mechanical, Electric, and Magnetic Effects of Homogeneous Crystals[2]

For crystals, many properties are generally anisotropic and thus often represented by tensors and vectors rather than scalers. For example, the relationships between the mechanical stress σ_{ij} and strain ε_{ij} are given by

$$\sigma_{ij} = C_{ijkl}\varepsilon_{kl}, \ \varepsilon_{ij} = s_{ijkl}\sigma_{kl} \tag{5.44}$$

where C_{ijkl} and s_{ijkl} are fourth-rank elastic stiffness and compliance tensors.

The relationships between electric field E_i and electric displacement D_i are

$$D_i = \epsilon_o E_i + P_i = \epsilon_o(\delta_{ij} + \chi_{ij})E_j = \epsilon_o K_{ij} E_j = k_{ij} E_j, \ P_i = \epsilon_o \chi_{ij} E_j \tag{5.45}$$

where ϵ_o, P_i, δ_{ij}, χ_{ij}, K_{ij}, and k_{ij} are vacuum permittivity $(8.854 \times 10^{-12} \text{ F/m, C/(Vm)})$, polarization (a vector quantity representing the electric dipole moment per unit volume, or polarization charge per unit area taken perpendicular to the direction of polarization), Kronecker delta function, dielectric susceptibility tensor, relative permittivity or the dielectric constant, and dielectric permittivity tensor, respectively.

The relationships between magnetic field H_i strength and magnetic induction B_i in a crystal are

---

[2] J. F. Nye, Physical Properties of Crystals, Their Representation by Tensors and Matrices, Oxford University Press, Oxford Science Publications 1995.

$$B_i = \mu_o H_i + I_i = \mu_o(\delta_{ij} + \psi_{ij})H_j = \mu_o M_{ij} H_j = \mu_{ij} H_j, \ I_i = k_o \psi_{ij} H_j,$$
$$(5.46)$$

where μ_o, I_i, ψ_{ij}, M_{ij}, and μ_{ij} are vacuum permeability $(4\pi/10^7 H/m, N/A^2)$, magnetization or magnetic moment per unit volume, magnetic susceptibility tensor, relative magnetic permeability tensor, and magnetic permeability tensor.

For a crystal involving electric, mechanical, magnetic, gravitational energy, the differential form for the internal energy is given by

$$dU = T dS + \mu dN + \phi dq + V \sigma_{ij} d\varepsilon_{ij} + V E_i dD_i + V H_i dB_i + gz dm \quad (5.47)$$

where U is internal energy, T temperature, S entropy, μ chemical potential, N number of moles, ϕ electrical potential, q electric charge, σ_{ij} and ε_{ij} are ijth components of stress and strain, E_i and D_i are the ith components of electrical field and electric displacement or flux density, H_i and B_i are the ith components of magnetic field and magnetization induction, g is the gravity, z is the height, and m is the mass. Repeating indices in the same term in Eqs. (5.44–5.47) above and equations follow below imply summation over those indices following the Einstein summation convention is adopted.

The pressure–volume term is assumed to be included in the stress–strain term. For example, if the stress is hydrostatic pressure,

$$\sigma_{ij} = -p\delta_{ij}. \quad (5.48)$$

where δ_{ij} is the Kronecker–Delta function which is 1 if $i = j$ and 0 otherwise. Therefore, in the case of hydrostatic pressure,

$$V\sigma_{ij}d\varepsilon_{ij} = -Vp\delta_{ij}d\varepsilon_{ij} = -pVd(\varepsilon_{11} + \varepsilon_{22} + \varepsilon_{33}) = -pdV. \quad (5.49)$$

The chemical, electric charge, and gravitational terms can be combined into a single term for a particular charged species i,

$$\mu_i dN + \phi \mathcal{F} z_i dN + gz M dN = (\mu_i + \phi \mathcal{F} z_i + gz M)dN, \quad (5.50)$$

where μ_i is chemical potential of species i, $\mathcal{F}$ is the Faraday constant, z_i is the valence of the species i, z is height, and M is molar mass. Therefore, if we understand the chemical potential in the differential form as the total potential which includes chemical, electric, and gravitational potentials, we can write down a slightly shorter version for the differential form,

$$dU = T dS + \mu dN + V \sigma_{ij} d\varepsilon_{ij} + V E_i dD_i + V H_i dB_i \quad (5.51)$$

The integrated form for the above equation is

$$U = TS + \mu N + V\sigma_{ij}\varepsilon_{ij} + VE_iD_i + VH_iB_i \tag{5.52}$$

The total energy of a system should include all the interaction energies with its surrounding, and the corresponding energy function has been given different names such as electric free energy or enthalpy, magnetic free energy, the grand free energy, etc.

For one mole of substance, Eq. (5.52) becomes

$$u = Ts + \mu + v\left(\sigma_{ij}\varepsilon_{ij} + E_iP_i + H_iB_i\right) \tag{5.53}$$

where u, s, μ, and v are molar internal energy, molar entropy, chemical potential, and molar volume, respectively.

For a unit volume, we can rewrite Eq. (5.52) in terms of densities,

$$u_v = Ts_v + \mu c + \sigma_{ij}\varepsilon_{ij} + E_iD_i + H_iB_i \tag{5.54}$$

$$du_v = Tds_v + \mu dc + \sigma_{ij}d\varepsilon_{ij} + E_idD_i + H_idB_i \tag{5.55}$$

where s_v and c are entropy per unit volume (entropy density) and number of moles per unit volume, respectively.

Below we will work with per unit volume basis, so we will drop the subscript v in the equations that follow.

$$u = Ts + \mu c + \sigma_{ij}\varepsilon_{ij} + E_iD_i + H_iB_i \tag{5.56}$$

$$du = Tds + \mu dc + \sigma_{ij}d\varepsilon_{ij} + E_idD_i + H_idB_i \tag{5.57}$$

The corresponding Helmholtz free energy density is given by

$$f = \mu c + \sigma_{ij}\varepsilon_{ij} + E_iD_i + H_iB_i \tag{5.58}$$

$$df = -sdT + \mu dc + \sigma_{ij}d\varepsilon_{ij} + E_idD_i + H_idB_i \tag{5.59}$$

where s is the entropy density.

If one chooses potentials and fields as the natural variables, the appropriate thermodynamic potential will be,

$$g = u - Ts - \sigma_{ij}\varepsilon_{ij} - E_iD_i - H_iB_i \tag{5.60}$$

$$dg = -sdT + \mu dc - \left(\varepsilon_{ij}d\sigma_{ij} + D_idE_i + B_idH_i\right) \tag{5.61}$$

where g is the Gibbs free energy density which is essentially the chemical energy density or chemical energy per unit volume.

5.5 Principal Properties

We can write down all the familiar definitions of properties using the following two versions of differential forms of fundamental equation of thermodynamics

$$df = -sdT + \mu dc + \sigma_{ij}d\varepsilon_{ij} + E_i dD_i + H_i dB_i \tag{5.62}$$

$$dg = -sdT + \mu dc - \left(\varepsilon_{ij}d\sigma_{ij} + D_i dE_i + B_i dH_i\right) \tag{5.63}$$

It is reminded that all the terms in the above two equations have the unit of energy per unit volume. To reduce the number of equations, let us assume the volume density of matter is a constant, so we have

$$df = -sdT + \sigma_{ij}d\varepsilon_{ij} + E_i dD_i + H_i dB_i \tag{5.64}$$

$$dg = -sdT - \left(\varepsilon_{ij}d\sigma_{ij} + D_i dE_i + B_i dH_i\right) \tag{5.65}$$

Immediately, we can write down the following definitions of stress, electric and magnetic fields based on the first derivatives of Helmholtz and Gibbs free energy densities,

$$\left(\frac{\partial f}{\partial \varepsilon_{ij}}\right)_{T,D,B_i} = \sigma_{ij}, \quad \left(\frac{\partial f}{\partial D_i}\right)_{T,\varepsilon,B_i,} = E_i, \quad \left(\frac{\partial f}{\partial B_i}\right)_{T,\varepsilon,D} = H_i \tag{5.66}$$

$$\left(\frac{\partial g}{\partial \sigma_{ij}}\right)_{T,E,H} = -\varepsilon_{ij}, \quad \left(\frac{\partial g}{\partial E_i}\right)_{T,\sigma,H} = -D_i, \quad \left(\frac{\partial g}{\partial H_i}\right)_{T,\sigma,E} = -B_i \tag{5.67}$$

The principal physical properties of a crystal are given by the second derivatives with respect to a particular thermodynamic variable; e.g., the second derivative of Gibbs free energy with respect to temperature is related to heat capacity at constant stress, electric, and magnetic fields.

Heat capacity

$$\left(\frac{\partial^2 g}{\partial T^2}\right)_{\sigma,E,H} = -\left(\frac{\partial s}{\partial T}\right)_{\sigma,E,H} = -\frac{c_p}{T} \tag{5.68}$$

It should be noted that the heat capacity here is on a per unit volume basis.

Elastic stiffness and compliance constants

The second derivative of Helmholtz free energy density with respect to strain is

$$\left(\frac{\partial^2 f}{\partial \varepsilon_{ij}\partial \varepsilon_{kl}}\right)_{T,D,B} = \left(\frac{\partial \sigma_{ij}}{\partial \varepsilon_{kl}}\right)_{T,D,B} = C_{ijkl} \tag{5.69}$$

where C_{ijkl} is a fourth-rank elastic stiffness modulus tensor which linearly connects second-rank elastic stress tensor, σ_{ij}, and second-rank elastic strain tensor, ε_{kl}.

The second derivative of Gibbs free energy density with respect to stress is

$$\left(\frac{\partial^2 g}{\partial \sigma_{ij} \partial \sigma_{kl}}\right)_{T,E,H} = -\left(\frac{\partial \varepsilon_{ij}}{\partial \sigma_{kl}}\right)_{T,E,H} = -C_{ijkl}^{-1} = -s_{ijkl} \qquad (5.70)$$

where C_{ijkl}^{-1} is the inverse of elastic stiffness tensor, and s_{ijkl} is the fourth-rank elastic compliance tensor.

Dielectric Permittivity Tensor

The second derivative of Gibbs free energy density with respect to electric field is

$$\left(\frac{\partial^2 g}{\partial E_i \partial E_j}\right)_{T,\sigma,H} = -\left(\frac{\partial D_i}{\partial E_j}\right)_{T,\sigma,H} = -\kappa_{ij} \qquad (5.71)$$

where κ_{ij} is the second-rank permittivity tensor.

The second derivative of Helmholtz free energy density with respect to the electric displacement is

$$\left(\frac{\partial^2 f}{\partial D_i \partial D_i}\right)_{T,\varepsilon,B} = \left(\frac{\partial E_i}{\partial D_j}\right)_{T,\varepsilon,B} = \kappa_{ij}^{-1} \qquad (5.72)$$

where the inverse to the second-rank dielectric permittivity tensor κ_{ij}^{-1} can also be called the second-rank dielectric stiffness tensor.

Sometimes, the electric polarization rather than electric displacement is used as an independent thermodynamic variable,

$$D_i = \kappa_{ij} E_j = \left(\delta_{ij} + \chi_{ij}\right)\epsilon_{\mathrm{o}} E_j = \epsilon_{\mathrm{o}} E_i + P_i \qquad (5.73)$$

where ϵ_{o} is the vacuum permittivity, and χ_{ij} is the second-rank dielectric susceptibility tensor. The dielectric permittivity tensor is related to the susceptibility tensor,

$$\kappa_{ij} = \left(\delta_{ij} + \chi_{ij}\right)\epsilon_{\mathrm{o}} \qquad (5.74)$$

where $\left(\delta_{ij} + \chi_{ij}\right)$ is the dielectric constant tensor.

It should be pointed out that either D_i or P_i or E_i can be chosen as an independent thermodynamic variable, but one cannot use two of them simultaneously as independent thermodynamic variables. The corresponding equations using electric polarization rather than electric displacement are

$$\mathrm{d}g = -s\mathrm{d}T - \varepsilon_{ij}\mathrm{d}\sigma_{ij} - P_i\mathrm{d}E_i - B_i\mathrm{d}H_i \qquad (5.75)$$

$$\left(\frac{\partial^2 g}{\partial E_i \partial E_j}\right)_{T,\sigma,H} = -\left(\frac{\partial P_i}{\partial E_j}\right)_{T,\sigma,H} = -\epsilon_0 \chi_{ij} \qquad (5.76)$$

Magnetic Permeability Tensor

The second derivative of Gibbs free energy density with respect to the magnetic field is the negative the second-rank magnetic permeability tensor,

$$\left(\frac{\partial^2 g}{\partial H_i \partial H_j}\right)_{T,\sigma,E} = -\left(\frac{\partial B_i}{\partial H_j}\right)_{T,\sigma,E} = -\mu_{ij} \qquad (5.77)$$

where μ_{ij} is the magnetic permeability tensor. Similarly, the second derivative of Helmholtz free energy density with respect to the magnetic inductance is

$$\left(\frac{\partial^2 f}{\partial B_i \partial B_i}\right)_{T,\varepsilon,D} = \left(\frac{\partial H_i}{\partial B_j}\right)_{T,\varepsilon,D} = \mu_{ij}^{-1} \qquad (5.78)$$

where the inverse to the second-rank magnetic permeability tensor μ_{ij}^{-1} can also be called the second-rank magnetic modulus tensor.

5.6   Coupled Properties

Thermoelastic Effects—Thermal Expansion Coefficients

The coupled physical properties of a crystal are given by the mixed second derivatives with respect to two thermodynamic variables; e.g., the second derivatives of Gibbs free energy density with respect to temperature and stress are

$$\frac{\partial^2 g}{\partial T \partial \sigma_{ij}} = -\left(\frac{\partial \varepsilon_{ij}}{\partial T}\right)_{\sigma,E,H} = -\alpha_{ij} \qquad (5.79)$$

$$\frac{\partial^2 g}{\partial \sigma_{ij} \partial T} = -\left(\frac{\partial s}{\partial \sigma_{ij}}\right)_{T,E,H} = -\alpha'_{ij} \qquad (5.80)$$

where α_{ij} is the second-rank thermal expansion coefficient tensor, and α'_{ij} is the second-rank elastocaloric tensor.

Pyroelectric Effects

The two mixed second derivatives of Gibbs free energy density with respect to temperature and electric field are

$$\frac{\partial^2 g}{\partial T \partial E_i} = -\left(\frac{\partial D_i}{\partial T}\right)_{\sigma,E,H} = -p_i \tag{5.81}$$

$$\frac{\partial^2 g}{\partial E_i \partial T} = -\left(\frac{\partial s}{\partial E_i}\right)_{T,\sigma,H} = -p_i' \tag{5.82}$$

where p_i is the pyroelectric vector, and p_i' is the electrocaloric vector.

Pyromagnetic Effects

The pyromagnetic effects can be expressed in terms of the mixed second derivatives of Gibbs free energy density with respect to temperature and magnetic fields,

$$\frac{\partial^2 g}{\partial T \partial H_i} = -\left(\frac{\partial B_i}{\partial T}\right)_{\sigma,E,H} = -\pi_i \tag{5.83}$$

$$\frac{\partial^2 g}{\partial H_i \partial T} = -\left(\frac{\partial s}{\partial H_i}\right)_{T,\sigma,E} = -\pi_i' \tag{5.84}$$

where π_i is the pyromagnetic vector, and π_i' is the magnetocaloric vector.

Piezoelectric effects

The mixed second derivatives of Gibbs free energy density with respect to stress and electric fields are

$$\frac{\partial^2 g}{\partial \sigma_{jk} \partial E_i} = -\left(\frac{\partial D_i}{\partial \sigma_{jk}}\right)_{T,E,H} = -d_{ijk} \tag{5.85}$$

$$\frac{\partial^2 g}{\partial E_k \partial \sigma_{ij}} = -\left(\frac{\partial \varepsilon_{ij}}{\partial E_k}\right)_{T,\sigma,H} = -d_{ijk}' \tag{5.86}$$

where d_{ijk} is the third-rank piezoelectric tensor, and d_{ijk}' is the third-rank converse piezoelectric tensor.

Piezomagnetic Effects

The piezomagnetic effects can be expressed in terms of the mixed second derivatives of Gibbs free energy density with respect to stress and magnetic fields,

$$\frac{\partial^2 g}{\partial \sigma_{jk} \partial H_i} = -\left(\frac{\partial B_i}{\partial \sigma_{jk}}\right)_H = -z_{ijk} \tag{5.87}$$

$$\frac{\partial^2 g}{\partial H_k \partial \sigma_{ij}} = -\left(\frac{\partial \varepsilon_{ij}}{\partial H_k}\right)_\sigma = -z_{ijk}' \tag{5.88}$$

where z_{ijk} is the third-rank piezomagnetic tensor, and z_{ijk}' is the third-rank converse piezomagnetic tensor.

Magnetoelectric Effects

The magnetoelectric effects can be expressed in terms of the mixed second derivatives of Gibbs free energy density with respect to electric and magnetic fields,

$$\frac{\partial^2 g}{\partial H_j \partial E_i} = -\left(\frac{\partial D_i}{\partial H_j}\right)_E = -\beta_{ij} \tag{5.89}$$

$$\frac{\partial^2 g}{\partial E_j \partial H_i} = -\left(\frac{\partial B_i}{\partial E_j}\right)_H = -\beta'_{ij} \tag{5.90}$$

where β_{ij} is the second-rank magnetoelectric tensor, and β'_{ij} is the second-rank converse magnetoelectric tensor.

5.7 Summary Comments on Relationships Between Derivatives and Properties.

Since the most common experimental conditions in materials science and engineering are constant temperature and constant pressure, Table 5.1 summarizes the definitions of most often used thermodynamic properties and their relations to the second derivatives of chemical potentials:

Here we summarize a few general observations on the relationships between derivatives of thermodynamic energy functions with respect to one or more thermodynamic variables and derivatives of one thermodynamic variable with respect to another.

- The first derivative of a fundamental thermodynamic energy function with respect to a thermodynamic variable is another thermodynamic variable within the same thermodynamic variable conjugate pair. For example,

$$T = \left(\frac{\partial U}{\partial S}\right)_{V,N} = \left(\frac{\partial H}{\partial S}\right)_{p,N}, \quad S = -\left(\frac{\partial F}{\partial T}\right)_{V,N} = -\left(\frac{\partial G}{\partial T}\right)_{p,N} = -\left(\frac{\partial \Xi}{\partial T}\right)_{V,\mu}$$

$$V = \left(\frac{\partial H}{\partial p}\right)_{S,N} = \left(\frac{\partial G}{\partial p}\right)_{T,N}, \quad p = -\left(\frac{\partial U}{\partial V}\right)_{S,N} = -\left(\frac{\partial F}{\partial V}\right)_{T,N} = -\left(\frac{\partial \Xi}{\partial V}\right)_{T,\mu}$$

Table 5.1 Thermodynamic properties defined by the first and second derivatives of chemical potential with respect to temperature and pressure

$\mu = \left(\frac{\partial G}{\partial N}\right)_{T,p}$	$d\mu = -sdT + vdp$	$c_p = \left(\frac{\partial h}{\partial T}\right)_p$	$\beta_T = -\frac{1}{v}\left(\frac{\partial v}{\partial p}\right)_T$	$\alpha = \frac{1}{v}\left(\frac{\partial v}{\partial T}\right)_p$
$s = -\left(\frac{\partial \mu}{\partial T}\right)_p$	$v = \left(\frac{\partial \mu}{\partial p}\right)_T$	$c_p = -T\left(\frac{\partial^2 \mu}{\partial T^2}\right)_p$	$\beta_T = -\frac{1}{v}\left(\frac{\partial^2 \mu}{\partial p^2}\right)_T$	$\alpha = \frac{1}{v}\frac{\partial^2 \mu}{\partial T \partial p}$

$$\mu = \left(\frac{\partial U}{\partial N}\right)_{S,V} = \left(\frac{\partial H}{\partial N}\right)_{S,p} = \left(\frac{\partial F}{\partial N}\right)_{T,V} = \left(\frac{\partial G}{\partial N}\right)_{T,p}, \ N = -\left(\frac{\partial \Xi}{\partial \mu}\right)_{T,V}$$

- The first derivatives of chemical potential with respect to its thermodynamic variables, temperature T and pressure p, are the corresponding molar thermodynamic quantities, molar entropy s and molar volume v,

$$s = -\left(\frac{\partial \mu}{\partial T}\right)_p, \ v = \left(\frac{\partial \mu}{\partial p}\right)_T$$

- The first derivative of a thermodynamic energy density function with respect to a component of a thermodynamic vector or a tensor variable is the corresponding component of another thermodynamic variable tensor within the same thermodynamic variable conjugate pair. For example,

$$\left(\frac{\partial f}{\partial \varepsilon_{ij}}\right)_{T,D,B} = \sigma_{ij}, \ \left(\frac{\partial f}{\partial D_i}\right)_{T,\varepsilon,B} = E_i, \ \left(\frac{\partial f}{\partial B_i}\right)_{T,\varepsilon,D} = H_i$$

$$\left(\frac{\partial g}{\partial \sigma_{ij}}\right)_{T,E,H} = -\varepsilon_{ij}, \ \left(\frac{\partial g}{\partial E_i}\right)_{T,\sigma,H} = -D_i, \ \left(\frac{\partial g}{\partial H_i}\right)_{T,\sigma,E} = -B_i$$

- The first derivatives of extensive thermodynamic variables (S, V, N) with respect to their corresponding intensive thermodynamic variables (T, p, μ) are thermodynamic properties measuring how susceptible a material is to external stimuli or how much matter can be stored per unit of potential change. For example,

$$\left(\frac{\partial S}{\partial T}\right)_{V,N} = -\left(\frac{\partial^2 F}{\partial T^2}\right)_{V,N} = \frac{C_V}{T}, \ \left(\frac{\partial S}{\partial T}\right)_{p,N} = -\left(\frac{\partial^2 G}{\partial T^2}\right)_{p,N} = \frac{C_p}{T},$$

$$\left(\frac{\partial V}{\partial p}\right)_{T,N} = \left(\frac{\partial^2 G}{\partial p^2}\right)_{T,N} = -V\beta_T, \ \left(\frac{\partial V}{\partial p}\right)_{S,N} = \left(\frac{\partial^2 H}{\partial p^2}\right)_{S,N} = -V\beta_S$$

- The first derivatives of intensive thermodynamic variables (T, p, μ) with respect to extensive thermodynamic variable (S, V, N) are materials moduli including thermal modulus, mechanical modulus (K_T), and chemical modulus. For example,

$$\left(\frac{\partial T}{\partial S}\right)_{p,N} = \left(\frac{\partial^2 H}{\partial S^2}\right)_{p,N} = \frac{T}{C_p}, \ \left(\frac{\partial p}{\partial V}\right)_{T,N} = -\left(\frac{\partial^2 F}{\partial V^2}\right)_{T,N} = -\frac{1}{V\beta_T} = -\frac{B_T}{V}$$

- The first derivatives of the densities of extensive thermodynamic variables $(\varepsilon_{ij}, D_i, B_i)$ with respect to their corresponding thermodynamic fields (σ_{ij}, E_i, H_i) are measures of materials susceptibility tensor properties such as elastic compliance tensor, dielectric permittivity tensor, and magnetic permeability tensor. For example, for a closed system,

$$\left(\frac{\partial \varepsilon_{ij}}{\partial \sigma_{kl}}\right)_{T,E,H} = -\left(\frac{\partial^2 g}{\partial \sigma_{ij}\partial \sigma_{kl}}\right)_{T,E,H} = s^E_{ijkl}, \left(\frac{\partial D_i}{\partial E_j}\right)_{T,\sigma,H}$$

$$= -\left(\frac{\partial^2 g}{\partial E_i \partial E_j}\right)_{T,\sigma,H} = \kappa^\sigma_{ij}$$

$$\left(\frac{\partial \varepsilon_{ij}}{\partial \sigma_{kl}}\right)_{T,D,H} = -\left(\frac{\partial^2 g}{\partial \sigma_{ij}\partial \sigma_{kl}}\right)_{T,D,H} = s^D_{ijkl}, \left(\frac{\partial D_i}{\partial E_j}\right)_{T,\varepsilon,H}$$

$$= -\left(\frac{\partial^2 g}{\partial E_i \partial E_j}\right)_{T,\varepsilon,H} = \kappa^\varepsilon_{ij}$$

- The first derivatives of thermodynamic fields (σ_{ij}, E, H) with respect to the densities of extensive thermodynamic variables $(\varepsilon_{ij}, D_i, B_i)$ are materials modulus tensors such as elastic stiffness or modulus tensor, dielectric stiffness or modulus tensor, and magnetic stiffness or modulus tensor. For example,

$$\left(\frac{\partial \sigma_{ij}}{\partial \varepsilon_{kl}}\right)_{T,E,H} = \left(\frac{\partial^2 f}{\partial \varepsilon_{ij}\partial \varepsilon_{kl}}\right)_{T,E,H} = C^E_{ijkl}, \left(\frac{\partial E_i}{\partial D_j}\right)_{T,\sigma,H}$$

$$= \left(\frac{\partial^2 f}{\partial D_i \partial D_j}\right)_{T,\sigma,H} = \left(\kappa^\sigma_{ij}\right)^{-1}$$

$$\left(\frac{\partial \sigma_{ij}}{\partial \varepsilon_{kl}}\right)_{T,D,H} = \left(\frac{\partial^2 f}{\partial \varepsilon_{ij}\partial \varepsilon_{kl}}\right)_{T,D,H} = C^D_{ijkl}, \left(\frac{\partial E_i}{\partial D_j}\right)_{T,\varepsilon,H}$$

$$= \left(\frac{\partial^2 f}{\partial D_i \partial D_j}\right)_{T,\varepsilon,H} = \left(\kappa^\varepsilon_{ij}\right)^{-1}$$

5.8 Examples

Example 1 The internal energy (u) per electron for the electrons in the conduction band can be approximately as $(3/2)k_B T + E_c$ where k_B is Boltzmann constant, T is temperature, and E_c is the conduction band edge energy or the minimum electron potential energy. What is the heat capacity of electrons at constant volume?

Solution
The constant volume heat capacity of free electrons in the conduction band is given by

$$c_v = \left(\frac{\partial u}{\partial T}\right)_v = \frac{3}{2}k_B, \text{ heat capacity per electron}$$

Example 2 The Gibbs free energy G of a monatomic ideal gas with N number of atoms as a function of temperature T and pressure p is given by

$$G(T, p, N) = -Nk_BT\left\{\ln\left[\frac{2k_BT}{p}\left(\frac{2\pi mk_BT}{h^2}\right)^{3/2}\right]\right\}$$

where k_B is the Boltzmann constant, m the effective mass of electrons, and h the Planck constant.

(a) Find the constant pressure heat capacity.
(b) Find the isothermal compressibility.

Solution

(a) The constant pressure heat capacity is given by

$$C_p = -T\left(\frac{\partial^2 G}{\partial T^2}\right)_{p,N} = \frac{5}{2}Nk_B$$

(b) The isothermal compressibility is given by

$$\beta_T = -\frac{1}{V}\left(\frac{\partial^2 G}{\partial p^2}\right)_{T,N} = \frac{1}{p}$$

Example 3 Consider a two-energy-level system with energy levels ε_o and ε_1, volume V, and N total number of independent particles, the Helmholtz free energy, entropy, and internal energy in terms of the energy levels, the number of particles and temperature are given by

$$F(T, V, N) = -Nk_BT\ln\left[e^{-\frac{\varepsilon_o}{k_BT}} + e^{-\frac{\varepsilon_1}{k_BT}}\right]$$

$$S(T, V, N) = Nk_B\ln\left[e^{-\frac{\varepsilon_o}{k_BT}} + e^{-\frac{\varepsilon_1}{k_BT}}\right] + \frac{N}{T}\frac{\varepsilon_oe^{-\frac{\varepsilon_o}{k_BT}} + \varepsilon_1e^{-\frac{\varepsilon_1}{k_BT}}}{\left[e^{-\frac{\varepsilon_o}{k_BT}} + e^{-\frac{\varepsilon_1}{k_BT}}\right]}$$

$$U(T, V, N) = N\frac{\varepsilon_oe^{-\frac{\varepsilon_o}{k_BT}} + \varepsilon_1e^{-\frac{\varepsilon_1}{k_BT}}}{\left[e^{-\frac{\varepsilon_o}{k_BT}} + e^{-\frac{\varepsilon_1}{k_BT}}\right]}$$

Determine the constant volume heat capacity from $F(T, V, N)$, $S(T, V, N)$, and $U(T, V, N)$.

Solution

The constant volume heat capacity is given by

$$C_V = \left(\frac{\partial U}{\partial T}\right)_{V,N} = T\left(\frac{\partial S}{\partial T}\right)_{V,N} = -T\left(\frac{\partial^2 F}{\partial T^2}\right)_{V,N} = \frac{N k_B \left(\frac{\varepsilon_1 - \varepsilon_0}{k_B T}\right)^2 e^{-\frac{\varepsilon_1 - \varepsilon_0}{k_B T}}}{\left(1 + e^{-\frac{\varepsilon_1 - \varepsilon_0}{k_B T}}\right)^2}$$

Example 4 The Helmholtz free energy of a blackbody radiation contained in a volume V at temperature T is given by

$$F(T, V) = -\frac{4\sigma}{3c} V T^4$$

where σ is the Stefan–Boltzmann constant, and c is the speed of light in vacuum.

(a) Derive an expression for the constant volume heat capacity as a function of T and V.

(b) Derive an expression for the isothermal bulk modulus as a function of T and V.

Solution

(a) The constant volume heat capacity is given by

$$C_V = -T\left(\frac{\partial^2 F}{\partial T^2}\right)_V = \frac{16\sigma}{c} V T^3$$

(b) The isothermal bulk modulus is given by

$$K_T = V\left(\frac{\partial^2 F}{\partial V^2}\right)_T = 0$$

Example 5 The electron density n at the conduction band edge of a semiconductor as a function of chemical potential of electrons is assumed to follow the Boltzmann distribution for electrons,

$$n = N_c \exp\left(\frac{E_f - E_c}{k_B T}\right)$$

where E_f is the Fermi level or the chemical potential of electrons, E_c is the conduction band edge energy, and N_c is the conduction band edge density of states. Obtain the chemical capacitance C_e of electrons per unit volume as a function of temperature and electron density.

Solution
The chemical capacitance of electrons per unit volume is given by

$$C_e = \left(\frac{\partial n}{\partial \mu} \right)_{T,V} = \left(\frac{\partial n}{\partial E_f} \right)_{T,V} = \frac{n}{k_B T}$$

5.9 Exercises

1. The internal energy of a monatomic ideal gas is only a function of temperature T and the number of moles N is given by

$$U(T, N) = \frac{3}{2} N R T$$

   Calculate the constant volume heat capacity C_V and the molar constant volume heat capacity c_v of the monatomic ideal gas.

2. The enthalpy of a monatomic ideal gas is only a function of temperature and is given by

$$H(T) = \frac{5}{2} N R T$$

   Calculate the constant pressure heat capacity C_p and the molar constant pressure heat capacity c_p of the monatomic ideal gas.

3. Calculate the difference between c_p and c_v for one mole of ideal gas at temperature T.

4. We consider a rubber at a certain temperature T as a polymer chain consisting of N rod-shaped monomers with length a. The Gibbs free energy G as a function of temperature T, force f, and number of monomers N is given by

$$G(T, f, N) = -N k_B T \ln \left[2 \cosh \left(\frac{af}{k_B T} \right) \right]$$

   where k_B is Boltzmann constant.

   (a) Find the heat capacity at constant f and N.
   (b) Find the thermal expansion coefficient at constant f and N.

5. Given the vibrational Helmholtz free energy of a crystal,

$$F(T) = Nu_o - 3Nk_B T \ln \frac{\exp(-\theta_E/T)}{1 - \exp(-\theta_E/T)}$$

where u_o and θ_E are constants independent of temperature, k_B is Boltzmann constant, N is the number of atoms, and T is temperature. Derive an expression for the constant volume heat capacity as a function of temperature T and number of atoms N.

6. The chemical potential of a monatomic ideal gas as a function of temperature T and pressure p is given by

$$\mu(T, p) = -RT \left\{ \ln \left[\frac{k_B T}{p} \left(\frac{2\pi m k_B T}{h^2} \right)^{3/2} \right] \right\}$$

where R is gas constant, m is atomic mass, k_B is the Boltzmann, and h is the Planck constant. Express the molar entropy of the monatomic ideal gas as a function of temperature and pressure.

(a) Express the molar volume of the monatomic ideal gas as a function of temperature and pressure.

(b) Express the molar constant pressure heat capacity of the monatomic ideal gas as a function of temperature and pressure.

(c) Express the molar constant volume heat capacity of the monatomic ideal gas as function of temperature and pressure

(d) Express the isothermal compressibility of the monatomic ideal gas as a function of temperature and pressure.

(e) Express the isothermal bulk modulus of the monatomic ideal gas as a function of temperature and pressure.

(f) Express the adiabatic compressibility of the monatomic ideal gas as a function of temperature and pressure.

(g) Express the adiabatic bulk modulus of the monatomic ideal gas as a function of temperature and pressure.

(h) Express the volume thermal expansion coefficient at constant pressure as a function of temperature and pressure.

7. Below are three different forms of fundamental equation of thermodynamics for an ideal gas:

$$\Xi(T, V, \mu) = -V k_B T \left(\frac{2\pi m k_B T}{h^2} \right)^{3/2} e^{\frac{\mu}{k_B T}}$$

$$\mu \left(T, \frac{V}{N} \right) = -k_B T \left\{ 1 + \ln \left[\frac{V}{N} \left(\frac{2\pi m k_B T}{h^2} \right)^{3/2} \right] \right\}$$

$$F(T, V, N) = -Nk_{\text{B}}T\left\{1 + \ln\left[\frac{V}{N}\left(\frac{2\pi mk_{\text{B}}T}{h^2}\right)^{3/2}\right]\right\}$$

where Ξ is grand potential energy, μ is chemical potential, F is Helmholtz free energy, N the number of atoms, T temperature, k_B the Boltzmann constant, V volume, m is atomic mass, and h is the Planck constant. Show that the chemical capacitance of the ideal gas at constant temperature and volume is given by

$$C_C = -\left(\frac{\partial^2 \Xi}{\partial \mu^2}\right)_{T,V} = \frac{1}{\left(\frac{\partial \mu}{\partial N}\right)_{T,V}} = \frac{1}{\left(\frac{\partial^2 F}{\partial N^2}\right)_{T,V}} = \frac{N}{k_{\text{B}}T}$$

8. The chemical potential of a material $\mu(G)$ as a function of temperature $T(K)$ and pressure p (bar) is approximated by

$$\mu(T, p) = A_0 + A_1 T + A_2 T\ln T + A_3 T^2 + A_4 p + A_5 Tp + A_6 p^2$$

where A_0, A_1, A_2, A_3, A_4, A_5, and A_6 are constants.

(a) Obtain an expression for the molar entropy of the material as a function of temperature and pressure.

(b) Obtain an expression for the molar enthalpy of the material as a function of temperature and pressure.

(c) Obtain an expression for the volume thermal expansion coefficient as a function of temperature and pressure.

(d) Obtain an expression for the isothermal compressibility of the material as a function of temperature and pressure.

(e) What is the amount of heat required to raise the temperature of one mole of the material by 100 K at 1 bar? (Express your result in terms of the constants A_0 to A_6 in the chemical potential expression)

(f) Obtain an expression for the molar constant volume heat capacity as a function of temperature and pressure.

9. The fundamental equation of thermodynamics of a monatomic Van der Waals gas with atomic mass m in terms of the Helmholtz free energy can be approximated as

$$F(T, V, N) = -\frac{aN^2}{V} - Nk_{\text{B}}T\left\{1 + \ln\left[\frac{(V - Nb)}{N}\left(\frac{2\pi mk_{\text{B}}T}{h^2}\right)^{3/2}\right]\right\}$$

where a and b are constants, N the number of atoms, T temperature, k_B the Boltzmann constant, V volume, and h the Planck constant.

(a) Express the constant volume heat capacity as a function of T, V, and N.

(b) Express the isothermal bulk modulus as a function of T, V, and N.

(c)   Express the chemical capacitance at constant temperature and volume of the monatomic Van der Waals gas as a function of T, V, and N.

10.   The chemical potentials of Si in diamond structure and liquid (l) Si as a function of temperature at 1 bar are given below using the enthalpy of Si in diamond (d) structure as the reference (from SGTE element database):

$$\mu_{Si}^{d} = -8162.6 + 137.23T - 22.832T \ln T - 1.9129 \times 10^{-3}T^{2}$$
$$- 0.003552 \times 10^{-6}T^{3} + 17667T^{-1} \quad G \ (298K < T < 1687K)$$

$$\mu_{Si}^{d} = -9457.6 + 167.28T - 27.196T \ln T$$
$$- 420.37 \times 10^{28}T^{-9} \quad G \ (1687K < T < 3600K)$$

$$\mu_{Si}^{l} = 42534 + 107.14T - 22.832T \ln T - 1.9129 \times 10^{-3}T^{2} - 0.003552 \times 10^{-6}T^{3}$$
$$+ 17667T^{-1} + 209.31 \times 10^{-23}T^{7} \quad G \ (298K < T < 1687K)$$

$$\mu_{Si}^{l} = 40371.0 + 137.72T - 27.196T \ln T \quad G \ (1687K < T < 3600K)$$

(a)   Derive the expressions for the molar enthalpy, molar entropy, and molar heat capacity of Si in diamond structure as functions of temperature.

(b)   Derive the expressions for the molar enthalpy, molar entropy, and molar heat capacity of liquid Si as functions of temperature.

Chapter 6
Relations Among Thermodynamic Properties

The thermodynamic properties obtained from the fundamental equations of thermodynamics through first and second derivatives of a fundamental equation of thermodynamics are mathematically related to each other and thus are not independent. The relations among different thermodynamic properties can be employed to obtain properties which are more difficult to experimentally measure or theoretically compute using those which are easier to measure or compute. The relations can also be used to examine if the properties obtained from different sources are consistent with each other. Therefore, they are extremely useful in thermodynamics. One of the most useful types of relations is the Maxwell relations which relate thermodynamic properties represented by mixed second derivatives of a fundamental energy or entropy function with respect to two variables belonging to different conjugate variable pairs, i.e., the coupling properties between thermal, mechanical, and chemical variables in a simple system. For example, thermal expansion coefficient and barocaloric effect are two properties which both involve the coupling between a mechanical variable and a thermal variable, and thus they must be related. For crystals, the coupled properties may involve tensorial mechanical, electric, and magnetic variables. Examples of coupled properties in a crystal include thermal expansion, thermoelectric, pyroelectric, piezoelectric, piezomagnetic, and magnetoelectric tensorial properties.

6.1 Maxwell Relations

Maxwell relations relate the mixed derivatives of a state function with respect to two of its variables. Let us start with the differential forms for the fundamental equations of thermodynamics. Since the most often-employed thermodynamic energy function is the Gibbs free energy function, we will start with the differential form for the Gibbs free energy as an example,

$$\mathrm{d}G = -S\mathrm{d}T + V\mathrm{d}p + \mu\mathrm{d}N \tag{6.1}$$

© Springer Nature Singapore Pte Ltd. 2022
L.-Q. Chen, *Thermodynamic Equilibrium and Stability of Materials*,
https://doi.org/10.1007/978-981-13-8691-6_6

From Eq. (6.1), one can write down the following first derivatives of G with respect to temperature T, pressure p, and the number of moles N,

$$\left(\frac{\partial G}{\partial T}\right)_{p,N} = -S, \quad \left(\frac{\partial G}{\partial p}\right)_{T,N} = V, \quad \left(\frac{\partial G}{\partial N}\right)_{T,p} = \mu \tag{6.2}$$

Now we take the mixed second derivatives of G with respect to temperature T and pressure p,

$$\left\{\frac{\partial}{\partial p}\left[\left(\frac{\partial G}{\partial T}\right)_{p,N}\right]\right\}_{T,N} = -\left(\frac{\partial S}{\partial p}\right)_{T,N}, \quad \left\{\frac{\partial}{\partial T}\left[\left(\frac{\partial G}{\partial p}\right)_{T,N}\right]\right\}_{p,N} = \left(\frac{\partial V}{\partial T}\right)_{p,N} \tag{6.3}$$

Since the Gibbs free energy is a property of a system, it is a state function. For a state function, the sequence of taking the first and second derivatives for a mixed second-order derivative does not matter, i.e.,

$$\left\{\frac{\partial}{\partial p}\left[\left(\frac{\partial G}{\partial T}\right)_{p,N}\right]\right\}_{T,N} = \left\{\frac{\partial}{\partial T}\left[\left(\frac{\partial G}{\partial p}\right)_{T,N}\right]\right\}_{p,N} \tag{6.4}$$

Therefore, we have a Maxwell relation

$$-\left(\frac{\partial S}{\partial p}\right)_{T,N} = \left(\frac{\partial V}{\partial T}\right)_{p,N} \tag{6.5}$$

Using the definition of volume thermal expansion coefficient α, we have

$$\left(\frac{\partial S}{\partial p}\right)_{T,N} = -\left(\frac{\partial V}{\partial T}\right)_{p,N} = -V\alpha \tag{6.6}$$

While the thermal expansion coefficient of a material is relatively easy to measure, we do not have a routine apparatus to directly measure the entropy change as a result of pressure change. However, we could use Eq. (6.6) to compute or estimate the change in entropy ΔS as a result of pressure change, e.g., from p_1 to p_2, using the information on the pressure dependence of thermal expansion coefficient as a function of pressure,

$$\Delta S = \int_{p_1}^{p_2}\left(\frac{\partial S}{\partial p}\right)_{T,N} dp = -\int_{p_1}^{p_2}\left(\frac{\partial V}{\partial T}\right)_{p,N} dp = -\int_{p_1}^{p_2} V\alpha \, dp \tag{6.7}$$

The entropy change as a result of pressure change at constant temperature $(\partial S/\partial p)_{T,N}$ is called the barocaloric effect.

Similarly, we can write down the mixed second derivatives of G with respect to T and N,

$$\left\{\frac{\partial}{\partial N}\left[\left(\frac{\partial G}{\partial T}\right)_{p,N}\right]\right\}_{T,p} = -\left[\frac{\partial S}{\partial N}\right]_{T,p}, \quad \left\{\frac{\partial}{\partial T}\left[\left(\frac{\partial G}{\partial N}\right)_{T,p}\right]\right\}_{p,N} = \left(\frac{\partial \mu}{\partial T}\right)_{p,N}$$

$$(6.8)$$

from which we have another Maxwell relation

$$-\left[\frac{\partial S}{\partial N}\right]_{T,p} = \left(\frac{\partial \mu}{\partial T}\right)_{p,N} \tag{6.9}$$

By definition, $(\partial S/\partial N)_{T,p}$ is the molar entropy s, and $(\partial \mu/\partial T)_{p,N}$ is $-s$, and hence we have

$$-\left[\frac{\partial S}{\partial N}\right]_{T,p} = \left(\frac{\partial \mu}{\partial T}\right)_{p,N} = -s \tag{6.10}$$

Therefore, the mixed derivatives of Gibbs free energy with respect to temperature and the number of moles do not generate a new relation since both mixed derivatives are definitions of negative of molar entropy.

We can also write down a final Maxwell relation based on Eq. (6.1) with respect to pressure and number of moles,

$$\left\{\frac{\partial}{\partial N}\left[\left(\frac{\partial G}{\partial p}\right)_{T,N}\right]\right\}_{T,p} = \left(\frac{\partial V}{\partial N}\right)_{T,p} = v,$$

$$\left\{\frac{\partial}{\partial p}\left[\left(\frac{\partial G}{\partial N}\right)_{T,p}\right]\right\}_{T,N} = \left(\frac{\partial \mu}{\partial p}\right)_{T,N} = v, \tag{6.11}$$

and hence,

$$\left(\frac{\partial V}{\partial N}\right)_{T,p} = \left(\frac{\partial \mu}{\partial p}\right)_{T,N} = v, \tag{6.12}$$

where v is the molar volume. The relation between mixed derivatives of Gibbs free energy with respect to pressure and the number of moles does not generate any new relation either.

From the above discussions, we can also see that it is sufficient to work with the chemical potential as a function of temperature and pressure rather than the Gibbs free energy of a material as a function of temperature, pressure, and the number of moles. The differential form of the fundamental equation of thermodynamics in terms of chemical potential as a function of temperature and pressure is essentially the Gibbs–Duhem relation for a single-component system or a multicomponent system

Table 6.1 Common Maxwell relations for simple systems

$\left(\frac{\partial S}{\partial p}\right)_T =$ $-\left(\frac{\partial V}{\partial T}\right)_p$	$\left(\frac{\partial T}{\partial V}\right)_S =$ $-\left(\frac{\partial p}{\partial S}\right)_V$	$\left(\frac{\partial S}{\partial V}\right)_T =$ $\left(\frac{\partial p}{\partial T}\right)_V$	$\left(\frac{\partial T}{\partial p}\right)_S =$ $\left(\frac{\partial V}{\partial S}\right)_p$
$\left(\frac{\partial s}{\partial p}\right)_T =$ $-\left(\frac{\partial v}{\partial T}\right)_p$	$\left(\frac{\partial T}{\partial v}\right)_s =$ $-\left(\frac{\partial p}{\partial s}\right)_v$	$\left(\frac{\partial s}{\partial v}\right)_T =$ $\left(\frac{\partial p}{\partial T}\right)_v$	$\left(\frac{\partial T}{\partial p}\right)_s =$ $\left(\frac{\partial v}{\partial s}\right)_p$

with a fixed chemical composition,

$$\mathrm{d}\mu = -s\mathrm{d}T + v\mathrm{d}p \tag{6.13}$$

where s and v are molar entropy and molar volume, respectively. We can immediately write down the Maxwell relation,

$$-\left(\frac{\partial s}{\partial p}\right)_T = \left(\frac{\partial v}{\partial T}\right)_p = v\alpha \tag{6.14}$$

We can write down similar sets of Maxwell relations based on the alternative expressions for the differential forms of the fundamental equation of thermodynamics in terms of internal energy, Helmholtz free energy, and enthalpy. For example,

$$\mathrm{d}U = T\mathrm{d}S - p\mathrm{d}V + \mu\mathrm{d}N, \quad \left(\frac{\partial T}{\partial V}\right)_{S,N} = -\left(\frac{\partial p}{\partial S}\right)_{V,N} \tag{6.15}$$

$$\mathrm{d}F = -S\mathrm{d}T - p\mathrm{d}V + \mu\mathrm{d}N, \quad \left(\frac{\partial S}{\partial V}\right)_{T,N} = \left(\frac{\partial p}{\partial T}\right)_{V,N} \tag{6.16}$$

$$\mathrm{d}H = T\mathrm{d}S + V\mathrm{d}p + \mu\mathrm{d}N, \quad \left(\frac{\partial T}{\partial p}\right)_{S,N} = \left(\frac{\partial V}{\partial S}\right)_{p,N} \tag{6.17}$$

Table 6.1 summarizes the most practically useful Maxwell relations for simple closed systems.

6.2 A Few Useful Strategies for Deriving Property Relations

To derive the relations between thermodynamic properties or to express thermodynamic derivatives in terms of well-known experimentally measurable thermodynamic properties, we often employ the following four tools:

(1)   The differential forms of the fundamental equations of thermodynamics.
(2)   The definitions of properties based on the first and second derivatives of thermodynamic energy functions.

(3) The Maxwell relations.
(4) The following two mathematical formulae for manipulating derivatives.

$$\left(\frac{\partial z}{\partial x}\right)_y = \frac{(\partial z/\partial w)_y}{(\partial x/\partial w)_y} \tag{6.18}$$

$$\left(\frac{\partial y}{\partial x}\right)_z = -\frac{(\partial z/\partial x)_y}{(\partial z/\partial y)_x} \tag{6.19}$$

Below we summarize six strategies to express derivatives in terms of thermodynamic properties and derive property relations.

- Strategy 1: For the derivative $(\partial z/\partial x)_y$, if either z or x is an extensive quantity while the other two are both intensive thermodynamic quantities, express the derivative in terms of a defined thermodynamic property such as heat capacity, compressibility, bulk modulus, or thermal expansion coefficient. If it cannot be directly expressed as a defined property, employ a Maxwell relation to convert it to another derivative which can be expressed as one of the defined, measurable thermodynamic properties.
- Strategy 2: For the derivative $(\partial z/\partial x)_y$, if z and x are both extensive thermodynamic quantities, either energy or amount of matter while y is an intensive thermodynamic potential quantity, choose another intensive parameter w (different from y), use Eq. (6.18) to express this derivative as a ratio of two derivatives, and then use Strategy 1 to express it in terms of a ratio of two defined properties.
- Strategy 3: For the derivative $(\partial y/\partial x)_z$, if either x or y or both are intensive potential quantities while z is an extensive variable representing a form of energy (or sometimes entropy), use Eq. (6.19) to express the derivative as a ratio of two derivatives, and then use Strategy 1 to express it in terms of a ratio of two defined properties.
- Strategy 4: For the derivative $(\partial z/\partial x)_y$, if z is a thermodynamic energy quantity while neither x nor y are a natural variable for z, start with the differential form for the fundamental equation of thermodynamics in terms of z, and use the above-described strategies if necessary to express the derivatives in terms of defined properties.
- Strategy 5: To derive an equation for the difference in a susceptibility or capacity, e.g., heat capacity, between two measurement conditions, e.g., constant volume or constant pressure, the first step is to write down the differential for the extensive thermodynamic variable, in this case, entropy S, involved in the definition of the capacity, e.g., the heat capacity, C_V, in terms of the corresponding intensive variable, in this case, temperature T, and the thermodynamic condition, in this case, constant volume V for a closed system with fixed number of moles N,

$$dS = \left(\frac{\partial S}{\partial T}\right)_V dT + \left(\frac{\partial S}{\partial V}\right)_T dV \tag{6.20}$$

- Strategy 6: To derive an equation for the difference in a susceptibility, e.g., elastic compliance tensor, between two measurement conditions, e.g., constant electric field (electrically free condition) or constant electric displacement (electrically clamped condition), the first step is to write down the differential for the density of the extensive thermodynamic variable, in this case, strain tensor ε_{ij}, involved in the definition of the susceptibility, e.g., the electrically clamped elastic compliance tensor, s_{ijkl}^{D}, in terms of the corresponding intensive variable, in this case, stress σ_{kl}, and the thermodynamic condition, in this case, constant electric displacement D_i, i.e., for a closed system,

$$d\varepsilon_{ij} = \left(\frac{\partial \varepsilon_{ij}}{\partial \sigma_{kl}}\right)_D d\sigma_{kl} + \left(\frac{\partial \varepsilon_{ij}}{\partial D_k}\right)_\sigma dD_k \qquad (6.21)$$

6.3 Examples of Applying Maxwell Relations

6.3.1 Barocaloric Effect

As mentioned above, one of the applications of Maxwell relations is to obtain properties that are difficult to measure experimentally using properties that are easier to measure. The barocaloric effect is a coupling property between mechanical and thermal contributions to the thermodynamics of a solid. It can be defined as the isothermal entropy change or the adiabatic temperature change of a solid as a result of an applied pressure, i.e.,

$$\left(\frac{\partial S}{\partial p}\right)_T \text{ or } \left(\frac{\partial T}{\partial p}\right)_S \qquad (6.22)$$

A direct measurement of entropy change as a result of pressure change is not possible although one could measure the amount of heat released as a result of the application of pressure to determine the entropy change. Since we cannot express the derivative of entropy with respect to pressure in terms of a defined property, we utilize a Maxwell relation to relate the isothermal entropy change with pressure to the isobaric volume change with temperature (Strategy 1), i.e.,

$$\left(\frac{\partial S}{\partial p}\right)_T = -\left(\frac{\partial V}{\partial T}\right)_p = -V\alpha \qquad (6.23)$$

where α is the volume thermal expansion coefficient which is reasonably easy to measure.

To relate the isothermal entropy change to adiabatic temperature change as a result of pressure change, we examine the change in entropy with temperature and pressure (Strategy 4). Under a reversible adiabatic condition, we have

$$dS = \left(\frac{\partial S}{\partial T}\right)_p dT + \left(\frac{\partial S}{\partial p}\right)_T dp = 0 \tag{6.24}$$

Therefore (Strategy 1),

$$dT = -\frac{(\partial S/\partial p)_T}{(\partial S/\partial T)_p} dp = -\frac{-V\alpha}{(C_p/T)} dp = \frac{TV\alpha}{C_p} dp \tag{6.25}$$

which shows that the barocaloric effect is proportional to the thermal expansion coefficient or the rate of volume change with temperature change.

We can also simply obtain the temperature change due to the barocaloric effect as

$$dT = \left(\frac{\partial T}{\partial p}\right)_S dp = -\frac{(\partial S/\partial p)_T}{(\partial S/\partial T)_p} dp = \frac{TV\alpha}{C_p} dp \tag{6.26}$$

where

$$\left(\frac{\partial T}{\partial p}\right)_S \tag{6.27}$$

is the practical measure of barocaloric effect, i.e., the rate of adiabatic temperature change due to the change in pressure. Since T, V, and C_p of a stable material are always positive, and the thermal expansion coefficient α can be positive or negative, the barocaloric effect can be either positive or negative.

6.3.2 Isothermal Volume Dependence of Entropy

It is generally believed that increasing volume of a material at constant temperature always increases the entropy of the material. Let us examine the isothermal volume dependence of entropy by writing down its thermodynamic expression,

$$\left(\frac{\partial S}{\partial V}\right)_T \tag{6.28}$$

which can be written as a ratio of two derivatives using Strategy 2,

$$\left(\frac{\partial S}{\partial V}\right)_T = \frac{(\partial S/\partial p)_T}{(\partial V/\partial p)_T} \tag{6.29}$$

We now use a Maxwell relation to express both derivatives on the right-hand side of Eq (6.29) in terms of measurable properties,

$$\left(\frac{\partial S}{\partial V}\right)_T = \frac{(\partial S/\partial p)_T}{(\partial V/\partial p)_T} = -\frac{(\partial V/\partial T)_p}{(\partial V/\partial p)_T} = \frac{\alpha}{\beta_T} \tag{6.30}$$

For a stable material, the isothermal compressibility is always positive. However, there are a small group of materials which have negative thermal expansion coefficients. For materials with negative thermal expansion coefficient, the entropy of the material decreases as volume increases at constant temperature, which is counterintuitive.

6.3.3 Joule Expansion Effect

In a Joule expansion process of a gas, a fixed amount of gas adiabatically expands into a vacuum, a process also called free expansion. This adiabatic expansion process is highly irreversible. Since the process is adiabatic, and there is no work interaction between the system and its surrounding due to the fact that the expansion is against a vacuum, i.e., zero pressure, the internal energy is constant during this process according to the first law of thermodynamics. Therefore, we could also call this process as an iso-energetic expansion process. Even though the process is highly irreversible, the temperature change is independent of the process path for the same volume change since temperature is a property of the system and thus a state function. Therefore, we can compute the temperature change assuming a reversible process. The Joule expansion effect can then be quantitatively defined as the iso-energetic temperature change as a result of volume change, i.e.,

$$\left(\frac{\partial T}{\partial V}\right)_U \tag{6.31}$$

To relate this effect to commonly measured thermodynamic properties, we use the mathematical relation [Eq. (6.19), Strategy 3] to rewrite the derivative [Expression (6.31)] as a ratio of two derivatives,

$$\left(\frac{\partial T}{\partial V}\right)_U = -\frac{(\partial U/\partial V)_T}{(\partial U/\partial T)_V} \tag{6.32}$$

We can immediately recognize that the denominator of the right-hand side of Eq. (6.31) is the constant volume heat capacity,

$$\left(\frac{\partial U}{\partial T}\right)_V = C_V \tag{6.33}$$

To obtain the isothermal internal energy change as a result of volume change, we examine the differential form of the fundamental equation of thermodynamics in terms of internal energy for a closed system since temperature T is not a natural variable for internal energy U (Strategy 4),

$$dU = TdS - pdV \tag{6.34}$$

Dividing both sides of the above equation by dV, we have

$$\frac{dU}{dV} = T\frac{dS}{dV} - p \tag{6.35}$$

Now let us keep temperature constant when we take the two derivatives in the above equation and use the mathematical relation (Eq. (6.18), Strategy 2) since both S and V are extensive quantities while T is an intensive quantity. We can rewrite Eq. (6.35) as

$$\left(\frac{\partial U}{\partial V}\right)_T = T\left(\frac{\partial S}{\partial V}\right)_T - p = T\frac{(\partial S/\partial p)_T}{(\partial V/\partial p)_T} - p \tag{6.36}$$

By using the Maxwell relation,

$$\left(\frac{\partial S}{\partial p}\right)_T = -\left(\frac{\partial V}{\partial T}\right)_p, \tag{6.37}$$

we have

$$\left(\frac{\partial U}{\partial V}\right)_T = T\frac{-(\partial V/\partial T)_p}{(\partial V/\partial p)_T} - p \tag{6.38}$$

Using the definitions for the isobaric volume thermal expansion coefficient α and the isothermal compressibility β_T, we have

$$\left(\frac{\partial U}{\partial V}\right)_T = T\frac{(1/V)(\partial V/\partial T)_p}{-(1/V)(\partial V/\partial p)_T} - p = \frac{T\alpha}{\beta_T} - p = T\alpha B_T - p \tag{6.39}$$

where B_T is the isothermal bulk modulus. Therefore, the Joule expansion effect or the rate of temperature change dT with respect to volume change dV at constant internal energy is then given by

$$\left(\frac{\partial T}{\partial V}\right)_U = -\frac{(\partial U/\partial V)_T}{(\partial U/\partial T)_V} = -\frac{1}{C_V}\left(\frac{T\alpha}{\beta_T} - p\right) = -\frac{1}{C_V}(T\alpha B_T - p) \tag{6.40}$$

For a finite volume change, the above equation will have to be integrated to obtain the finite temperature change for the free expansion process, which requires the volume dependences of the quantities on the right-hand side of Eq. (6.40). Using Eq. (6.40), it is easy to show that the Joule expansion effect is zero for ideal gases since for ideal gases $T\alpha = 1$ and $\beta_T p = 1$.

6.3.4   Joule–Thomson Expansion Effect

The Joule–Thomson expansion effect is the temperature change of a fluid as a result of pressure change, while the enthalpy of the fluid is constant. It has a number of applications including refrigeration and liquification of gases. The magnitude of the Joule–Thomson expansion effect of a fluid is thermodynamically quantified as

$$\left(\frac{\partial T}{\partial p}\right)_H \tag{6.41}$$

Using the mathematical relation [Eq. (6.19), strategy 3] since T and p are intensive, and H is an extensive energy quantity, we can rewrite the above derivative [Expression Eq. (6.41)] as

$$\left(\frac{\partial T}{\partial p}\right)_H = -\frac{(\partial H/\partial p)_T}{(\partial H/\partial T)_p} \tag{6.42}$$

We can immediately recognize that the denominator of the right-hand side of Eq. (6.42) is the constant pressure heat capacity,

$$\left(\frac{\partial H}{\partial T}\right)_p = C_p \tag{6.43}$$

To relate the partial derivative in the numerator of Eq. (6.42) to familiar thermodynamic properties, we use the differential form of the fundamental equation of thermodynamics in terms of enthalpy since temperature T is not a natural variable for H (Strategy 4),

$$dH = T\,dS + V\,dp \tag{6.44}$$

Dividing both sides of the above equation by dp, we have

$$\frac{dH}{dp} = T\frac{dS}{dp} + V \tag{6.45}$$

Let us keep temperature constant for the two derivatives of the above equation, we can rewrite Eq. (6.45) as

$$\left(\frac{\partial H}{\partial p}\right)_T = T\left(\frac{\partial S}{\partial p}\right)_T + V \tag{6.46}$$

By using the Maxwell relation (Strategy 1),

$$\left(\frac{\partial S}{\partial p}\right)_T = -\left(\frac{\partial V}{\partial T}\right)_p, \tag{6.47}$$

we have

$$\left(\frac{\partial H}{\partial p}\right)_T = -T\left(\frac{\partial V}{\partial T}\right)_p + V \tag{6.48}$$

Using the definitions for the isobaric volume thermal expansion coefficient α, we have

$$\left(\frac{\partial H}{\partial p}\right)_T = V(1 - T\alpha) \tag{6.49}$$

Therefore, the magnitude of the Joule–Thomson effect of a fluid is given by

$$\left(\frac{\partial T}{\partial p}\right)_H = -\frac{(\partial H/\partial p)_T}{(\partial H/\partial T)_p} = \frac{V(T\alpha - 1)}{C_p} \tag{6.50}$$

As one can easily see that this effect is zero for ideal gases since for ideal gases, $T\alpha = 1$. However, it is nonzero for real gases and liquids.

6.3.5   Grüneisen Parameter

The Grüneisen parameter represents the temperature increase of a material as the material is adiabatically, reversibly compressed. For an adiabatic reversible process, the entropy of the material is constant, while the degree of compression is measured by the relative volume change dV/V. Therefore, the dimensionless Grüneisen parameter Υ is the relative temperature change (dT/T) as a result of relative volume change at constant entropy, i.e.,

$$\Upsilon = -\left[\frac{(\partial T/T)}{(\partial V/V)}\right]_S = -\left(\frac{\partial \ln T}{\partial \ln V}\right)_S = -\frac{V}{T}\left[\frac{\partial T}{\partial V}\right]_S = V\left(\frac{\partial p}{\partial U}\right)_V \tag{6.51}$$

where the negative sign is used to ensure that Υ is positive since during an isentropic compression, the temperature increases, while the volume decreases.

We can express the Grüneisen parameter using experimentally measurable properties by employing Maxwell relations and mathematical manipulations. For example, using the mathematical relation [Eq. (6.19), Strategy 3] gives

$$\left(\frac{\partial T}{\partial V}\right)_S = -\frac{(\partial S/\partial V)_T}{(\partial S/\partial T)_V} \tag{6.52}$$

Now we can rewrite Eq. (6.51) as

$$\Upsilon = -\frac{V}{T}\left[\frac{\partial T}{\partial V}\right]_S = \frac{V}{T}\frac{(\partial S/\partial V)_T}{(\partial S/\partial T)_V} \tag{6.53}$$

Then by using following mathematical relation [Eq. (6.18), Strategy 2],

$$\left(\frac{\partial S}{\partial V}\right)_T = \frac{(\partial S/\partial p)_T}{(\partial V/\partial p)_T} \tag{6.54}$$

we can rewrite Eq. (6.53) as

$$\Upsilon = \frac{V}{T}\frac{(\partial S/\partial p)_T}{(\partial S/\partial T)_V(\partial V/\partial p)_T} \tag{6.55}$$

Now using the following Maxwell relation (Strategy 1),

$$-\left(\frac{\partial S}{\partial p}\right)_T = \left(\frac{\partial V}{\partial T}\right)_p \tag{6.56}$$

we can rewrite Eq. (6.55) as

$$\Upsilon = \frac{V}{T}\frac{-(\partial V/\partial T)_p}{(\partial S/\partial T)_V(\partial V/\partial p)_T} \tag{6.57}$$

We recall the definitions of thermal expansion coefficient α, molar constant volume heat capacity c_v, and isothermal compressibility β_T,

$$\Upsilon = \frac{V\alpha}{Nc_v\beta_T} = \frac{v\alpha}{c_v\beta_T} = \frac{\alpha}{cc_v\beta_T}, \tag{6.58}$$

where v is molar volume (V/N), and $c = N/V$. Therefore, the Grüneisen parameter is zero for harmonic crystals in which the thermal expansion coefficient is zero.

Based on the relations among isothermal and adiabatic compressibility and bulk moduli and between constant volume and constant pressure heat capacity, we can also write the Gruneisen parameter as

$$\Upsilon = \frac{\alpha}{cc_p\beta_S} = \frac{\alpha B_T}{cc_v} = \frac{\alpha B_S}{cc_p} \tag{6.59}$$

where β_S is the adiabatic compressibility, B_T is the isothermal bulk modulus, B_S is the adiabatic bulk modulus, and c_p is the constant pressure heat capacity.

For a monoatomic ideal gas, $\alpha_p = 1/T$, $c_v = 3R/2$, $\beta_T = 1/p$, and hence

$$\Upsilon = \frac{\alpha}{cc_v\beta_T} = \frac{1/T}{c(3R/2)(1/p)} = \frac{pV}{NT(3R/2)} = \frac{2}{3} \tag{6.60}$$

6.3.6 Property Relations Between Coupled Properties in Crystals

We consider possible thermal, mechanical, electric, and magnetic contributions to thermodynamics of a crystal. The coupled physical properties of a crystal are given by the mixed second derivatives of a thermodynamic energy function with respect to two thermodynamic variables. The coupled properties are related through Maxwell relations.

6.3.6.1 Thermal Strain and Elastocaloric Effect

The thermal strain and elastocaloric effects involve thermomechanical coupling. The differential form for the Gibbs free energy per unit volume g including only the thermal and mechanical contributions is

$$dg = -sdT - \varepsilon_{ij}d\sigma_{ij} \tag{6.61}$$

where s is the entropy per unit volume (Note: We omit the subscript v for labeling a density quantity on a per unit volume basis in this section). The Maxwell relation is

$$\left(\frac{\partial \varepsilon_{ij}}{\partial T}\right)_{\sigma,E,H} = \alpha_{ij} = \left(\frac{\partial s}{\partial \sigma_{ij}}\right)_{T,E,H} = \alpha'_{ij} \tag{6.62}$$

where α_{ij} is the second-rank thermal expansion coefficient tensor, and α'_{ij} is the second-rank elastocaloric coefficient tensor.

6.3.6.2 Pyroelectric and Electrocaloric Effects

The pyroelectric and electrocaloric effects involve coupling between thermal and dielectric contributions. The differential form for the Gibbs free energy density with only thermal and dielectric contributions is

$$dg = -sdT - D_i dE_i \tag{6.63}$$

The Maxwell relation is

$$\left(\frac{\partial D_i}{\partial T}\right)_{\sigma,E,H} = p_i = \left(\frac{\partial s}{\partial E_i}\right)_{T,\sigma,H} = p_i' \tag{6.64}$$

where p_i is the pyroelectric coefficient vector, and p_i' is the electrocaloric coefficient vector.

6.3.6.3 Pyromagnetic and Magnetocaloric Effects

The pyromagnetic and magnetocaloric effects involve coupling between thermal and magnetic contributions. The differential form for the Gibbs free energy density with only thermal and magnetic contributions is

$$dg = -s dT - B_i dH_i \tag{6.65}$$

The Maxwell relation is

$$\left(\frac{\partial B_i}{\partial T}\right)_{\sigma,E,H} = \pi_i = \left(\frac{\partial s}{\partial H_i}\right)_{T,\sigma,E} = \pi_i' \tag{6.66}$$

where π_i is the pyromagnetic coefficient vector, and π_i' is the magnetocaloric coefficient vector.

6.3.6.4 Piezoelectric Effects

The direct and converse piezoelectric effects involve coupling between mechanical and dielectric contributions. The differential form for the Gibbs free energy density with only mechanical and dielectric contributions is

$$dg = -\varepsilon_{ij} d\sigma_{ij} - D_i dE_i \tag{6.67}$$

The Maxwell relation is

$$\left(\frac{\partial D_i}{\partial \sigma_{jk}}\right)_{T,E,H} = d_{ijk} = \left(\frac{\partial \varepsilon_{ij}}{\partial E_k}\right)_{T,\sigma,H} = d_{ijk}' \tag{6.68}$$

where d_{ijk} is the third-rank direct piezoelectric constant tensor, and d_{ijk}' is the third-rank converse piezoelectric constant tensor.

6.3.6.5 Piezomagnetic Effects

The direct and converse piezomagnetic effects involve coupling between mechanical and magnetic contributions. The differential form for the Gibbs free energy density with only mechanical and magnetic contributions is

$$dg = -\varepsilon_{ij} d\sigma_{ij} - B_i dH_i \tag{6.69}$$

The Maxwell relation is

$$\left(\frac{\partial B_i}{\partial \sigma_{jk}} \right)_H = z_{ijk} = \left(\frac{\partial \varepsilon_{ij}}{\partial H_k} \right)_\sigma = z'_{ijk} \tag{6.70}$$

where z_{ijk} is the third-rank direct piezomagnetic constant tensor, and z'_{ijk} is the third-rank converse piezomagnetic constant tensor.

6.3.6.6 Magnetoelectric Effects

The magnetoelectric effects involve coupling between magnetic and dielectric contributions. The differential form for the Gibbs free energy density with only magnetic and dielectric contributions is

$$dg = -D_i dE_i - BdH_i \tag{6.71}$$

The Maxwell relation is

$$\left(\frac{\partial D_i}{\partial H_j} \right)_E = \beta_{ij} = \left(\frac{\partial B_i}{\partial E_i} \right)_H = \beta'_{ij} \tag{6.72}$$

where β_{ij} is the second-rank magnetoelectric constant tensor, and β'_{ij} is the second-rank electromagnetic tensor.

6.4 Summary Comments on Relationships Between Coupled Properties

Based on the discussions in the last few sections, we can summarize a few general observations on the relationships between coupled properties and relationships between derivatives and properties:

- The first derivative of an extensive variable or a density of an extensive variable with respect to an intensive variable or a thermodynamic field belonging to another thermodynamic variable pair is a coupled property. Two corresponding coupled properties are related by a Maxwell relation. For example,

$$\left(\frac{\partial V}{\partial T}\right)_p = \left(\frac{\partial^2 G}{\partial T \partial p}\right) = V\alpha, \quad -\left(\frac{\partial S}{\partial p}\right)_T$$

$$= \left(\frac{\partial^2 G}{\partial p \partial T}\right), \quad \left(\frac{\partial V}{\partial T}\right)_p = -\left(\frac{\partial S}{\partial p}\right)_T$$

$$\left(\frac{\partial D_i}{\partial \sigma_{jk}}\right)_E = -\left(\frac{\partial^2 g}{\partial \sigma_{jk} \partial E_i}\right) = d_{ijk}, \quad \left(\frac{\partial \varepsilon_{ij}}{\partial E_k}\right)_\sigma$$

$$= -\left(\frac{\partial^2 g}{\partial E_k \partial \sigma_{ij}}\right) = d'_{ijk}, \quad d_{ijk} = d'_{ijk}$$

- The first derivative of an extensive variable or density of an extensive variable with respect to another extensive variable or density of an extensive variable is a ratio of two thermodynamic properties (See Strategy 2 and 1 above). For example,

$$\left(\frac{\partial S}{\partial V}\right)_T = \frac{\left(\frac{\partial S}{\partial p}\right)_T}{\left(\frac{\partial V}{\partial p}\right)_T} = \frac{-\left(\frac{\partial V}{\partial T}\right)_p}{\left(\frac{\partial V}{\partial p}\right)_T} = \frac{a}{\beta_T}$$

$$\left(\frac{\partial D_i}{\partial \varepsilon_{jk}}\right)_E = \left(\frac{\partial D_i}{\partial \sigma_{lm}}\right)_E \left[\left(\frac{\partial \varepsilon_{jk}}{\partial \sigma_{lm}}\right)_E\right]^{-1} = d_{ilm}^E \left(S_{jklm}^E\right)^{-1}$$

$$\left(\frac{\partial \varepsilon_{ij}}{\partial D_k}\right)_\sigma = \left(\frac{\partial \varepsilon_{ij}}{\partial E_l}\right)_\sigma \left[\left(\frac{\partial D_k}{\partial E_l}\right)_\sigma\right]^{-1} = d_{ijl}^\sigma \left(\kappa_{kl}^\sigma\right)^{-1}$$

- The first derivative of an intensive variable or a thermodynamic field with respect to another intensive variable or another field is a ratio of two thermodynamic properties (See Strategy 3 above and 1). For example,

$$\left(\frac{\partial T}{\partial p}\right)_S = -\frac{\left(\frac{\partial S}{\partial p}\right)_T}{\left(\frac{\partial S}{\partial T}\right)_p} = -\frac{-\left(\frac{\partial V}{\partial T}\right)_p}{\frac{C_p}{T}} = \frac{TVa}{C_p} = \frac{Tva}{c_p}$$

$$\left(\frac{\partial E_i}{\partial \sigma_{jk}}\right)_D = -\left(\frac{\partial D_l}{\partial \sigma_{jk}}\right)_E \left[\left(\frac{\partial D_l}{\partial E_i}\right)_\sigma\right]^{-1} = -d_{ljk}^E \left(k_{li}^\sigma\right)^{-1}$$

6.5   Examples

Example 1 Express the rate of enthalpy change with respect to volume change during an adiabatic, free volume expansion.

Solution

It is an adiabatic, free expansion, so there is neither heat transfer nor work interaction, and hence the internal energy is constant during the process. Therefore, the enthalpy change with respect to volume at constant internal energy is given by

$$\left(\frac{\partial H}{\partial V}\right)_U = \left[\frac{\partial (U + pV)}{\partial V}\right]_U = p + V\left(\frac{\partial p}{\partial V}\right)_U = p - V\frac{\left(\frac{\partial U}{\partial V}\right)_p}{\left(\frac{\partial U}{\partial p}\right)_V}$$

$$= p - V\frac{T\left(\frac{\partial S}{\partial V}\right)_p - p}{T\left(\frac{\partial S}{\partial p}\right)_V}$$

Expressing the derivatives in the above expression in terms of measurable properties, we have

$$\left(\frac{\partial H}{\partial V}\right)_U = p - V\frac{\frac{C_p}{V\alpha} - p}{\frac{C_V\beta_T}{\alpha}} = p - \frac{C_p - pV\alpha}{C_V\beta_T} = p - \frac{C_p - pv\alpha}{c_v\beta_T}$$

Example 2 Using the Maxwell relation relating the two mixed second derivatives of internal energy U with respect to temperature T and volume V to prove that the volume dependence of constant volume heat capacity at constant temperature is given by.

$$\left(\frac{\partial C_V}{\partial V}\right)_T = T\left(\frac{\partial^2 p}{\partial T^2}\right)_V$$

Solution

$$\left(\frac{\partial C_V}{\partial V}\right)_T = \left[\frac{\partial}{\partial V}\left(\frac{\partial U}{\partial T}\right)_V\right]_T = \left[\frac{\partial}{\partial T}\left(\frac{\partial U}{\partial V}\right)_T\right]_V = \left[\frac{\partial}{\partial T}\left(T\left(\frac{\partial S}{\partial V}\right)_T - p\right)\right]_V$$

Therefore,

$$\left(\frac{\partial C_V}{\partial V}\right)_T = \left[\frac{\partial}{\partial T}\left(T\left(\frac{\partial p}{\partial T}\right)_V - p\right)\right]_V = T\left(\frac{\partial^2 p}{\partial T^2}\right)_V$$

Example 3 Show that the difference between the constant pressure and constant volume heat capacity, $c_p - c_v$, is given by

$$c_p - c_v = vT\frac{\alpha^2}{\beta_T}$$

where v is molar volume, T is temperature, α is volume thermal expansion coefficient, and β_T is the isothermal compressibility.

Solution

Since heat capacity involves the molar entropy change ds with respect to temperature change dT, and we would like to determine the change of entropy with respect to temperature at either constant volume or constant pressure. We first express the change in molar entropy ds as a result of temperature change dT and molar volume change dv (See Strategy 5)

$$ds = \left(\frac{\partial s}{\partial T} \right)_v dT + \left(\frac{\partial s}{\partial v} \right)_T dv$$

Dividing both sides of the above equation by dT while keeping pressure p constant and then multiply both sides by T, we have

$$T \left(\frac{\partial s}{\partial T} \right)_p = T \left(\frac{\partial s}{\partial T} \right)_v + T \left(\frac{\partial s}{\partial v} \right)_T \left(\frac{\partial v}{\partial T} \right)_p$$

According to the definitions of constant volume and constant pressure heat capacities in terms of molar entropy (See Strategy 1 above), we can rewrite the above equation as

$$c_p = c_v + T \left(\frac{\partial s}{\partial v} \right)_T \left(\frac{\partial v}{\partial T} \right)_p$$

We can then use the mathematical manipulations (Strategy 2) and a Maxwell relation as discussed above (See Strategy 1), and the definitions of thermal expansion coefficient and isothermal compressibility to rewrite the above equation,

$$c_p = c_v + T \left(\frac{\partial s/\partial p}{\partial v/\partial p} \right)_T \left(\frac{\partial v}{\partial T} \right)_p = c_v - T \frac{(\partial v/\partial T)_p}{(\partial v/\partial p)_T} \left(\frac{\partial v}{\partial T} \right)_p$$

or

$$c_p - c_v = vT \frac{\alpha^2}{\beta_T}$$

which can also be expressed as

$$c_p = c_v [1 + \Upsilon \alpha T]$$

where Υ is the Grüneisen parameter [see Eq. (6.58)].

Example 4 Show that the difference between mechanically free (constant stress or stress-free) dielectric permittivity κ_{ij}^σ and mechanically clamped (constant strain) dielectric permittivity κ_{ij}^ε at constant temperature is given by

$$\kappa_{ij}^{\sigma} - \kappa_{ij}^{\varepsilon} = C_{mnkl}^{E} d_{imn} d_{jkl}$$

where d_{imn} and d_{jkl} are piezoelectric coefficient tensors, and C_{mnkl}^{E} is the elastic stiffness tensor measured at constant electric field (or electrically free stiffness tensor).

Solution

We start with expressing the change in the electric displacement vector dD_i as a result of the change in the electric field dE_j and change in strain $d\varepsilon_{kl}$ (See Strategy 6),

$$dD_i = \left(\frac{\partial D_i}{\partial E_j}\right)_{\varepsilon} dE_j + \left(\frac{\partial D_i}{\partial \varepsilon_{kl}}\right)_{E} d\varepsilon_{kl}$$

Dividing both sides of the above equation by dE_j while keeping stress σ_{ij} constant, we have

$$\left(\frac{\partial D_i}{\partial E_j}\right)_{\sigma} = \left(\frac{\partial D_i}{\partial E_j}\right)_{\varepsilon} + \left(\frac{\partial D_i}{\partial \varepsilon_{kl}}\right)_{E}\left(\frac{\partial \varepsilon_{kl}}{\partial E_j}\right)_{\sigma}$$

which can be rewritten as (Strategy 2)

$$\left(\frac{\partial D_i}{\partial E_j}\right)_{\sigma} = \left(\frac{\partial D_i}{\partial E_j}\right)_{\varepsilon} + \left(\frac{\partial D_i/\partial \sigma_{mn}}{\partial \varepsilon_{kl}/\partial \sigma_{mn}}\right)_{E}\left(\frac{\partial \varepsilon_{kl}}{\partial E_j}\right)_{\sigma}$$

or

$$\left(\frac{\partial D_i}{\partial E_j}\right)_{\sigma} = \left(\frac{\partial D_i}{\partial E_j}\right)_{\varepsilon} + \left(\frac{\partial \sigma_{mn}}{\partial \varepsilon_{kl}}\right)_{E}\left(\frac{\partial D_i}{\partial \sigma_{mn}}\right)_{E}\left(\frac{\partial \varepsilon_{jk}}{\partial E_l}\right)_{\sigma}$$

Therefore

$$\left(\frac{\partial D_i}{\partial E_j}\right)_{\sigma} - \left(\frac{\partial D_i}{\partial E_j}\right)_{\varepsilon} = \left(\frac{\partial \sigma_{mn}}{\partial \varepsilon_{kl}}\right)_{E}\left(\frac{\partial D_i}{\partial \sigma_{mn}}\right)_{E}\left(\frac{\partial \varepsilon_{jk}}{\partial E_l}\right)_{\sigma}$$

Using the definitions of dielectric permittivity, elastic modulus, and piezoelectric coefficients (See Strategy 1), we have

$$\kappa_{ij}^{\sigma} - \kappa_{ij}^{\varepsilon} = C_{mnkl}^{E} d_{imn} d_{jkl}$$

6.6 Exercises

1. Given the following properties of Si solid at room temperature: molar constant pressure heat capacity $c_p = 19.789$ J/(mol K), density $\rho = 2.3290$ g/cm^3, linear thermal expansion coefficient, $\alpha_l = 2.6 \times 10^{-6}$/K, and the isothermal bulk modulus $B_T = 97.6$ GPa, calculate the molar constant volume heat capacity c_v of Si at room temperature.

2. Using data from exercise 1 for Si, calculate the molar entropy change of Si at room temperature and 1 bar after it is isothermally compressed to 10 MPa using the following two different assumptions:

    (a) Assuming that the thermal expansion coefficient is independent of pressure and Si is incompressible.

    (b) Assuming that the thermal expansion coefficient and the compressibility (or bulk modulus) are both constant and are independent of pressure.

3. Show that the relative difference between constant volume and constant pressure heat capacity can also be expressed as

$$\frac{c_p - c_v}{c_v} = \Upsilon \alpha T$$

where Υ is the Grüneisen parameter, a is volume thermal expansion coefficient, and T is temperature.

4. Show that if

$$\left(\frac{\partial c_p}{\partial p}\right)_T = 0,$$

the volume thermal expansion coefficient is given by (Hint: start with Strategy 1)

$$\alpha = \frac{1}{T}$$

5. Prove that for a closed system (Hint: start with Strategy 4),

$$\left(\frac{\partial U}{\partial V}\right)_T = \frac{T\alpha}{\beta_T} - p$$

6. Prove that for a closed system (Hint: The best approach is to start with an entropic representation for the fundamental equation of thermodynamics with natural variables $(1/T)$ and V),

$$\left(\frac{\partial U}{\partial V}\right)_T = T^2\left[\frac{\partial}{\partial T}\left(\frac{p}{T}\right)\right]_V$$

7. Prove that for a closed system (Hint: Start with the Maxwell relation from the differential form for the fundamental equation of thermodynamics for a simple closed system and recognize that $TdS = dU$ under the condition of constant volume V),

$$V\left(\frac{\partial p}{\partial U}\right)_V = -\frac{V}{T}\left[\frac{\partial T}{\partial V}\right]_S$$

8. Using the Maxwell relation, which relates the two mixed second derivatives of enthalpy H with respect to temperature T and pressure p, to show that

$$\left(\frac{\partial C_p}{\partial p}\right)_T = -TV\left[\alpha^2 + \left(\frac{\partial \alpha}{\partial T}\right)_p\right]$$

where α is the volume thermal expansion coefficient.

9. Show that the difference between the isothermal compressibility β_T and adiabatic compressibility β_S is given by (Hint: Start with Strategy 5)

$$\beta_T = \beta_S + vT\frac{\alpha^2}{c_p}$$

where v is molar volume, T is temperature, α is volume thermal expansion coefficient, and c_p is molar constant pressure heat capacity.

10. Express molar volume v of a material as a function of temperature T and pressure p in terms of c_p, α, and β_T assuming c_p, α, β_T are constants independent of temperature and pressure. (Hint: Start with a differential form for v as a function of T and p)

11. Show that the difference between the isothermal elastic compliance tensor s_{ij}^T and the adiabatic elastic compliance s_{ij}^S at constant electric field is given by

$$s_{ijkl}^T - s_{ijkl}^S = \frac{T\alpha_{ij}\alpha_{kl}}{c_\sigma}$$

where α_{ij} is thermal expansion coefficient tensor, and c_σ is heat capacity per unit volume measured at constant stress. (Hint: Start with Strategy 5).

12. Show that the difference between the electrically free elastic compliance tensor s_{ij}^E and the electrically clamped elastic compliance s_{ij}^D at constant temperature is given by

$$s^{E}_{ijkl} - s^{D}_{ijkl} = \left(\kappa^{-1}_{mn}\right)^{\sigma} d_{mij} d_{nkl}$$

where d_{mij} and d_{nkl} are piezoelectric tensors, and κ^{-1}_{mn} is the inverse to the dielectric permittivity tensor κ_{mn}. (Hint: Start with Strategy 6).

13. Show that the difference between the heat capacity $C_{V,N}$ measured at constant volume V and constant number of moles N and the heat capacity $C_{V,\mu}$ measured at constant volume V and constant chemical potential μ is given by (Hint: Start with Strategy 5 or 6)

$$C_{V,N} = C_{V,\mu} - T \left(\frac{\partial N}{\partial \mu}\right)_{T,V} \left(\frac{\partial \mu}{\partial T}\right)^{2}_{V,N}$$

Chapter 7
Equilibrium and Stability

This chapter discusses the equilibrium states of a material system and their stability criteria. An equilibrium state is a stationary state at which the properties of a system no longer change with time, and there are no fluxes of matter, i.e., no mass diffusion, no heat transport, and no electric current, going through the system.

A stable equilibrium state is at the global minimum in the potential energy function with respect to an extensive thermodynamic variable of a system and is stable against any arbitrary fluctuations. A state that is time-stationary and stable with respect to small perturbations but unstable with respect to some large perturbations that would allow the system to reach a more stable state is a metastable state (see Fig. 7.1). A metastable equilibrium state is located at a local minimum in the potential energy landscape. An unstable equilibrium state is at a local maximum of the potential energy landscape with a negative local curvature.

In this chapter, we discuss the thermodynamic principles to determine when a system reaches thermodynamic equilibrium, the equilibrium conditions for the potentials, and the criteria to evaluate whether an equilibrium state is stable. The initial non-equilibrium state typically arises from the changes in the external conditions that lead to the changes in certain thermodynamic parameters in the system. There are two equivalent basic principles that can be used to determine the thermodynamic equilibrium of a system: one is the maximum entropy principle, and the other is the minimum energy principle.

7.1 Maximum Entropy Principle

We first consider the condition of equilibrium for an isolated system with an initial amount of entropy S and chemical matter N enclosed in a volume V with rigid walls that are impermeable to any chemical species and non-conducting to heat or entropy transfer, i.e.,

© Springer Nature Singapore Pte Ltd. 2022
L.-Q. Chen, *Thermodynamic Equilibrium and Stability of Materials*,
https://doi.org/10.1007/978-981-13-8691-6_7

Fig. 7.1 Illustration of
stable, metastable, and
unstable equilibrium states

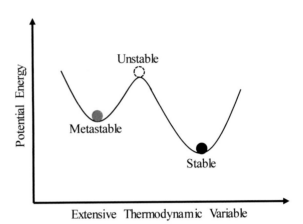

$$dS^e = dV^e = dN^e = 0 \tag{7.1}$$

where dS^e, dV^e, and dN^e are the infinitesimal amounts of exchanges in entropy,
volume, and substance between the system and its surrounding. Since there are no
interactions between the system and its surrounding, the internal energy must be a
constant ($dU = 0$) according to the first law of thermodynamics.

During the evolution of the isolated system from an initially non-equilibrium state
to the final equilibrium state, the second law of thermodynamics states that

$$dS = dS^e + dS^{ir} = dS^{ir} = \frac{D}{T}d\xi > 0 \tag{7.2}$$

i.e., the entropy of the system is increasing due to the entropy production or dissipation of thermodynamic driving force D during irreversible processes for the system
evolving from the initially non-equilibrium state ($D > 0$, $\xi = 0$) to the final equilibrium state ($D = 0$, $\xi = \xi_{eq}$, where ξ_{eq} is the equilibrium extent of the irreversible
process that maximizes the entropy). After the system reaches equilibrium, there is
no more entropy creation

$$dS = dS^{ir} = \frac{D}{T}d\xi = 0 \tag{7.3}$$

i.e., the entropy is maximized after the system reaches equilibrium. Therefore, the
equilibrium condition can be obtained using the entropy maximum principle:

- *For an isolated system at thermodynamic equilibrium, its entropy S is maximized
  at constant energy* $dU = 0$, *constant volume* $dV = 0$, *and constant amount of
  substance* $dN = 0$. Expressing this maximum entropy principle using calculus,

$$(dS)_{U,V,N} = 0 \tag{7.4}$$

$$\left(\delta^2 S\right)_{U,V,N} < 0 \tag{7.5}$$

where $\left(\delta^2 S\right)_{U,V,N}$ is the second-order variation of S at constant U, V, and N. Equation (7.4) ensures that the entropy of the system is at an extremum (maximum or minimum) while Eq. (7.5) guarantees that the extremum is a maximum, i.e., any perturbation to the equilibrium state while keeping the U, V, and N of the system constant will decrease the entropy of the system.

7.2 Minimum Internal Energy Principle

The above maximum entropy principle formulated based on an isolated system can be employed to derive a minimum internal energy principle by combining the first and second laws of thermodynamics. Under constant volume and constant number of moles,

$$dV = dV^e = 0, dN = dN^e = 0$$

According to the first law of thermodynamics, we have

$$dU = T dS^e \tag{7.6}$$

Now the entropy change dS for the system is equal to the sum of the entropy transferred dS^e from the surrounding to the system and the entropy produced dS^{ir}, i.e.,

$$dS = dS^e + dS^{ir} \tag{7.7}$$

Therefore,

$$dU = T dS^e = T dS - T dS^{ir} \tag{7.8}$$

In the maximum entropy principle, the entropy keeps increasing during the irreversible evolution process from the initially non-equilibrium state to the equilibrium state until the entropy is maximized while the internal energy is constant, which is the case for an isolated system. However, if we now assume that we could maintain a constant entropy for the system,

$$dS = 0$$

Equation (7.8) becomes

$$dU = -T dS^{ir} = -D d\xi \leq 0 \tag{7.9}$$

i.e., the internal energy must keep decreasing during the irreversible process if the entropy of the system remains constant. However, in order for the system to maintain constant entropy, entropy must be removed from the system during an irreversible process which produces entropy. Therefore, in this case, the system is not isolated as entropy is removed from the system to maintain constant entropy, and the internal energy of the system decreases due to the removal of entropy during the process.

After the system establishes thermodynamic equilibrium, all the driving force D is consumed, the entropy production stops, and internal energy U no longer changes with time and reaches minimum. Expressing internal energy minimization using mathematics, we have

$$(dU)_{S,V,N} = 0 \tag{7.10}$$

i.e., the internal energy of a macroscopic system at equilibrium is at an extreme value at a given set of constant thermodynamic variables: entropy, volume, and number of moles.

For the internal energy to be at minimum at equilibrium, it requires that

$$\left(\delta^2 U\right)_{S,V,N} > 0 \tag{7.11}$$

where $\left(\delta^2 U\right)_{S,V,N}$ is the second-order variation of U at constant S, V, and N.

Equations (7.10) and (7.11) define the energy minimum principle and the equilibrium states of a system: *The internal energy U of a system is minimized at equilibrium at constant entropy S, volume V, and amount of substance N, and any perturbation to the equilibrium state while keeping S, V, and N of the system constant increases the internal energy U of the system.*

7.3 Minimum Enthalpy Principle

Under constant pressure and constant number of moles,

$$dV = dV^e, dN = dN^e = 0$$

Hence, the first law of thermodynamics becomes

$$dU = TdS^e - pdV \tag{7.12}$$

Now applying the second law of thermodynamics by taking into account the entropy produced dS^{ir}, the combined first and second law of thermodynamics becomes

$$dU = TdS^e - pdV = TdS - TdS^{ir} - pdV \tag{7.13}$$

Moving the $p\,dV$ term in Eq. (7.13) to the left-hand side, we have

$$dU + p\,dV = T\,dS^e = T\,dS - T\,dS^{ir} \tag{7.14}$$

Under constant entropy and pressure,

$$dU + p\,dV = d(U + pV) = dH = -T\,dS^{ir} = -D\,d\xi \leq 0 \tag{7.15}$$

i.e., enthalpy must keep decreasing during an irreversible process if the pressure and entropy of the system are kept constant. After the system establishes thermodynamic equilibrium, enthalpy H reaches minimum. At equilibrium,

$$(dH)_{S,p,N} = 0 \tag{7.16}$$

i.e., the enthalpy of a macroscopic system at equilibrium is at an extreme value at a given set of constant thermodynamic variables: entropy, pressure, and number of moles.

Minimum enthalpy at equilibrium requires that

$$\left(\delta^2 H\right)_{S,p,N} > 0 \tag{7.17}$$

where $\left(\delta^2 H\right)_{S,p,N}$ is the second-order variation of H at constant S, p, and N.

Therefore, at equilibrium, the enthalpy H of a system is minimized at constant entropy S, constant pressure p, and constant amount of substance N, and any perturbation to the equilibrium state while keeping the S, p, and N of the system constant will increase the enthalpy H of the system.

7.4 Minimum Helmholtz Free Energy Principle

The combined first and second law of thermodynamics for a closed system at constant volume is

$$dU = T\,dS^e = T\left(dS - dS^{ir}\right) = T\,dS - T\,dS^{ir} \tag{7.18}$$

Moving the $T\,dS$ term in Eq. (7.18) to the left-hand side, we have

$$dU - T\,dS = -T\,dS^{ir} \tag{7.19}$$

If we assume that the temperature of the system is constant, we get

$$dU - T\,dS = d(U - TS) = dF = -T\,dS^{ir} = -D\,d\xi \leq 0 \tag{7.20}$$

i.e., the Helmholtz free energy must keep decreasing during an irreversible process if the temperature and volume of the system remain constant. After the system establishes thermodynamic equilibrium, the Helmholtz free energy reaches minimum. At equilibrium,

$$(dF)_{T,V,N} = 0 \tag{7.21}$$

i.e., the Helmholtz free energy of a macroscopic system at equilibrium is at an extreme value at a given set of constant thermodynamic variables: temperature, volume, and number of moles.

For the Helmholtz free energy to be at minimum at equilibrium, it requires that

$$\left(\delta^2 F\right)_{T,V,N} > 0 \tag{7.22}$$

where $\left(\delta^2 F\right)_{T,V,N}$ is the second-order variation of F at constant T, V, and N.

Therefore, at equilibrium, the Helmholtz free energy F of a system is minimized at constant temperature T, constant volume V, and constant amount of substance N, and any perturbation to the equilibrium state, while keeping T, V, and N of the system constant, will increase the Helmholtz free energy F of the system.

7.5   Minimum Gibbs Free Energy Principle

The combined first and second law of thermodynamics for a closed system is

$$dU = T\left(dS - dS^{ir}\right) - pdV = TdS - TdS^{ir} - pdV \tag{7.23}$$

Moving the TdS and $-pdV$ terms in Eq. (7.23) to the left-hand side, we have

$$dU - TdS + pdV = -TdS^{ir} \tag{7.24}$$

If we assume that the temperature and pressure of the system are constant, we have

$$dU - TdS + pdV = d(U - TS + pV) = dG = -TdS^{ir} = -Dd\xi \leq 0 \tag{7.25}$$

i.e., the Gibbs free energy must keep decreasing during an irreversible process if the temperature and pressure of the system are maintained constant. After the system establishes thermodynamic equilibrium, the Gibbs free energy reaches minimum. At equilibrium,

$$(dG)_{T,p,N} = 0 \tag{7.26}$$

i.e., the Gibbs free energy of a macroscopic system at equilibrium is at an extreme value at a given set of constant thermodynamic variables: temperature, pressure, and number of moles.

For the Gibbs free energy to be at minimum at equilibrium, it requires that

$$\left(\delta^2 G\right)_{T,p,N} > 0 \tag{7.27}$$

where $\left(\delta^2 G\right)_{T,p,N}$ is the second-order variation of G at constant T, p, and N.

According to above Eqs. (7.26) and (7.27), *the Gibbs free energy or free enthalpy G of a system is minimized at equilibrium at constant temperature T, constant pressure p, and constant amount of substance N,* and any perturbation to the equilibrium state while keeping T, p, and overall N of the system constant increases the Gibbs free energy G of the system.

7.6 Equilibrium Conditions for Potentials

At thermodynamic equilibrium, the entropy of the isolated system is maximized at constant internal energy, volume, and number of moles, or an energy function is minimized under a specified set of thermodynamic conditions. Now we examine the consequence of the equilibrium condition to the thermodynamic potentials T, p, and μ in the system.

Example 1 Equilibrium condition for temperature T

Consider an isolated system containing two hypothetical subsystems separated by a wall which is fixed and impermeable to chemical species, but is diathermal, allowing entropy or heat transfer (see Fig. 7.2). According to the above maximum entropy principle, the total entropy of the isolated composite system is maximized at equilibrium at the fixed total energy, $dU_1 + dU_2 = 0$, total volume, $dV_1 + dV_2 = 0$, and total number of moles, $dN_1 + dN_2 = 0$, i.e.,

$$dS_{tot} = dS_1 + dS_2 = 0 \tag{7.28}$$

Since the wall separating the two subsystems is fixed, $dV_1 = dV_2 = 0$, and impermeable $dN_1 = dN_2 = 0$, we have

Fig. 7.2 Thermal equilibrium between two subsystems

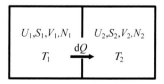

| U_1,S_1,V_1,N_1 | U_2,S_2,V_2,N_2 |
| T_1 | $\xrightarrow{dQ}$ | T_2 |

$$dS_{\text{tot}} = dS_1 + dS_2 = \frac{dU_1}{T_1} + \frac{dU_2}{T_2} = \left(\frac{1}{T_2} - \frac{1}{T_1}\right)dQ = 0 \qquad (7.29)$$

Furthermore, the amount of heat transfer dQ is arbitrary, and thus the maximization of entropy requires that

$$\frac{1}{T_2} - \frac{1}{T_1} = 0$$

or

$$T_1 = T_2 \qquad (7.30)$$

i.e., the temperature of an isolated system at thermodynamic equilibrium is uniform throughout the system.

Example 2 Equilibrium condition for pressure p

Now we consider a composite system at a fixed temperature $dT = 0$, fixed total volume $dV = 0$, and fixed total amount of substance $dN = 0$ (see Fig. 7.3). The wall separating the two subsystems is now movable but still impermeable $dN_1 = dN_2 = 0$. In this case, it is convenient to use the Helmholtz free energy F. The differential form for F is

$$dF = -SdT - pdV + \mu dN$$

Therefore,

$$dF_{\text{tot}} = dF_1 + dF_2 = -p_1 dV_1 - p_2 dV_2 \qquad (7.31)$$

However, the total volume of the combined composite system is fixed, i.e., $dV_1 + dV_2 = 0$. Hence at equilibrium,

$$dF_{\text{tot}} = (p_2 - p_1)dV_1 = 0 \qquad (7.32)$$

Since the change dV_1 is arbitrary, *the pressure has to be uniform at equilibrium,*

$$p_1 = p_2 \qquad (7.33)$$

Fig. 7.3 Mechanical equilibrium between two subsystems

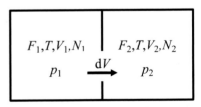

Fig. 7.4 Chemical equilibrium between two subsystems

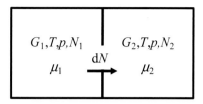

Example 3 Equilibrium condition for chemical potential μ

The differential form for the Gibbs free energy is

$$dG = -SdT + Vdp + \mu dN$$

Now consider two subsystems with chemical potentials μ_1 and μ_2 and the same temperature and pressure as shown in Fig. 7.4. Since T and p are uniform,

$$dG = \mu dN \tag{7.34}$$

If the wall separating the two subsystems is allowed mass transfer, then

$$dG_{tot} = dG_1 + dG_2 = -\mu_1 dN + \mu_2 dN \tag{7.35}$$

At equilibrium,

$$dG_{tot} = 0 = (\mu_2 - \mu_1)dN \Rightarrow \mu_2 = \mu_1 = \mu \tag{7.36}$$

i.e., at equilibrium, it requires that the chemical potential be uniform in the system.

7.7 General Equilibrium Conditions for Potentials

In general, all the potentials in a simple system at equilibrium are uniform,

$$T = \text{uniform}, \ p = \text{uniform}, \ \mu = \text{uniform}$$

If any of the potentials is not uniform, i.e., if there is a spatial gradient in any of the potentials, the system is not at equilibrium, and there will be corresponding transport of thermal, mechanical, or chemical matter. A thermal potential or temperature gradient leads to thermal transport or heat transfer or entropy transfer; a pressure gradient leads to volume flow; and chemical potential gradients of chemical species lead to mass transfer.

If there are differences in potentials at the same location, corresponding kinetic processes are expected to take place at the location. For example, there can be phase transitions or chemical reactions if they reduce the chemical potential of a component or a combination of several components at a given spatial location. All these processes dissipate chemical energy and create entropy, essentially converting chemical energy to thermal energy.

It should be cautioned that the uniform potential condition is derived assuming no internal constraints or heterogeneity that prevent the transport of corresponding matter within the system. For example, in a multiphase system, if one of the chemical species is not soluble in one of the phases or in one spatial region, it implies the chemical potential of that species in that phase or region must be higher than the rest regions of the system. This is analogous to the scenario for the water levels of a lake with many small islands. The water level in the lake is uniform at equilibrium, but the water levels on those islands would be higher if there is water (or lower if there are wells whose water levels are lower than that of the lake, but water is not allowed to transport between the wells and the lake).

7.8 Thermodynamic Fields

We define a potential as the amount of specific form of energy per unit amount of the corresponding matter, i.e., the energy intensity. On the other hand, a field can be defined as the negative gradient of a potential. For example, the electric field, $\vec{E}$, is the negative gradient of electric potential, $-\nabla\phi$. The chemical field for a chemical species i is the negative chemical potential gradient of species i, $-\nabla\mu_i$, or the negative difference between a final state and initial state, $D = -\Delta\mu = \mu^{\text{initial}} - \mu^{\text{final}}$, the thermodynamic driving force.

If the chemical potential of a species is not uniform in a system, the species will move from a spatial region with high chemical potential to another region with low chemical potential until equilibrium is achieved with uniform chemical potential, or the chemical potential of a species or a combination of several chemical species will decrease until it reaches its minimum at a given location. Chemical potential gradient is the driving force if a process is taking place due to the chemical potential difference of a species at two different spatial locations, leading to chemical diffusion or flow. This is analogous to an object moving down a slope driven by the height difference under the gravitational field.

Chemical potential change is the driving force if a process is taking place at the same spatial location, temperature, pressure, and composition, e.g., for phase transitions without composition changes, or chemical reactions or phase transitions with composition changes (decomposition or precipitation reactions), but the overall composition of each individual elemental species is maintained constant. This is analogous to the height change of a body at the same location under a gravitational potential field.

If the temperature is not uniform, a thermal field exists. The thermal field is the negative temperature gradient, $-\nabla T$. Heat or entropy will flow from a spatial region with high temperature to other regions with lower temperature until equilibrium is achieved with uniform temperature.

If the pressure or stress is not uniform in a material, there exists a mechanical field. A mechanical field is the pressure gradient, ∇p, or negative stress gradient, $-\nabla \sigma_{ij}$. Volume is expected to flow from a low- to high-pressure region, or a region at high pressure will expand against another region with low pressure until equilibrium is achieved with uniform pressure.

7.9 Thermodynamic Stability Criteria

Thermodynamic equilibrium states can be stable, metastable, or unstable. Here, we further discuss the criteria that can be employed to determine if a thermodynamic state is intrinsically stable, metastable, or unstable. Unstable systems under a given thermodynamic condition will spontaneously evolve, with small perturbations, to a different equilibrium state which is stable or is at least metastable under the same thermodynamic condition.

When a system is at a stable equilibrium state, any perturbation, large or small, to the equilibrium state, e.g., spatial perturbations in any of the thermodynamic quantities such as entropy, temperature, volume, pressure, the number of moles, and chemical potential of a chemical substance in a system under a given set of thermodynamic conditions will spontaneously fall back to the equilibrium state. A thermodynamically metastable equilibrium state is stable against small perturbations to the equilibrium state, e.g., spatial variations with very small amplitudes to any of the thermodynamic quantities, but unstable against sufficiently large perturbations.

There are several approaches that we can employ to derive the same set of stability criteria. One is to use the entropic representation in which the entropy of a stable equilibrium state of an isolated system is at a maximum, and any perturbation to the stable equilibrium state, while maintaining the same total internal energy, volume, and amount of substance, decreases the entropy. A second approach is to use one of the energy representations, e.g., internal energy at constant total amount of entropy, volume, and number of moles, or enthalpy at constant entropy, pressure, and number of moles, or Helmholtz free energy at constant temperature, volume, and number of moles, or Gibbs free energy at constant temperature, pressure, and number of moles. Any perturbation to a stable equilibrium state increases internal energy, or enthalpy, or Helmholtz free energy, or Gibbs free energy depending on the thermodynamic conditions. The third approach is to consider the amount of entropy produced when an equilibrium state is subjected to a perturbation. The amount of entropy produced arising from perturbing a stable equilibrium state is negative.

According to the second law of thermodynamics, small changes in entropy dS, internal energy dU, enthalpy dH, Helmholtz free energy dF, and Gibbs free energy dG arising from small perturbations to a stable equilibrium state can all be directly related to the amount of entropy produced dS^{ir} under different thermodynamic conditions, i.e.,

Fig. 7.5 A small perturbation to an initially equilibrium state leading to two systems with the same total internal energy $2U$, volume $2V$, and number of moles $2N$ as the initial state

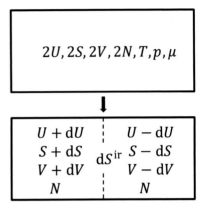

$$dS^{ir} = Dd\xi / T$$

$$= \begin{bmatrix} dS & \text{constant internal energy, volume, and number of moles} \\ -dU/T & \text{constant entropy, volume, and number of moles} \\ -dH/T & \text{constant entropy, pressure, and number of moles} \\ -dF/T & \text{constant temperature, volume, and number of moles} \\ -dG/T & \text{constant temperature, pressure, and number of moles} \end{bmatrix}$$

Therefore, the stability criteria for different thermodynamic conditions can be formulated using the amount of entropy produced when a stable equilibrium state is subject to small perturbations in any or a combination of thermodynamic variables. For simplicity, we first only consider the perturbations to the thermal and mechanical variables (entropy S, temperature T, volume V, and pressure p). We will discuss the stability of equilibrium states with regard to chemical variables (chemical potentials and the number of moles of each species or phases) in Chapter 12 on binary or multicomponent solutions.

Let us consider a simple stable equilibrium system with total internal energy $2U$, entropy $2S$, volume $2V$, and number of moles $2N$ hypothetically separated into two identical systems with each having energy U, entropy S, and volume V plus small perturbations $\pm dU, \pm dS$, and $\pm dV$ while keeping their overall total entropy, volume, and the number of moles at the same $2U$, $2V$, and $2N$ as the original state (Fig. 7.5). It should be pointed out that since the thermodynamic variables are coupled to each other, the small perturbations in internal energy, entropy, and volume also lead to small perturbations in temperature and pressure. Following Kondepudi and Prigogine,[1] the entropy produced ΔS^{ir} from small perturbations dS, dT, dV, and dp to a stable equilibrium state can be obtained by a Taylor series expansion of the entropy of the perturbed system with respect to the original stable state to second order in entropy $\delta^2 S$, i.e.,

---

[1] Dilip Kondepudi and Ilya Prigogine, Modern Thermodynamics: From Heat Engines to Dissipative Structures, John Wiley & Sons, England 1998.

$$\Delta S^{\text{ir}} = \delta^2 S = -\frac{1}{T}(\mathrm{d}S\mathrm{d}T - \mathrm{d}V\mathrm{d}p) < 0 \tag{7.37}$$

It is reminded that since the original state is an equilibrium state, all the thermo-dynamic variables are uniform, and the first-order variation in entropy $\delta S = 0$ at the initial uniform internal energy, volume, and number of moles.

Now we consider different energetic representations under different thermody-namic conditions. For example, the increase in the internal energy due to the pertur-bations in entropy $\mathrm{d}S$ and volume $\mathrm{d}V$ in the initially equilibrium state with energy U, entropy S, and volume V while maintaining constant overall entropy S and volume V is given by

$$\Delta U = -T\Delta S^{\text{ir}} = \mathrm{d}S\mathrm{d}T - \mathrm{d}V\mathrm{d}p > 0 \tag{7.38}$$

The perturbations in temperature and pressure can be represented in terms of perturbations in entropy and volume,

$$\mathrm{d}T = \left(\frac{\partial T}{\partial S}\right)_V \mathrm{d}S + \left(\frac{\partial T}{\partial V}\right)_S \mathrm{d}V \tag{7.39}$$

$$\mathrm{d}p = \left(\frac{\partial p}{\partial S}\right)_V \mathrm{d}S + \left(\frac{\partial p}{\partial V}\right)_S \mathrm{d}V \tag{7.40}$$

Substituting Eqs. (7.39) and (7.40) into Eq. (7.38), we have

$$\Delta U = \mathrm{d}S\left[\left(\frac{\partial T}{\partial S}\right)_V \mathrm{d}S + \left(\frac{\partial T}{\partial V}\right)_S \mathrm{d}V\right] - \mathrm{d}V\left[\left(\frac{\partial p}{\partial S}\right)_V \mathrm{d}S + \left(\frac{\partial p}{\partial V}\right)_S \mathrm{d}V\right] > 0 \tag{7.41}$$

or

$$\Delta U = \left(\frac{\partial T}{\partial S}\right)_V (\mathrm{d}S)^2 + \left(\frac{\partial T}{\partial V}\right)_S \mathrm{d}S\mathrm{d}V - \left(\frac{\partial p}{\partial S}\right)_V \mathrm{d}V\mathrm{d}S - \left(\frac{\partial p}{\partial V}\right)_S (\mathrm{d}V)^2 > 0 \tag{7.42}$$

If the original equilibrium state is stable or metastable, any infinitesimal pertur-bation should increase the internal energy of the system, $\Delta U > 0$. For example, if we consider only perturbation in entropy, i.e., $\mathrm{d}V = 0$ in Eq. (7.42), we have

$$\Delta U = \left(\frac{\partial T}{\partial S}\right)_V (\mathrm{d}S)^2 = \left(\frac{\partial^2 U}{\partial S^2}\right)_V (\mathrm{d}S)^2 > 0 \tag{7.43}$$

Since $(\mathrm{d}S)^2$ is always positive, we have the following conclusion that for a stable or metastable equilibrium state,

$$\left(\frac{\partial^2 U}{\partial S^2}\right)_V > 0 \tag{7.44}$$

Similarly, we may consider only a small perturbation to the volume, and we reach a similar conclusion that for a stable or metastable equilibrium state,

$$\Delta U = -\left(\frac{\partial p}{\partial V}\right)_S (dV)^2 = \left(\frac{\partial^2 U}{\partial V^2}\right)_S (dV)^2 > 0 \tag{7.45}$$

Since $(dV)^2$ is always positive, we have the following criterion for a stable or metastable equilibrium state,

$$\left(\frac{\partial^2 U}{\partial V^2}\right)_S > 0 \tag{7.46}$$

If we consider simultaneous small perturbations in entropy and volume, we have

$$\Delta U = \left(\frac{\partial T}{\partial S}\right)_V (dS)^2 + \left(\frac{\partial T}{\partial V}\right)_S dSdV - \left(\frac{\partial p}{\partial S}\right)_V dVdS - \left(\frac{\partial p}{\partial V}\right)_S (dV)^2 > 0 \tag{7.47}$$

Taking into the Maxwell relation between $(\partial T/\partial V)_S$ and $-(\partial p/\partial S)_V$, we can rewrite the above equation as

$$\Delta U = \left(\frac{\partial^2 U}{\partial S^2}\right)_V (dS)^2 + 2\frac{\partial^2 U}{\partial S\partial V}dSdV + \left(\frac{\partial^2 U}{\partial V^2}\right)_S (dV)^2 > 0 \tag{7.48}$$

Equation (7.48) can be rearranged as

$$\Delta U = \left(\frac{\partial^2 U}{\partial S^2}\right)_V \left[dS + \frac{\frac{\partial^2 U}{\partial S\partial V}}{\left(\frac{\partial^2 U}{\partial S^2}\right)_V}dV\right]^2 + \left[\left(\frac{\partial^2 U}{\partial V^2}\right)_S - \frac{\left(\frac{\partial^2 U}{\partial S\partial V}\right)^2}{\left(\frac{\partial^2 U}{\partial S^2}\right)_V}\right](dV)^2 > 0 \tag{7.49}$$

Since $(\partial^2 U/\partial S^2)_V > 0$ for a stable equilibrium state, to guarantee that a thermodynamic equilibrium state is a stable or metastable state requires

$$\left[\left(\frac{\partial^2 U}{\partial V^2}\right)_S - \frac{\left(\frac{\partial^2 U}{\partial S\partial V}\right)^2}{\left(\frac{\partial^2 U}{\partial S^2}\right)_V}\right] = \left[\frac{\left(\frac{\partial^2 U}{\partial S^2}\right)_V \left(\frac{\partial^2 U}{\partial V^2}\right)_S - \left(\frac{\partial^2 U}{\partial S\partial V}\right)^2}{\left(\frac{\partial^2 U}{\partial S^2}\right)_V}\right] > 0 \tag{7.50}$$

or

$$\left(\frac{\partial^2 U}{\partial S^2}\right)_V \left(\frac{\partial^2 U}{\partial V^2}\right)_S - \left(\frac{\partial^2 U}{\partial S \partial V}\right)^2 = \begin{vmatrix} \left(\frac{\partial^2 U}{\partial S^2}\right)_V & \frac{\partial^2 U}{\partial S \partial V} \\ \frac{\partial^2 U}{\partial S \partial V} & \left(\frac{\partial^2 U}{\partial V^2}\right)_S \end{vmatrix} > 0 \qquad (7.51)$$

We can express the above stability criteria using measurable or computable thermodynamic properties. For example,

$$\left(\frac{\partial^2 U}{\partial S^2}\right)_V = \left(\frac{\partial T}{\partial S}\right)_{V,N} = \frac{T}{C_V} > 0 \Rightarrow C_V > 0 \qquad (7.52)$$

i.e., for a thermodynamically stable or metastable equilibrium state, the constant volume heat capacity has to be positive. For the second criterion,

$$\left(\frac{\partial^2 U}{\partial V^2}\right)_S = -\left(\frac{\partial p}{\partial V}\right)_S = \frac{1}{V \beta_S} > 0 \Rightarrow \beta_S > 0 \qquad (7.53)$$

i.e., for a thermodynamically stable or metastable equilibrium state, the adiabatic compressibility has to be positive.

For the third stability criterion, we can rewrite the criterion (Eq. 7.50) as[2]

$$\left[\left(\frac{\partial^2 U}{\partial V^2}\right)_S - \frac{\left(\frac{\partial^2 U}{\partial S \partial V}\right)^2}{\left(\frac{\partial^2 U}{\partial S^2}\right)_V}\right] = \left(\frac{\partial^2 F}{\partial V^2}\right)_T > 0 \qquad (7.54)$$

where F is the Helmholtz free energy. Therefore, the third criterion implies

$$\left(\frac{\partial^2 F}{\partial V^2}\right)_{T,N} = -\left(\frac{\partial p}{\partial V}\right)_{T,N} = \frac{1}{V \beta_T} > 0 \Rightarrow \beta_T > 0 \qquad (7.55)$$

i.e., for a thermodynamically stable or metastable equilibrium state, the isothermal compressibility has to be positive.

Furthermore, we can rewrite the inequality (Eq. 7.54) as

$$\left(\frac{\partial^2 F}{\partial V^2}\right)_T - \left(\frac{\partial^2 U}{\partial V^2}\right)_S = -\frac{\left(\frac{\partial^2 U}{\partial S \partial V}\right)^2}{\left(\frac{\partial^2 U}{\partial S^2}\right)_V} < 0 \qquad (7.56)$$

Based on Eq. (7.56), it can be deduced that the derivative

$$\left(\frac{\partial^2 F}{\partial V^2}\right)_T$$

---

[2] Michael Modell and Robert C. Reid, Thermodynamics and Its Applications, Second Edition, Prentice-Hall International, 1983.

will reach zero before

$$\left(\frac{\partial^2 U}{\partial V^2}\right)_S$$

reaches zero as a material approaches instability. Therefore, the third criterion will be violated first before the second criterion is violated as a material approaches instability. As a matter of fact, the third stability criterion Eq. (7.55) is sufficient to guarantee that all three criteria (Eqs. 7.52, 7.53, and 7.55) are satisfied.

Similarly, we may employ the perturbations to a stable equilibrium state under the condition of constant entropy and pressure,

$$\Delta H = -T\Delta S^{\text{ir}} = \mathrm{d}S\mathrm{d}T - \mathrm{d}V\mathrm{d}p > 0 \qquad (7.57)$$

In this case, we express the perturbations in temperature and volume in terms of perturbations in entropy and pressure,

$$\mathrm{d}T = \left(\frac{\partial T}{\partial S}\right)_p \mathrm{d}S + \left(\frac{\partial T}{\partial p}\right)_S \mathrm{d}p \qquad (7.58)$$

$$\mathrm{d}V = \left(\frac{\partial V}{\partial S}\right)_p \mathrm{d}S + \left(\frac{\partial V}{\partial p}\right)_S \mathrm{d}p \qquad (7.59)$$

Substituting Eqs. (7.58) and (7.59) into Eq. (7.57), we have

$$\Delta H = -T\Delta S^{\text{ir}} = \left(\frac{\partial T}{\partial S}\right)_p (\mathrm{d}S)^2 - \left(\frac{\partial V}{\partial p}\right)_S (\mathrm{d}p)^2 > 0 \qquad (7.60)$$

or

$$\Delta H = -T\Delta S^{\text{ir}} = \left(\frac{\partial^2 H}{\partial S^2}\right)_p (\mathrm{d}S)^2 - \left(\frac{\partial^2 H}{\partial p^2}\right)_S (\mathrm{d}p)^2 > 0 \qquad (7.61)$$

Therefore, for a thermodynamically stable state,

$$\left(\frac{\partial^2 H}{\partial S^2}\right)_p > 0, \quad \left(\frac{\partial^2 H}{\partial p^2}\right)_S < 0 \qquad (7.62)$$

which implies that for a stable equilibrium state.

$$\left(\frac{\partial^2 H}{\partial S^2}\right)_p = \frac{T}{C_p} > 0 \Rightarrow C_p > 0, \quad \left(\frac{\partial^2 H}{\partial p^2}\right)_S = -V\beta_S < 0 \Rightarrow \beta_S > 0 \quad (7.63)$$

For small perturbations to a stable equilibrium state under the condition of constant temperature and volume,

$$\Delta F = -T\Delta S^{\text{ir}} = dSdT - dVdp > 0 \tag{7.64}$$

In this case, we express the perturbations in entropy and pressure in terms of perturbations in temperature and volume,

$$dS = \left(\frac{\partial S}{\partial T}\right)_V dT + \left(\frac{\partial S}{\partial V}\right)_T dV \tag{7.65}$$

$$dp = \left(\frac{\partial p}{\partial T}\right)_V dT + \left(\frac{\partial p}{\partial V}\right)_T dV \tag{7.66}$$

Substituting Eqs. (7.65) and (7.66) into Eq. (7.64), we have

$$\Delta F = -T\Delta S^{\text{ir}} = \left(\frac{\partial S}{\partial T}\right)_V (dT)^2 - \left(\frac{\partial p}{\partial V}\right)_T (dV)^2 > 0 \tag{7.67}$$

or

$$\Delta F = -T\Delta S^{\text{ir}} = -\left(\frac{\partial^2 F}{\partial T^2}\right)_V (dT)^2 + \left(\frac{\partial^2 F}{\partial V^2}\right)_T (dV)^2 > 0 \tag{7.68}$$

Therefore, for a thermodynamically stable state,

$$\left(\frac{\partial^2 F}{\partial T^2}\right)_V < 0, \quad \left(\frac{\partial^2 F}{\partial V^2}\right)_T > 0 \tag{7.69}$$

which implies that for a stable equilibrium state,

$$\left(\frac{\partial^2 F}{\partial T^2}\right)_V = -\frac{C_V}{T} < 0 \Rightarrow C_V > 0, \quad \left(\frac{\partial^2 F}{\partial V^2}\right)_T = \frac{1}{V\beta_T} > 0 \Rightarrow \beta_T > 0 \tag{7.70}$$

For small perturbations to a stable equilibrium state under the condition of constant temperature and pressure,

$$\Delta G = -T\Delta S^{\text{ir}} = dSdT - dVdp > 0 \tag{7.71}$$

In this case, we express the perturbations in entropy and volume in terms of perturbations in temperature and pressure,

$$dS = \left(\frac{\partial S}{\partial T}\right)_p dT + \left(\frac{\partial S}{\partial p}\right)_T dp \tag{7.72}$$

$$dV = \left(\frac{\partial V}{\partial T}\right)_p dT + \left(\frac{\partial V}{\partial p}\right)_T dp \tag{7.73}$$

Substituting Eqs. (7.72) and (7.73) into Eq. (7.71), we have

$$\Delta G = \left(\frac{\partial S}{\partial T}\right)_p (dT)^2 + \left(\frac{\partial S}{\partial p}\right)_T dT\,dp - \left(\frac{\partial V}{\partial T}\right)_p dT\,dp - \left(\frac{\partial V}{\partial p}\right)_T (dp)^2 > 0 \tag{7.74}$$

or

$$\Delta G = -\left(\frac{\partial^2 G}{\partial T^2}\right)_p (dT)^2 - \frac{\partial^2 G}{\partial p\partial T} dT\,dp - \frac{\partial^2 G}{\partial T\partial p} dT\,dp - \left(\frac{\partial^2 G}{\partial p^2}\right)_T (dp)^2 > 0 \tag{7.75}$$

Therefore, for a thermodynamically stable state,

$$\left(\frac{\partial^2 G}{\partial T^2}\right)_p < 0, \quad \left(\frac{\partial^2 G}{\partial p^2}\right)_T < 0 \tag{7.76}$$

$$\begin{vmatrix} -\left(\frac{\partial^2 G}{\partial T^2}\right)_p & -\frac{\partial^2 G}{\partial T\partial p} \\ -\frac{\partial^2 G}{\partial p\partial T} & -\left(\frac{\partial^2 G}{\partial p^2}\right)_T \end{vmatrix} = \left(\frac{\partial^2 G}{\partial T^2}\right)_p \left(\frac{\partial^2 G}{\partial p^2}\right)_T - \left(\frac{\partial^2 G}{\partial T\partial p}\right)^2 > 0 \tag{7.77}$$

which implies that for a stable equilibrium state.

$$\left(\frac{\partial^2 G}{\partial T^2}\right)_p = -\frac{C_p}{T} < 0 \Rightarrow C_p > 0, \quad \left(\frac{\partial^2 G}{\partial p^2}\right)_T = -V\beta_T < 0 \Rightarrow \beta_T > 0 \tag{7.78}$$

$$\left(\frac{\partial^2 G}{\partial T^2}\right)_p \left(\frac{\partial^2 G}{\partial p^2}\right)_T - \left(\frac{\partial^2 G}{\partial T\partial p}\right)^2 = \frac{C_p V\beta_T}{T} - V^2\alpha^2 = \frac{V C_p \beta_S}{T} = \frac{V C_V \beta_T}{T} > 0 \tag{7.79}$$

Therefore, from the above discussions, we can conclude that for a system at stable or metastable equilibrium, the thermodynamic energy functions are convex functions of their extensive variables and concave functions of their intensive variables.

$$\left(\frac{\partial^2 U}{\partial S^2}\right)_V > 0, \quad \left(\frac{\partial^2 U}{\partial V^2}\right)_S > 0, \quad \left(\frac{\partial^2 H}{\partial S^2}\right)_p > 0, \quad \left(\frac{\partial^2 F}{\partial V^2}\right)_T > 0 \tag{7.80}$$

$$\left(\frac{\partial^2 H}{\partial p^2}\right)_S < 0, \quad \left(\frac{\partial^2 F}{\partial T^2}\right)_V < 0, \quad \left(\frac{\partial^2 G}{\partial T^2}\right)_p < 0, \quad \left(\frac{\partial^2 G}{\partial p^2}\right)_T < 0 \tag{7.81}$$

If we translate the stability conditions from in terms of derivatives to in terms of properties, we can show that all the physical properties defined by the second derivatives of an energy function with respect to a thermodynamic variable should be positive for a stable or metastable system. For example,

$$\left(\frac{\partial^2 H}{\partial p^2}\right)_S = \left(\frac{\partial V}{\partial p}\right)_S = -V\beta_S < 0 \Rightarrow \beta_S > 0 \tag{7.82}$$

$$\left(\frac{\partial^2 F}{\partial T^2}\right)_V = -\left(\frac{\partial S}{\partial T}\right)_V = -\frac{C_V}{T} < 0 \Rightarrow c_v > 0 \tag{7.83}$$

$$\left(\frac{\partial^2 G}{\partial T^2}\right)_p = -\left(\frac{\partial S}{\partial T}\right)_p = -\frac{C_p}{T} < 0 \Rightarrow c_p > 0 \tag{7.84}$$

$$\left(\frac{\partial^2 G}{\partial p^2}\right)_T = \left(\frac{\partial V}{\partial p}\right)_p = -V\beta_T < 0 \Rightarrow \beta_T > 0 \tag{7.85}$$

If any of the properties become negative, it signals the instability of a material with respect its phase transition to another state. It should be noted that properties defined by the mixed second derivatives of an energy function with respect to two variables, e.g., thermal expansion coefficient, are not necessarily always positive for a stable material.

For one mole of a material, the stability criteria can be written in terms of molar properties such as molar enthalpy h, molar Helmholtz free energy f, molar entropy s, molar volume v, and chemical potential μ. For example,

$$\left(\frac{\partial^2 h}{\partial p^2}\right)_S = \left(\frac{\partial v}{\partial p}\right)_S = -v\beta_s < 0 \Rightarrow \beta_s > 0 \tag{7.86}$$

$$\left(\frac{\partial^2 f}{\partial T^2}\right)_v = -\left(\frac{\partial s}{\partial T}\right)_v = -\frac{c_v}{T} < 0 \Rightarrow c_v > 0 \tag{7.87}$$

$$\left(\frac{\partial^2 \mu}{\partial T^2}\right)_p = -\left(\frac{\partial s}{\partial T}\right)_p = -\frac{c_p}{T} < 0 \Rightarrow c_p > 0 \tag{7.88}$$

$$\left(\frac{\partial^2 \mu}{\partial p^2}\right)_T = \left(\frac{\partial v}{\partial p}\right)_p = -v\beta_T < 0 \Rightarrow \beta_T > 0 \tag{7.89}$$

Finally, we have shown previously that the constant pressure heat capacity is greater than the constant volume heat capacity, and the isothermal compressibility is greater than the adiabatic compressibility, i.e.,

$$c_p - c_v > 0 \tag{7.90}$$

and

$$\beta_T - \beta_s > 0 \tag{7.91}$$

Therefore, for a material to be stable,

$$c_p > c_v > 0, \quad \beta_T > \beta_s > 0 \tag{7.92}$$

An intrinsic thermodynamic instability develops when any one of the thermodynamic moduli (thermal (c_v, c_p), mechanical (β_s, β_T), or chemical) becomes negative.

It should be pointed out that real crystalline solids may undergo other types of material instability, e.g., with respect to crystallographic shear, electric polar instability, etc., and hence additional stability criteria will be required to determine the thermodynamic stability of a material. These additional criteria can be established by including additional thermodynamic contributions, e.g., strain energy, dielectric energy, etc., to the thermodynamic equilibrium and stability.

Example: Thermodynamic Stability of van der Waals Fluids

The Helmholtz free energy of a van der Waals fluid as a function of temperature T, volume V, and the number of moles N is given by

$$F(T, V, N) = -\frac{aN^2}{V} - Nk_BT\left\{1 + \ln\left[\frac{(V - Nb)}{N}\left(\frac{2\pi mk_BT}{h^2}\right)^{3/2}\right]\right\}$$

where a and b are the van der Waals parameters measuring the attraction between molecules and the molecular volume, respectively, m is the mass of a molecule, k_B is the Boltzmann constant, and h is the Planck constant. The critical point of a van der Waals fluid describes a thermodynamic state at which a homogeneous state starts to lose its stability with respect to its separation to two phases with different fluid density. At the critical point, $T = T_c$, $p = p_c$, and $V = V_c$, both the second and third derivatives of the Helmholtz free energy with respect to volume at constant temperature and number of moles become zero. Therefore, the critical temperature, pressure, and volume can be obtained in terms of the van der Waals parameters a and b and the number of moles of N:

$$T_c = \frac{8a}{27k_Bb}, \quad p_c = \frac{a}{27b^2}, \quad V_c = 3Nb,$$

We can then express the Helmholtz free energy (normalized by $3Nk_BT_c/8$) using the normalized temperature ($\tau = T/T_c$) and volume ($v = V/V_c$),

$$f(\tau, v) = -\frac{3}{v} - \frac{8\tau}{3}\left\{1 + \ln(3v - 1) + \frac{3}{2}\ln\tau + \ln\left[\left(\frac{2\pi mk_BT_c}{h^2}\right)^{3/2}\frac{k_BT_c}{8p_c}\right]\right\}$$

(a)   Derive the normalized pressure as a function of normalized temperature and volume.

(b) Derive the normalized bulk modulus as a function of normalized temperature and volume.

(c) Plot the normalized bulk modulus as a function of normalized volume at several temperatures from $0.75T_c$ to $1.25T_c$ and indicate the range of volumes for each temperature that a homogeneous fluid is thermodynamically unstable.

Solution

(a) The normalized pressure can be obtained from the normalized Helmholtz free energy as

$$p = -\left(\frac{\partial f}{\partial v}\right)_T = -\frac{3}{v^2} + \frac{8\tau}{(3v-1)}$$

(b) The normalized bulk modulus can be obtained from the normalized pressure as a function of normalized volume as

$$B_T = -v\left(\frac{\partial p}{\partial v}\right)_T = 6\left[\frac{4\tau v}{(3v-1)^2} - \frac{1}{v^2}\right]$$

(c) The normalized bulk modulus as a function of normalized volume for a series of normalized temperatures is plotted in Fig. 7.6. The ranges of volumes at each temperature from $0.75T_c$ to T_c that give rise to a negative bulk modulus indicate that a single homogeneous fluid is thermodynamically unstable within these ranges of temperatures and volumes. The line represented by the dotted line is the normalized bulk modulus as a function of normalized volume at the critical temperature $T_c (\tau = 1.0)$. The normalized bulk modulus becomes zero at the normalized temperature T_c and normalized volume V_c ($v = 1$) represented by the open circle in the figure.

Fig. 7.6 Normalized bulk modulus as a function of normalized volume

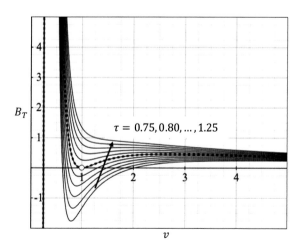

7.10   Exercises

1.   If one doubles the size of a substance, the chemical potential of the substance is also doubled: True or false?

2.   For an ice–water mixture with 80% volume fraction of ice at 0 °C and 1 bar, the chemical potential of ice is 4 times the chemical potential of water, true or false?

3.   Mixing 1 mol of ice at -10.0 °C and 1 mol of water at 25 °C and then enclosing it with a flexible adiabatic wall, what is the final temperature of the system, and what are the amounts of ice and water? The molar constant pressure heat capacity of ice is approximately 38 J/(mol K), and the molar constant pressure heat capacity of water is approximately 75 J/(mol K). The molar heat of melting of ice is 6007 J/mol.

4.   The curvature of a Helmholtz free energy versus temperature at constant volume plot for a substance is always negative, true or false?

5.   The curvature of a Helmholtz free energy versus volume at constant temperature plot for a substance is always negative, true or false?

6.   Show that

$$\left(\frac{\partial^2 U}{\partial S^2}\right)(dS)^2 + 2\left(\frac{\partial^2 U}{\partial S \partial V}\right)dSdV + \left(\frac{\partial^2 U}{\partial V^2}\right)(dV)^2 > 0$$

can be written as

$$\left[\left(\frac{\partial^2 U}{\partial S^2}\right)dS + \left(\frac{\partial^2 U}{\partial S \partial V}\right)dV\right]^2 + \left[\left(\frac{\partial^2 U}{\partial S^2}\right)\left(\frac{\partial^2 U}{\partial V^2}\right) - \left(\frac{\partial^2 U}{\partial S \partial V}\right)^2\right](dV)^2 > 0$$

7.   Given the following expression,

$$\left(\frac{\partial^2 U}{\partial S^2}\right)\left(\frac{\partial^2 U}{\partial V^2}\right) - \left(\frac{\partial^2 U}{\partial S \partial V}\right)^2$$

(a)   Express the above expression in terms of materials properties such as heat capacity, volume thermal expansion coefficient, compressibility, and temperature.

(b)   Based on the materials properties and the stability criteria discussed in this chapter, is the value for the above expression positive or negative for a stable material?

8.   Show that for a stable material, its constant pressure heat capacity is greater than the constant volume heat capacity.

9.   Show that for a stable material, the isothermal compressibility is greater than the adiabatic compressibility.

10.   Using the fundamental equation of thermodynamics for monatomic ideal gases, show that an ideal gas is a stable state, i.e., it satisfies all the thermodynamic stability criteria.

11.   Given the following expression,

$$\left(\frac{\partial^2 G}{\partial T^2}\right)_p \left(\frac{\partial^2 G}{\partial p^2}\right)_T - \left(\frac{\partial^2 G}{\partial T \partial p}\right)^2$$

(iii)   Express the above expression in terms of materials properties such as heat capacity, volume thermal expansion coefficient, compressibility, and temperature.

(iv)   Based on the materials properties and the stability criteria discussed in this chapter, is the value for the above expression positive or negative for a stable material?

Chapter 8
Thermodynamic Calculations of Materials Processes

Materials processes refer to the changes of a material from one thermodynamic state to another. Generally speaking, there are three types of materials processes: (a) a change in thermodynamic state of a material due to the change in thermodynamic variables such as temperature and pressure (or volume in theoretical calculations, or under an external field such as a stress, electric, and magnetic field); (b) phase transitions at a given temperature and pressure from one physical state such as solid, liquid, gas to another, or from one crystal structure to another, or from a single-phase state to multiphase state and vice versa; (c) chemical reactions at a given temperature and pressure resulting in changes in chemical bonding among different chemical species and thus changes in chemical components and compositions. This chapter discusses how to compute the changes in different forms of thermodynamic energy and entropy functions for these materials processes, given the knowledge of thermodynamic properties such as heat capacity, mechanical compressibility or modulus, thermal expansion coefficient, enthalpy changes or heat of phase transitions at equilibrium transition temperatures and pressures, as well as thermodynamic properties at a reference state, e.g., room temperature 298 K and ambient pressure 1 bar. From these calculations, we can determine the thermodynamic driving forces and thus directions for phase transitions and chemical reactions as thermodynamic conditions change, as well as the heat or entropy exchanges or work interactions between a material system and its surrounding during a materials process.

It is usually more convenient to control intensive potentials such as temperature and pressure (or fields such as stress, electric, and magnetic fields) to conduct experimental measurements. However, computationally, it is easier to perform calculations at fixed volumes. For example, the interatomic interaction potential energy of a perfect crystal relative to individual atoms in a fixed volume is the most straightforward quantity to be computed using density function theory (DFT) calculations. The lattice vibrational contributions to energy and entropy and thus Helmholtz free energy can also be obtained at fixed volumes using a combination of DFT calculations, lattice dynamics as well as statistical thermodynamics. As a result, the molar

© Springer Nature Singapore Pte Ltd. 2022
L.-Q. Chen, *Thermodynamic Equilibrium and Stability of Materials*,
https://doi.org/10.1007/978-981-13-8691-6_8

enthalpy and chemical potential (molar Gibbs free energy) as a function of temperature and pressure are more accessible through experiments whereas molar internal energy, entropy, and Helmholtz free energy as a function of temperature and volume are typically obtained by computations. The relations between thermodynamic properties allow us to connect the experimentally measured and theoretically computed properties. Therefore, in order to be able to relate the computed internal energy, enthalpy, and Helmholtz free energy to the experimentally measured quantities such as chemical potential, it is important to be able to compute the changes for the different energy functions and entropy as thermodynamic conditions are altered. In this chapter, we first outline the procedures and derive the equations for the changes in energy functions and entropy as the temperature, pressure, and volume vary in terms of experimentally measured or theoretically computed properties such as heat capacity, compressibility, and thermal expansion coefficient. We will then discuss the calculations of changes in enthalpy, entropy, and, chemical poential or Gibbs free energy due to phase transitions and chemical reactions at a given temperature and pressure.

8.1 Changes in Thermodynamic Properties with Temperature at Constant Volume

To calculate the changes in the molar internal energy Δu, enthalpy Δh, entropy Δs, Helmholtz free energy Δf, and chemical potential $\Delta \mu$ of a material as a result of temperature change ΔT at constant molar volume v, we first express their rates of change with respect to a change in temperature at constant molar volume using mathematical derivatives,

$$\left(\frac{\partial u}{\partial T}\right)_v, \left(\frac{\partial h}{\partial T}\right)_v, \left(\frac{\partial s}{\partial T}\right)_v, \left(\frac{\partial f}{\partial T}\right)_v, \left(\frac{\partial \mu}{\partial T}\right)_v \tag{8.1}$$

The corresponding infinitesimal changes in internal energy du, enthalpy dh, entropy ds, Helmholtz free energy df, and chemical potential $d\mu$ are then given by

$$du = \left(\frac{\partial u}{\partial T}\right)_v dT, \, dh = \left(\frac{\partial h}{\partial T}\right)_v dT, \, ds = \left(\frac{\partial s}{\partial T}\right)_v dT, \tag{8.2}$$

$$df = \left(\frac{\partial f}{\partial T}\right)_v dT, \, d\mu = \left(\frac{\partial \mu}{\partial T}\right)_v dT \tag{8.3}$$

The next step is to express the above rates of changes using the definitions of measurable properties, and if that is not possible, use the differential forms of u, h, s, f, and μ to rewrite the partial derivatives. For example,

$$\left(\frac{\partial u}{\partial T}\right)_v = c_v \tag{8.4}$$

$$\left(\frac{\partial h}{\partial T}\right)_v = \left(\frac{T\,\mathrm{d}s + v\,\mathrm{d}p}{\mathrm{d}T}\right)_v = T\left(\frac{\mathrm{d}s}{\mathrm{d}T}\right)_v + v\left(\frac{\mathrm{d}p}{\mathrm{d}T}\right)_v = c_v + v\left(\frac{\mathrm{d}p}{\mathrm{d}T}\right)_v \tag{8.5}$$

$$\left(\frac{\partial s}{\partial T}\right)_v = \frac{c_v}{T} \tag{8.6}$$

$$\left(\frac{\partial f}{\partial T}\right)_v = -s \tag{8.7}$$

$$\left(\frac{\partial \mu}{\partial T}\right)_v = \left(\frac{-s\,\mathrm{d}T + v\,\mathrm{d}p}{\mathrm{d}T}\right)_v = -s + v\left(\frac{\mathrm{d}p}{\mathrm{d}T}\right)_v \tag{8.8}$$

where c_v is the molar constant volume heat capacity defined as

$$c_v = \left(\frac{\partial u}{\partial T}\right)_v = T\left(\frac{\mathrm{d}s}{\mathrm{d}T}\right)_v \tag{8.9}$$

We then use a mathematical equality[1] to convert the remaining partial derivative on the right-hand sides of Eqs. (8.5) and (8.8) to a ratio of two partial derivatives that can be expressed as a ratio of two measurable thermodynamic properties,

$$\left(\frac{\mathrm{d}p}{\mathrm{d}T}\right)_v = -\frac{\left(\frac{\mathrm{d}v}{\mathrm{d}T}\right)_p}{\left(\frac{\mathrm{d}v}{\mathrm{d}p}\right)_T} = \frac{\alpha}{\beta_T} \tag{8.10}$$

where α is volume thermal expansion coefficient, and β_T is isothermal compressibility defined as

$$\alpha = \frac{1}{v}\left(\frac{\partial v}{\partial T}\right)_p, \beta_T = -\frac{1}{v}\left(\frac{\partial v}{\partial p}\right)_T \tag{8.11}$$

We can now express all the rates of changes in u, h, s, f, and μ with respect to the temperature change at constant volume in terms of measurable properties except the molar entropy on the right-hand sides of Eqs. (8.7) and (8.8), which is not directly measurable:

$$\left(\frac{\partial u}{\partial T}\right)_v = c_v \tag{8.12}$$

---

[1] $\left(\frac{\partial x}{\partial y}\right)_z = -\frac{\left(\frac{\partial z}{\partial y}\right)_x}{\left(\frac{\partial z}{\partial x}\right)_y}$

$$\left(\frac{\partial h}{\partial T}\right)_v = c_v + \frac{v\alpha}{\beta_T} \tag{8.13}$$

$$\left(\frac{\partial s}{\partial T}\right)_v = \frac{c_v}{T} \tag{8.14}$$

$$\left(\frac{\partial f}{\partial T}\right)_v = -s \tag{8.15}$$

$$\left(\frac{\partial \mu}{\partial T}\right)_v = -s + \frac{v\alpha}{\beta_T} \tag{8.16}$$

The finite changes in u, h, s, f, and μ due to a finite temperature change are then given by

$$\Delta u = u(T, v_0) - u(T_0, v_0) = \int_{T_0}^{T} \left(\frac{\partial u}{\partial T}\right)_v dT = \int_{T_0}^{T} c_v dT \tag{8.17}$$

$$\Delta h = h(T, v_0) - h(T_0, v_0) = \int_{T_0}^{T} \left(\frac{\partial h}{\partial T}\right)_v dT = \int_{T_0}^{T} \left(c_v + \frac{v_0\alpha}{\beta_T}\right) dT \tag{8.18}$$

$$\Delta s = s(T, v_0) - s(T_0, v_0) = \int_{T_0}^{T} \left(\frac{\partial s}{\partial T}\right)_v dT = \int_{T_0}^{T} \frac{c_v}{T} dT \tag{8.19}$$

$$\Delta f = f(T, v_0) - f(T_0, v_0) = \int_{T_0}^{T} \left(\frac{\partial f}{\partial T}\right)_v dT = -\int_{T_0}^{T} s dT \tag{8.20}$$

$$\Delta \mu = \mu(T, v_0) - \mu(T_0, v_0) = \int_{T_0}^{T} \left(\frac{\partial \mu}{\partial T}\right)_v dT = \int_{T_0}^{T} \left(-s + \frac{v_0\alpha}{\beta_T}\right) dT \tag{8.21}$$

where T_0 and v_0 are the temperature and molar volume of the initial state. It is noted that the calculations of changes in the molar Helmholtz free energy (Eq. 8.20) and chemical potential (Eq. 8.21) with respect to temperature change at constant volume require the knowledge of molar entropy s as a function of temperature. The molar entropy can be obtained from Eq. (8.19),

$$s(T, v_0) = s(T_0, v_0) + \Delta s = s(T_0, v_0) + \int_{T_0}^{T} \frac{c_v}{T} dT$$

which requires the information on the molar entropy of the initial state $s(T_o, v_o)$. Using the above equation for the molar entropy, we can express the molar Helmholtz free energy and chemical potential in terms of measurable properties and the molar entropy of the initial state,

$$\Delta f = -\int_{T_o}^{T}\left[s(T_o, v_o) + \int_{T_o}^{T} \frac{c_v}{T}\mathrm{d}T \right]\mathrm{d}T \tag{8.22}$$

$$\Delta \mu = \int_{T_o}^{T}\left\{ -\left[s(T_o, v_o) + \int_{T_o}^{T} \frac{c_v}{T}\mathrm{d}T \right] + \frac{v_o \alpha}{\beta_T} \right\}\mathrm{d}T \tag{8.23}$$

By performing integration by parts[2] of the above two equations, we obtain

$$\Delta f = -s(T_o, v_o)(T - T_o) + \int_{T_o}^{T} c_v \mathrm{d}T - T\int_{T_o}^{T} \frac{c_v}{T}\mathrm{d}T \tag{8.24}$$

$$\Delta \mu = -s(T_o, v_o)(T - T_o) + \int_{T_o}^{T}\left(c_v + \frac{v_o \alpha}{\beta_T}\right)\mathrm{d}T - T\int_{T_o}^{T} \frac{c_v}{T}\mathrm{d}T \tag{8.25}$$

Therefore, the general equations for the finite changes in $u, h, s, f,$ and μ due to a finite temperature change at constant volume in terms of measurable or computable properties are:

$$\Delta u = \int_{T_o}^{T} c_v \mathrm{d}T \tag{8.26}$$

$$\Delta h = \int_{T_o}^{T}\left(c_v + \frac{v_o \alpha}{\beta_T}\right)\mathrm{d}T \tag{8.27}$$

$$\Delta s = \int_{T_o}^{T} \frac{c_v}{T}\mathrm{d}T \tag{8.28}$$

$$\Delta f = -s(T_o, v_o)(T - T_o) + \int_{T_o}^{T} c_v \mathrm{d}T - T\int_{T_o}^{T} \frac{c_v}{T}\mathrm{d}T \tag{8.29}$$

---

[2] $\int_{T_o}^{T}\left(\int_{T_o}^{T} \frac{c_v}{T}\mathrm{d}T \right)\mathrm{d}T = T\int_{T_o}^{T} \frac{c_v}{T}\mathrm{d}T - \int_{T_o}^{T} c_v \mathrm{d}T$

$$\Delta \mu = -s(T_0, v_0)(T - T_0) + \int_{T_0}^{T} \left(c_v + \frac{v_0 \alpha}{\beta_T} \right) dT - T \int_{T_0}^{T} \frac{c_v}{T} dT \qquad (8.30)$$

The changes in the molar thermal energy Δu_T, molar mechanical energy Δu_M, and molar chemical energy Δu_C of the system as a result of temperature change from T_0 to T at constant volume are:

$$\Delta u_T = \int_{T_0}^{T} d(Ts) = \int_{T_0}^{T} s \, dT + \int_{T_0}^{T} T \, ds = -\Delta f + \Delta u = -\Delta f + Q \qquad (8.31)$$

$$\Delta u_M = -\int_{T_0}^{T} d(pv) = -v_0 \int_{T_0}^{T} \left(\frac{\partial p}{\partial T} \right)_v dT = -v_0 \int_{T_0}^{T} \frac{\alpha}{\beta_T} dT = \Delta f - \Delta \mu \quad (8.32)$$

$$\Delta u_C = \Delta \mu \qquad (8.33)$$

where Q is the amount of heat absorbed from the surrounding during the process.

The heat capacity c_v, isothermal compressibility β_T, and thermal expansion coefficient α are generally temperature-dependent, and thus one has to perform the integrations to obtain all the changes. However, if the heat capacity c_v, isothermal compressibility β_T, and thermal expansion coefficient α are assumed to be constant independent of temperature, the changes in u, h, s, f, and μ due to a finite temperature change at constant volume in terms of measurable properties can be simplified:

$$\Delta u = c_v(T - T_0) \qquad (8.34)$$

$$\Delta h = \left(c_v + \frac{v_0 \alpha}{\beta_T} \right)(T - T_0) \qquad (8.35)$$

$$\Delta s = c_v \ln(T/T_0) \qquad (8.36)$$

$$\Delta f = [c_v - s(T_0, v_0)](T - T_0) - c_v T \ln(T/T_0) \qquad (8.37)$$

$$\Delta \mu = \left[c_v + \frac{v_0 \alpha}{\beta_T} - s(T_0, v_0) \right](T - T_0) - c_v T \ln \frac{T}{T_0} \qquad (8.38)$$

For ideal gases, although the heat capacity is constant independent of temperature, the compressibility and thermal expansion coefficient are not constant. For example, for monatomic ideal gases, $c_v = 3R/2$, $\alpha = 1/T$, $\beta_T = 1/p$, and thus $(v_0 \alpha)/\beta_T = R$, where R is the gas constant. Therefore, the changes in u, h, s, f, and μ due to a finite temperature change at constant volume for ideal gases are given by

$$\Delta u = c_v(T - T_o) \tag{8.39}$$

$$\Delta h = c_p(T - T_o) \tag{8.40}$$

$$\Delta s = c_v \ln(T/T_o) \tag{8.41}$$

$$\Delta f = [c_v - s(T_o, v_o)](T - T_o) - c_v T \ln(T/T_o) \tag{8.42}$$

$$\Delta \mu = \left[c_p - s(T_o, v_o)\right](T - T_o) - c_v T \ln(T/T_o) \tag{8.43}$$

8.2 Changes in Thermodynamic Properties with Temperature at Constant Pressure

To calculate the changes in the molar internal energy Δu, enthalpy Δh, entropy Δs, Helmholtz free energy Δf, and chemical potential $\Delta \mu$ of a material as a result of temperature change ΔT at constant pressure p, we again first express their rates of change with respect to change in temperature at constant pressure as mathematical derivatives,

$$\left(\frac{\partial u}{\partial T}\right)_p, \left(\frac{\partial h}{\partial T}\right)_p, \left(\frac{\partial s}{\partial T}\right)_p, \left(\frac{\partial f}{\partial T}\right)_p, \left(\frac{\partial \mu}{\partial T}\right)_p \tag{8.44}$$

The infinitesimal changes in internal energy du, enthalpy dh, entropy ds, Helmholtz free energy df, and chemical potential $d\mu$ are then

$$du = \left(\frac{\partial u}{\partial T}\right)_p dT, dh = \left(\frac{\partial h}{\partial T}\right)_p dT, ds = \left(\frac{\partial s}{\partial T}\right)_p dT, \tag{8.45}$$

$$df = \left(\frac{\partial f}{\partial T}\right)_p dT, d\mu = \left(\frac{\partial \mu}{\partial T}\right)_p dT \tag{8.46}$$

We express the rates of changes using the definitions of properties:

$$\left(\frac{\partial u}{\partial T}\right)_p = \left(\frac{T ds - p dv}{dT}\right)_p = c_p - p\left(\frac{\partial v}{\partial T}\right)_p = c_p - p v \alpha \tag{8.47}$$

$$\left(\frac{\partial h}{\partial T}\right)_p = c_p \tag{8.48}$$

$$\left(\frac{\partial s}{\partial T}\right)_p = \frac{c_p}{T} \tag{8.49}$$

$$\left(\frac{\partial f}{\partial T}\right)_p = \left(\frac{-s dT - p dv}{dT}\right)_p = -s - p\left(\frac{\partial v}{\partial T}\right)_p = -s - pv\alpha \qquad (8.50)$$

$$\left(\frac{\partial \mu}{\partial T}\right)_p = -s \qquad (8.51)$$

where c_p is the molar constant pressure heat capacity,

$$c_p = \left(\frac{\partial h}{\partial T}\right)_p = T\left(\frac{\partial s}{\partial T}\right)_p \qquad (8.52)$$

The finite changes in u, h, s, f, and μ due to a finite temperature change at constant pressure are given by

$$\Delta u = u(T, p_0) - u(T_0, p_0) = \int_{T_0}^{T} \left(\frac{\partial u}{\partial T}\right)_p dT = \int_{T_0}^{T} (c_p - p_0 v\alpha) dT \qquad (8.53)$$

$$\Delta h = h(T, p_0) - h(T_0, p_0) = \int_{T_0}^{T} \left(\frac{\partial h}{\partial T}\right)_p dT = \int_{T_0}^{T} c_p dT \qquad (8.54)$$

$$\Delta s = s(T, p_0) - s(T_0, p_0) = \int_{T_0}^{T} \left(\frac{\partial s}{\partial T}\right)_p dT = \int_{T_0}^{T} \frac{c_p}{T} dT \qquad (8.55)$$

$$\Delta f = f(T, p_0) - f(T_0, p_0) = \int_{T_0}^{T} \left(\frac{\partial f}{\partial T}\right)_p dT = -\int_{T_0}^{T} (s + p_0 v\alpha) dT \qquad (8.56)$$

$$\Delta \mu = \mu(T, p_0) - \mu(T_0, p_0) = \int_{T_0}^{T} \left(\frac{\partial \mu}{\partial T}\right)_p dT = -\int_{T_0}^{T} s dT \qquad (8.57)$$

where p_0 is the pressure of the initial state. The entropy as a function of temperature at constant pressure in Eqs. (8.56) and (8.57) can be obtained from Eq. (8.55), i.e.,

$$s(T, p_0) = s(T_0, p_0) + \Delta s = s(T_0, p_0) + \int_{T_0}^{T} \frac{c_p}{T} dT$$

where $s(T_0, p_0)$ is the molar entropy of the initial state. Using the molar entropy as a function of temperature, we can rewrite the changes in molar Helmholtz free energy and chemical potential as

$$\Delta f = -\int_{T_0}^{T} (s + p_0 v\alpha)\,dT = -\int_{T_0}^{T} \left[s(T_0, p_0) + \int_{T_0}^{T} \frac{c_p}{T}\,dT + p_0 v\alpha \right] dT \quad (8.58)$$

$$\Delta \mu = \int_{T_0}^{T} \left\{ -\left[s(T_0, p_0) + \int_{T_0}^{T} \frac{c_p}{T}\,dT \right] \right\} dT \quad (8.59)$$

By performing integration by parts[3] of the above two equations, we obtain

$$\Delta f = -s(T_0, p_0)(T - T_0) + \int_{T_0}^{T} \left(c_p - p_0 v\alpha \right)\,dT - T\int_{T_0}^{T} \frac{c_p}{T}\,dT \quad (8.60)$$

$$\Delta \mu = -s(T_0, p_0)(T - T_0) + \int_{T_0}^{T} c_p\,dT - T\int_{T_0}^{T} \frac{c_p}{T}\,dT \quad (8.61)$$

Therefore, the general expressions for the finite changes in u, h, s, f, and μ due to a finite temperature change at constant pressure in terms of measurable properties are given by

$$\Delta u = \int_{T_0}^{T} \left(c_p - p_0 v\alpha \right)\,dT \quad (8.62)$$

$$\Delta h = \int_{T_0}^{T} c_p\,dT \quad (8.63)$$

$$\Delta s = \int_{T_0}^{T} \frac{c_p}{T}\,dT \quad (8.64)$$

$$\Delta f = -s(T_0, p_0)(T - T_0) + \int_{T_0}^{T} \left(c_p - p_0 v\alpha \right)\,dT - T\int_{T_0}^{T} \frac{c_p}{T}\,dT \quad (8.65)$$

$$\Delta \mu = -s(T_0, p_0)(T - T_0) + \int_{T_0}^{T} c_p\,dT - T\int_{T_0}^{T} \frac{c_p}{T}\,dT \quad (8.66)$$

---

[3] $\int_{T_0}^{T} \left(\int_{T_0}^{T} \frac{c_p}{T}\,dT \right) dT = T\int_{T_0}^{T} \frac{c_p}{T}\,dT - \int_{T_0}^{T} c_p\,dT$

The changes in the molar thermal energy Δu_T, molar mechanical energy Δu_M, and molar chemical energy Δu_C of the system as a result of a temperature change from T_0 to T at constant pressure are:

$$\Delta u_\mathrm{T} = \int_{T_0}^{T} \mathrm{d}(Ts) = \int_{T_0}^{T} s\,\mathrm{d}T + \int_{T_0}^{T} T\,\mathrm{d}s = -\Delta\mu + \Delta h = -\Delta\mu + Q \qquad (8.67)$$

$$\Delta u_\mathrm{M} = -\int_{T_0}^{T} \mathrm{d}(pv) = -p_0 \int_{T_0}^{T} \left(\frac{\partial v}{\partial T}\right)_p \mathrm{d}T = -p_0 \int_{T_0}^{T} v\alpha\,\mathrm{d}T = -p_0(v - v_0) = W$$

$$(8.68)$$

$$\Delta u_\mathrm{C} = \Delta\mu \qquad (8.69)$$

where Q is the amount of heat absorbed from the surrounding during the process, and W is the reversible work done on the system by the surrounding during the process.

If the heat capacity c_p and thermal expansion coefficient α are assumed to be constant independent of temperature, we can express volume v in Eqs. (8.62) and (8.65) as a function of temperature at constant pressure using the definition of thermal expansion coefficient,

$$\alpha = \frac{1}{v(T, p_0)} \frac{\mathrm{d}v(T, p_0)}{\mathrm{d}T} \qquad (8.70)$$

or

$$d\ln v(T, p_0) = \alpha\,\mathrm{d}T \qquad (8.71)$$

Integrating the above equation for both sides from T_0 to T, we have

$$\int_{T_0}^{T} d\ln v(T, p_0) = \int_{T_0}^{T} \alpha\,\mathrm{d}T \qquad (8.72)$$

and

$$\ln \frac{v(T, p_0)}{v_0(T_0, p_0)} = \alpha(T - T_0) \qquad (8.73)$$

where $v_0(T_0, p_0)$ is the molar volume of the initial state. Equation (8.73) can be rewritten as

$$v(T, p_0) = v_0(T_0, p_0)e^{\alpha(T - T_0)} = v_0 e^{\alpha(T - T_0)} \qquad (8.74)$$

Substituting the above expression into Eqs. (8.62) and (8.65), we have

$$\Delta u = c_p(T - T_o) - p_o \alpha v_o \int_{T_o}^{T} e^{\alpha(T - T_o)} dT \qquad (8.75)$$

$$\Delta f = \left[c_p - s(T_o, p_o) \right](T - T_o) - p_o \alpha v_o \int_{T_o}^{T} e^{\alpha(T - T_o)} dT - c_p T \ln \frac{T}{T_o} \qquad (8.76)$$

Carrying out the integrations in the above two equations, we get

$$\Delta u = c_p(T - T_o) - p_o v_o \left[e^{\alpha(T - T_o)} - 1 \right] \qquad (8.77)$$

and

$$\Delta f = \left[c_p - s(T_o, p_o) \right](T - T_o) - p_o v_o \left[e^{\alpha(T - T_o)} - 1 \right] - c_p T \ln(T/T_o) \qquad (8.78)$$

Therefore, if heat capacity c_p and thermal expansion coefficient α are both assumed to be constant independent of temperature, the changes in u, h, s, f, and μ due to a finite temperature change at constant pressure in terms of measurable properties can be rewritten as

$$\Delta u = c_p(T - T_o) - p_o v_o \left[e^{\alpha(T - T_o)} - 1 \right] \qquad (8.79)$$

$$\Delta h = c_p(T - T_o) \qquad (8.80)$$

$$\Delta s = c_p \ln(T/T_o) \qquad (8.81)$$

$$\Delta f = \left[c_p - s(T_o, p_o) \right](T - T_o) - p_o v_o \left[e^{\alpha(T - T_o)} - 1 \right] - c_p T \ln(T/T_o) \qquad (8.82)$$

$$\Delta \mu = \left[c_p - s(T_o, p_o) \right](T - T_o) - c_p T \ln(T/T_o) \qquad (8.83)$$

The thermal expansion coefficient α in solids is generally small, and we can usually make the following approximation[4]

$$e^{\alpha(T - T_o)} - 1 \approx \alpha(T - T_o),$$

With this approximation, the changes in the molar internal energy and molar Helmholtz free energy can be rewritten as

---

[4] $e^x \approx 1 + x$ for $x \ll 1$

$$\Delta u \approx \left[c_p - p_0 v_0 \alpha \right] (T - T_0) \tag{8.84}$$

and

$$\Delta f \approx \left[c_p - p_0 v_0 \alpha - s(T_0, p_0) \right] (T - T_0) - c_p T \ln(T/T_0) \tag{8.85}$$

For ideal gases, c_p is constant, and $\alpha = 1/T$, and thus the changes in u, h, s, f, and μ due to a finite temperature change at constant pressure are given by

$$\Delta u = c_v (T - T_0) \tag{8.86}$$

$$\Delta h = c_p (T - T_0) \tag{8.87}$$

$$\Delta s = c_p \ln(T/T_0) \tag{8.88}$$

$$\Delta f = [c_v - s(T_0, p_0)](T - T_0) - c_p T \ln(T/T_0) \tag{8.89}$$

$$\Delta \mu = \left[c_p - s(T_0, p_0) \right] (T - T_0) - c_p T \ln(T/T_0) \tag{8.90}$$

8.3   Changes in Thermodynamic Properties with Volume at Constant Temperature

To obtain the changes in the molar internal energy Δu, enthalpy Δh, entropy Δs, Helmholtz free energy Δf, and chemical potential $\Delta \mu$ of a material due to volume change Δv at constant temperature T, we first write down their rates of change with respect to a change in volume at constant temperature,

$$\left(\frac{\partial u}{\partial v} \right)_T, \left(\frac{\partial h}{\partial v} \right)_T, \left(\frac{\partial s}{\partial v} \right)_T, \left(\frac{\partial f}{\partial v} \right)_T, \left(\frac{\partial \mu}{\partial v} \right)_T \tag{8.91}$$

The infinitesimal changes in internal energy du, enthalpy dh, entropy ds, Helmholtz free energy df, and chemical potential $d\mu$ are then

$$du = \left(\frac{\partial u}{\partial v} \right)_T dv, \; dh = \left(\frac{\partial h}{\partial v} \right)_T dv, \; ds = \left(\frac{\partial s}{\partial v} \right)_T dv, \tag{8.92}$$

$$df = \left(\frac{\partial f}{\partial v} \right)_T dv, \; d\mu = \left(\frac{\partial \mu}{\partial v} \right)_T dv \tag{8.93}$$

The rates of their changes in u, h, s, f, and μ with respect to volume at constant temperature in terms of measurable properties are given by

$$\left(\frac{\partial u}{\partial v}\right)_T = T\left(\frac{\partial s}{\partial v}\right)_T - p = \frac{T\alpha}{\beta_T} - p \tag{8.94}$$

$$\left(\frac{\partial h}{\partial v}\right)_T = T\left(\frac{\partial s}{\partial v}\right)_T + v\left(\frac{\partial p}{\partial v}\right)_T = \frac{T\alpha}{\beta_T} - \frac{1}{\beta_T} = \frac{1}{\beta_T}(T\alpha - 1) \tag{8.95}$$

$$\left(\frac{\partial s}{\partial v}\right)_T = \frac{\alpha}{\beta_T} \tag{8.96}$$

$$\left(\frac{\partial f}{\partial v}\right)_T = -p \tag{8.97}$$

$$\left(\frac{\partial \mu}{\partial v}\right)_T = v\left(\frac{\partial p}{\partial v}\right)_T = -\frac{1}{\beta_T} \tag{8.98}$$

The general expressions for the finite changes in u, h, s, f, and μ due to a finite volume change at constant temperature are then given by

$$\Delta u = u(T_o, v) - u(T_o, v_o) = \int_{v_o}^{v}\left(\frac{\partial u}{\partial v}\right)_T dv = \int_{v_o}^{v}\left(\frac{T\alpha}{\beta_T} - p\right)dv \tag{8.99}$$

$$\Delta h = h(T_o, v) - h(T_o, v_o) = \int_{v_o}^{v}\left(\frac{\partial h}{\partial v}\right)_T dv = \int_{v_o}^{v}\frac{1}{\beta_T}(T\alpha - 1)dv \tag{8.100}$$

$$\Delta s = s(T_o, v) - s(T_o, v_o) = \int_{v_o}^{v}\left(\frac{\partial s}{\partial v}\right)_T dv = \int_{v_o}^{v}\frac{\alpha}{\beta_T}dv \tag{8.101}$$

$$\Delta f = f(T_o, v) - f(T_o, v_o) = \int_{v_o}^{v}\left(\frac{\partial f}{\partial v}\right)_T dv = -\int_{v_o}^{v}pdv \tag{8.102}$$

$$\Delta \mu = \mu(T_o, v) - \mu(T_o, v_o) = \int_{v_o}^{v}\left(\frac{\partial \mu}{\partial v}\right)_T dv = -\int_{v_o}^{v}\frac{1}{\beta_T}dv \tag{8.103}$$

The changes in the molar thermal energy Δu_T, molar mechanical energy Δu_M, and molar chemical energy Δu_C of the system due to a volume change at constant temperature are:

$$\Delta u_T = T_o\Delta s = Q \tag{8.104}$$

$$\Delta u_M = -\int_{v_o}^{v} \left[\frac{\partial(pv)}{\partial v}\right]_T dv = -\int_{v_o}^{v} p\,dv + \int_{v_o}^{v} \frac{1}{\beta_T} dv = W - \Delta\mu \qquad (8.105)$$

$$\Delta u_C = \Delta\mu \qquad (8.106)$$

where Q is the amount of heat absorbed from the surrounding during the process, and W is the reversible work done on the system by the surrounding during the process.

If we assume the thermal expansion coefficient and the isothermal compressibility are both constant independent of molar volume, we can express p as a function of v. The isothermal compressibility is given by

$$\beta_T = -\frac{1}{v}\left(\frac{\partial v}{\partial p}\right)_T = -\left(\frac{\partial \ln v}{\partial p}\right)_T \qquad (8.107)$$

Integrating the above equation from v_o to v for v, and p_o to p for p, we have

$$v(T_o, p) = v_o(T_o, p_o)e^{-\beta_T(p-p_o)} = v_o e^{-\beta_T(p-p_o)} \qquad (8.108)$$

Solving p from the above equation, we have

$$p = p_o - \frac{1}{\beta_T}\ln\frac{v}{v_o} \qquad (8.109)$$

Using the above volume dependence of pressure, we can evaluate the integral,

$$\int_{v_o}^{v} p\,dv = \int_{v_o}^{v}\left(p_o - \frac{1}{\beta_T}\ln\frac{v}{v_o}\right)dv \qquad (8.110)$$

Integrating the above equation using integration by parts,[5] we have

$$\int_{v_o}^{v} p\,dv = \left(\frac{1}{\beta_T} + p_o\right)(v - v_o) - \frac{1}{\beta_T}\left[v\ln\frac{v}{v_o}\right] \qquad (8.111)$$

Therefore, for constant thermal expansion coefficient and isothermal compressibility, the finite changes in u, h, s, f, and μ due to a volume change at constant temperature are given by

$$\Delta u = \int_{v_o}^{v}\left(\frac{T_o\alpha}{\beta_T} - p\right)dv = \left(\frac{T_o\alpha - 1}{\beta_T} - p_o\right)(v - v_o) + \frac{1}{\beta_T}\left[v\ln\frac{v}{v_o}\right] \qquad (8.112)$$

---

[5] $\int \ln x\,dx = x\ln x - x$

$$\Delta h = \int_{v_0}^{v} \frac{1}{\beta_T}(T_o\alpha - 1)dv = \frac{1}{\beta_T}(T_o\alpha - 1)(v - v_0) \qquad (8.113)$$

$$\Delta s = \int_{v_0}^{v} \frac{\alpha}{\beta_T}dv = \frac{\alpha}{\beta_T}(v - v_0) \qquad (8.114)$$

$$\Delta f = -\int_{v_0}^{v} pdv = -\left(\frac{1}{\beta_T} + p_o\right)(v - v_0) + \frac{1}{\beta_T}\left[v\ln\frac{v}{v_0}\right] \qquad (8.115)$$

$$\Delta\mu = -\int_{v_0}^{v} \frac{1}{\beta_T}dv = -\frac{1}{\beta_T}(v - v_0) \qquad (8.116)$$

It is noted that the entropy of a material increases with volume at constant temperature only if the thermal expansion coefficient is positive but decreases with volume at constant temperature if the thermal expansion coefficient is negative since the isothermal compressibility is always positive for a thermodynamically stable material.

For ideal gases, $\alpha = 1/T$ and $\beta_T = 1/p$, the finite changes in u, h, s, f, and μ due to a finite volume change at constant temperature can be easily computed:

$$\Delta u = \int_{v_0}^{v} \left(\frac{T\alpha}{\beta_T} - p\right)dv = 0 \qquad (8.117)$$

$$\Delta h = \int_{v_0}^{v} \frac{1}{\beta_T}(T\alpha - 1)dv = 0 \qquad (8.118)$$

$$\Delta s = \int_{v_0}^{v} \frac{\alpha}{\beta_T}dv = \int_{v_0}^{v} \frac{p}{T}dv = \int_{v_0}^{v} \frac{R}{v}dv = R\ln\frac{v}{v_0} \qquad (8.119)$$

$$\Delta f = -\int_{v_0}^{v} pdv = -RT_o\ln\frac{v}{v_0} \qquad (8.120)$$

$$\Delta\mu = -\int_{v_0}^{v} \frac{1}{\beta_T}dv = -\int_{v_0}^{v} \frac{1}{\beta_T}dv = -RT_o\ln\frac{v}{v_0} \qquad (8.121)$$

8.4 Changes in Thermodynamic Properties with Pressure at Constant Temperature

To determine the changes in the molar internal energy Δu, enthalpy Δh, entropy Δs, Helmholtz free energy Δf, and chemical potential $\Delta \mu$ of a material as a result of pressure change Δp at constant temperature T, we first determine their rates of change with respect to change in pressure at constant temperature,

$$\left(\frac{\partial u}{\partial p}\right)_T, \left(\frac{\partial h}{\partial p}\right)_T, \left(\frac{\partial s}{\partial p}\right)_T, \left(\frac{\partial f}{\partial p}\right)_T, \left(\frac{\partial \mu}{\partial p}\right)_T \tag{8.122}$$

The infinitesimal changes in internal energy du, enthalpy dh, entropy ds, Helmholtz free energy df, and chemical potential $d\mu$ due to infinitesimal pressure change dp at constant temperature T are then

$$du = \left(\frac{\partial u}{\partial p}\right)_T dp, \, dh = \left(\frac{\partial h}{\partial p}\right)_T dp, \, ds = \left(\frac{\partial s}{\partial p}\right)_T dp, \tag{8.123}$$

$$df = \left(\frac{\partial f}{\partial p}\right)_T dp, \, d\mu = \left(\frac{\partial \mu}{\partial p}\right)_T dp \tag{8.124}$$

We then express the rates of changes using the definitions of measurable properties:

$$\left(\frac{\partial u}{\partial p}\right)_T = \left(\frac{Tds - pdv}{dp}\right)_T = T\left(\frac{\partial s}{\partial p}\right)_T - p\left(\frac{\partial v}{\partial p}\right)_T = -Tv\alpha + pv\beta_T \tag{8.125}$$

$$\left(\frac{\partial h}{\partial p}\right)_T = \left(\frac{Tds + vdp}{dp}\right)_T = T\left(\frac{\partial s}{\partial p}\right)_T + v = -Tv\alpha + v = v(1 - T\alpha) \tag{8.126}$$

$$\left(\frac{\partial s}{\partial p}\right)_T = -v\alpha \tag{8.127}$$

$$\left(\frac{\partial f}{\partial p}\right)_T = \left(\frac{-sdT - pdv}{dp}\right)_T = pv\beta_T \tag{8.128}$$

$$\left(\frac{\partial \mu}{\partial p}\right)_T = v \tag{8.129}$$

The general expressions for the finite changes in u, h, s, f, and μ due to a finite pressure change at constant temperature are given by

$$\Delta u = u(T_0, p) - u(T_0, p_0) = \int_{p_0}^{p} \left(\frac{\partial u}{\partial p} \right)_T dp = \int_{p_0}^{p} v(p\beta_T - T_0\alpha) dp \qquad (8.130)$$

$$\Delta h = h(T_0, p) - h(T_0, p_0) = \int_{p_0}^{p} \left(\frac{\partial h}{\partial p} \right)_T dp = \int_{p_0}^{p} v(1 - T_0\alpha) dp \qquad (8.131)$$

$$\Delta s = s(T_0, p) - s(T_0, p_0) = \int_{p_0}^{p} \left(\frac{\partial s}{\partial p} \right)_T dp = -\int_{p_0}^{p} v\alpha dp \qquad (8.132)$$

$$\Delta f = f(T_0, p) - f(T_0, p_0) = \int_{p_0}^{p} \left(\frac{\partial f}{\partial p} \right)_T dp = \int_{p_0}^{p} pv\beta_T dp \qquad (8.133)$$

$$\Delta \mu = \mu(T_0, p) - \mu(T_0, p_0) = \int_{p_0}^{p} \left(\frac{\partial \mu}{\partial p} \right)_T dp = \int_{p_0}^{p} v dp \qquad (8.134)$$

The changes in the molar thermal energy Δu_T, molar mechanical energy Δu_M, and molar chemical energy Δu_C of the system due to a pressure change at constant temperature are:

$$\Delta u_T = T_0 \Delta s = Q \qquad (8.135)$$

$$\Delta u_M = -\int_{p_0}^{p} \left[\frac{\partial (pv)}{\partial p} \right]_T dp = -\int_{p_0}^{p} v dp + \int_{p_0}^{p} pv\beta_T dp = -\Delta \mu + W \qquad (8.136)$$

$$\Delta u_C = \Delta \mu \qquad (8.137)$$

where Q is the amount of heat absorbed from the surrounding during the process, and W is the reversible work done on the system by the surrounding during the process.

For constant isothermal compressibility independent of pressure, we have shown that

$$v(T_0, p) = v_0(T_0, p_0)e^{-\beta_T(p-p_0)} = v_0 e^{-\beta_T(p-p_0)}$$

Therefore,

$$\int_{p_0}^{p} v dp = v_0 \int_{p_0}^{p} e^{-\beta_T(p-p_0)} dp = -\frac{v_0}{\beta_T} \left[e^{-\beta_T(p-p_0)} - 1 \right]$$

and

$$\int_{p_0}^{p} v p \beta_T \mathrm{d}p = \beta_T v_0 \int_{p_0}^{p} p e^{-\beta_T (p - p_0)} \mathrm{d}p \tag{8.138}$$

Performing the integration by parts[6] of the right-hand side of the above equation, we have

$$\beta_T \int_{p_0}^{p} v p \mathrm{d}p = -v_0 \left(p e^{-\beta_T (p - p_0)} - p_0 \right) - \frac{v_0}{\beta_T} \left[e^{-\beta_T (p - p_0)} - 1 \right] \tag{8.139}$$

If we assume both the thermal expansion coefficient and the isothermal compressibility are constant independent of pressure, we have the finite changes in $u, h, s, f,$ and μ due to a finite pressure change at constant temperature given by

$$\Delta u = v_0 \left\{ \left[p_0 - \frac{(T_0 \alpha - 1)}{\beta_T} \right] - \left[p - \frac{(T_0 \alpha - 1)}{\beta_T} \right] e^{-\beta_T (p - p_0)} \right\} \tag{8.140}$$

$$\Delta h = \frac{(T_0 \alpha - 1) v_0}{\beta_T} \left[e^{-\beta_T (p - p_0)} - 1 \right] \tag{8.141}$$

$$\Delta s = \frac{\alpha v_0}{\beta_T} \left[e^{-\beta_T (p - p_0)} - 1 \right] \tag{8.142}$$

$$\Delta f = -v_0 \left\{ \left(p e^{-\beta_T (p - p_0)} - p_0 \right) + \frac{1}{\beta_T} \left[e^{-\beta_T (p - p_0)} - 1 \right] \right\} \tag{8.143}$$

$$\Delta \mu = -\frac{v_0}{\beta_T} \left[e^{-\beta_T (p - p_0)} - 1 \right] \tag{8.144}$$

If we assume that the thermal expansion coefficient is constant and the isothermal compressibility also constant and very small, $e^{-\beta_T (p - p_0)} \approx 1 - \beta_T (p - p_0) + \frac{1}{2} [\beta_T (p - p_0)]^2$, we can rewrite the finite changes in $u, h, s, f,$ and μ due to a finite pressure change at constant temperature as

$$\Delta u = v_0 \left\{ \left[p_0 + \frac{(1 - T\alpha)}{\beta_T} \right] - \left[p + \frac{(1 - T\alpha)}{\beta_T} \right] \right.$$
$$\left. \left[1 - \beta_T (p - p_0) + \frac{1}{2} [\beta_T (p - p_0)]^2 \right] \right\} \tag{8.145}$$

$$\Delta h = v_0 (1 - T\alpha) \left[(p - p_0) - \frac{\beta_T}{2} (p - p_0)^2 \right] \tag{8.146}$$

---

[6] $\int x e^x \mathrm{d}x = (x - 1)e^x$; $\int x e^{ax} \mathrm{d}x = \frac{e^{ax}}{a^2}(ax - 1)$

$$\Delta s = -v_o \alpha \left[(p - p_o) - \frac{\beta_T}{2}(p - p_o)^2 \right] \tag{8.147}$$

$$\Delta f = v_o \beta_T \left\{ p(p - p_o) - \frac{1}{2}(p - p_o)^2 - \frac{\beta_T}{2}p(p - p_o)^2 \right\} \tag{8.148}$$

$$\Delta \mu = v_o \left[(p - p_o) - \frac{\beta_T}{2}(p - p_o)^2 \right] \tag{8.149}$$

If a condensed phase is assumed to be incompressible, i.e., $\beta_T = 0$, and the thermal expansion coefficient is a constant, we can further simplify the expressions for the changes in u, h, s, f, and μ due to a finite pressure change at constant temperature as

$$\Delta u = -T_o \alpha v_o(p - p_o) \tag{8.150}$$

$$\Delta h = (1 - T_o \alpha)v_o(p - p_o) \tag{8.151}$$

$$\Delta s = -\alpha v_o(p - p_o) \tag{8.152}$$

$$\Delta f = 0 \tag{8.153}$$

$$\Delta \mu = v_o(p - p_o) \tag{8.154}$$

For ideal gases, the changes in u, h, s, f, and μ due to a finite pressure change at constant temperature are given by

$$\Delta u = \int_{p_o}^{p} v(p\beta_T - T_o \alpha)\mathrm{d}p = 0 \tag{8.155}$$

$$\Delta h = \int_{p_o}^{p} v(1 - T_o \alpha)\mathrm{d}p = 0 \tag{8.156}$$

$$\Delta s = -\int_{p_o}^{p} v\alpha \mathrm{d}p = -R \ln \frac{p}{p_o} \tag{8.157}$$

$$\Delta f = \int_{p_o}^{p} pv\beta_T \mathrm{d}p = RT \ln \frac{p}{p_o} \tag{8.158}$$

$$\Delta \mu = \int_{p_0}^{p} v \mathrm{d}p = RT \ln \frac{p}{p_0} \tag{8.159}$$

8.5 Changes in Thermodynamic Properties with Both Temperature and Volume

The changes in molar internal energy Δu, molar enthalpy Δh, entropy Δs, Helmholtz free energy Δf, and chemical potential $\Delta \mu$ of a system due to simultaneous changes in temperature and volume from T_0 and v_0 to T and v can be mathematically expressed as

$$\Delta u = u(T, v) - u(T_0, v_0) = \int_{T_0, v_0}^{T, v} \left[\left(\frac{\partial u}{\partial T} \right)_v \mathrm{d}T + \left(\frac{\partial u}{\partial v} \right)_T \mathrm{d}v \right] \tag{8.160}$$

$$\Delta h = h(T, v) - h(T_0, v_0) = \int_{T_0, v_0}^{T, v} \left[\left(\frac{\partial h}{\partial T} \right)_v \mathrm{d}T + \left(\frac{\partial h}{\partial v} \right)_T \mathrm{d}v \right] \tag{8.161}$$

$$\Delta s = s(T, v) - s(T_0, v_0) = \int_{T_0, v_0}^{T, v} \left[\left(\frac{\partial s}{\partial T} \right)_v \mathrm{d}T + \left(\frac{\partial s}{\partial v} \right)_T \mathrm{d}v \right] \tag{8.162}$$

$$\Delta f = f(T, v) - f(T_0, v_0) = \int_{T_0, v_0}^{T, v} \left[\left(\frac{\partial f}{\partial T} \right)_v \mathrm{d}T + \left(\frac{\partial f}{\partial v} \right)_T \mathrm{d}v \right] \tag{8.163}$$

$$\Delta \mu = \mu(T, v) - \mu(T_0, v_0) = \int_{T_0, v_0}^{T, v} \left[\left(\frac{\partial \mu}{\partial T} \right)_v \mathrm{d}T + \left(\frac{\partial \mu}{\partial v} \right)_T \mathrm{d}v \right] \tag{8.164}$$

Combining changes in u, h, s, f, and μ due to temperature change at constant volume and those due to volume change at constant temperature, we have the general expressions:

$$\Delta u = \int_{T_0, v_0}^{T, v_0} c_v \mathrm{d}T + \int_{T, v_0}^{T, v} \left(\frac{T\alpha}{\beta_T} - p \right) \mathrm{d}v \tag{8.165}$$

$$\Delta h = \int_{T_0, v_0}^{T, v_0} \left(c_v + v_0 \frac{\alpha}{\beta_T} \right) \mathrm{d}T + \int_{T_0, v_0}^{T, v} \frac{1}{\beta_T} (T\alpha - 1) \mathrm{d}v \tag{8.166}$$

$$\Delta s = \int\limits_{T_o,v_o}^{T,v_o} \frac{c_v}{T} dT + \int\limits_{T_o,v_o}^{T,v} \frac{\alpha}{\beta_T} dv \tag{8.167}$$

$$\Delta f = \int\limits_{T_o,v_o}^{T,v_o} [c_v - s(T_o, v_o)] dT - T \int\limits_{T_o,v_o}^{T,v_o} \frac{c_v}{T} dT - \int\limits_{T,v_o}^{T,v} p dv \tag{8.168}$$

$$\Delta \mu = \int\limits_{T_o,v_o}^{T,v_o} \left[c_v + \frac{v_o \alpha}{\beta_T} - s(T_o, v_o) \right] dT - T \int\limits_{T_o,v_o}^{T,v_o} \frac{c_v}{T} dT - \int\limits_{T,v_o}^{T,v} \frac{1}{\beta_T} dv \tag{8.169}$$

If c_v, β_T, and α are constant independent of temperature and volume, we have

$$\Delta u = c_v(T - T_o) + \left(\frac{T\alpha - 1}{\beta_T} - p_o \right)(v - v_o) + \frac{1}{\beta_T} \left[v \ln \frac{v}{v_o} \right] \tag{8.170}$$

$$\Delta h = \left(c_v + \frac{v_o \alpha}{\beta_T} \right)(T - T_o) + \frac{1}{\beta_T}(T\alpha - 1)(v - v_o) \tag{8.171}$$

$$\Delta s = c_v ln \frac{T}{T_o} + \frac{\alpha}{\beta_T}(v - v_o) \tag{8.172}$$

$$\Delta f = [c_v - s(T_o, v_o)](T - T_o)$$
$$- c_v T \ln \frac{T}{T_o} - \left(\frac{1}{\beta_T} + p_o \right)(v - v_o) + \frac{1}{\beta_T} \left[v \ln \frac{v}{v_o} \right] \tag{8.173}$$

$$\Delta \mu = \left[c_v + \frac{v_o \alpha}{\beta_T} - s(T_o, v_o) \right](T - T_o) - c_v T \ln \frac{T}{T_o} - \frac{1}{\beta_T}(v - v_o) \tag{8.174}$$

For ideal gases, we have

$$\Delta u = c_v(T - T_o) \tag{8.175}$$

$$\Delta h = c_p(T - T_o) \tag{8.176}$$

$$\Delta s = c_v \ln \frac{T}{T_o} + R \ln \frac{v}{v_o} \tag{8.177}$$

$$\Delta f = [c_v - s^o(T_o, v_o)](T - T_o) - c_v T \ln \frac{T}{T_o} - RT \ln \frac{v}{v_o} \tag{8.178}$$

$$\Delta \mu = [c_p - s^o(T_o, v_o)](T - T_o) - c_v T \ln \frac{T}{T_o} - RT \ln \frac{v}{v_o} \tag{8.179}$$

8.6 Changes in Thermodynamic Properties with Both Temperature and Pressure

The changes in molar internal energy Δu, enthalpy Δh, entropy Δs, Helmholtz free energy Δf, and chemical potential $\Delta \mu$ of a system due to simultaneous temperature and pressure changes can be expressed as

$$\Delta u = u(T, p) - u(T_0, p_0) = \int_{T_0, p_0}^{T, p} \left[\left(\frac{\partial u}{\partial T} \right)_p dT + \left(\frac{\partial u}{\partial p} \right)_T dp \right] \tag{8.180}$$

$$\Delta h = h(T, p) - h(T_0, p_0) = \int_{T_0, p_0}^{T, p} \left[\left(\frac{\partial h}{\partial T} \right)_p dT + \left(\frac{\partial h}{\partial p} \right)_T dp \right] \tag{8.181}$$

$$\Delta s = s(T, p) - s(T_0, p_0) = \int_{T_0, p_0}^{T, p} \left[\left(\frac{\partial s}{\partial T} \right)_p dT + \left(\frac{\partial s}{\partial p} \right)_T dp \right] \tag{8.182}$$

$$\Delta f = f(T, p) - f(T_0, p_0) = \int_{T_0, p_0}^{T, p} \left[\left(\frac{\partial f}{\partial T} \right)_p dT + \left(\frac{\partial f}{\partial p} \right)_T dp \right] \tag{8.183}$$

$$\Delta \mu = \mu(T, p) - \mu(T_0, p_0) = \int_{T_0, p_0}^{T, p} \left[\left(\frac{\partial \mu}{\partial T} \right)_p dT + \left(\frac{\partial \mu}{\partial p} \right)_T dp \right] \tag{8.184}$$

Using the results above (Sects. (8.2) and (8.4)) for the temperature and pressure dependences of Δu, Δh, Δs, Δf, and $\Delta \mu$, we obtain their finite changes due to simultaneous changes in temperature and pressure,

$$\Delta u = \int_{T_0, p_0}^{T, p_0} (c_p - pv\alpha) dT + \int_{T, p_0}^{T, p} v(p\beta_T - T\alpha) dp \tag{8.185}$$

$$\Delta h = \int_{T_0, p_0}^{T, p_0} c_p dT + \int_{T, p_0}^{T, p} v(1 - T\alpha) dp \tag{8.186}$$

$$\Delta s = \int_{T_0, p_0}^{T, p_0} \frac{c_p}{T} dT - \int_{T, p_0}^{T, p} v\alpha dp \tag{8.187}$$

$$\Delta f = \int_{T_o,p_o}^{T,p_o} \left[c_p - s(T_o, p_o) - pv\alpha \right] dT - T \int_{T_o,p_o}^{T,p_o} \frac{c_p}{T} dT + \int_{T,p_o}^{T,p} pv\beta_T dp \qquad (8.188)$$

$$\Delta \mu = \int_{T_o,p_o}^{T,p_o} \left[c_p - s(T_o, p_o) \right] dT - T \int_{T_o,p_o}^{T,p_o} \frac{c_p}{T} dT + \int_{T,p_o}^{T,p} v \, dp \qquad (8.189)$$

As it has been shown above, if the thermal expansion coefficient α is a constant independent of temperature, we can express volume $v(T, p_o)$ as a function of temperature using the definition of thermal expansion coefficient,

$$v(T, p_o) = v_o(T_o, p_o)e^{\alpha(T-T_o)} = v_o e^{\alpha(T-T_o)} \qquad (8.190)$$

If we also assume that the isothermal compressibility β_T is constant independent of pressure, we can express v as a function of p,

$$v(T, p) = v(T, p_o)e^{-\beta_T(p-p_o)} \qquad (8.191)$$

Combining the effects of temperature and pressure on volume, we have

$$v(T, p) = v_o(T_o, p_o)e^{\alpha(T-T_o)-\beta_T(p-p_o)} = v_o e^{\alpha(T-T_o)-\beta_T(p-p_o)} \qquad (8.192)$$

Therefore, the general expressions for the changes in u, h, s, f, and μ due to the simultaneous changes in temperature and pressure assuming constant heat capacity, constant thermal expansion coefficient, and constant compressibility using T_o and p_o as the initial state are given by

$$\Delta u = c_p(T - T_o) - p_o v_o \left[e^{\alpha(T-T_o)} - 1 \right] + v_o e^{\alpha(T-T_o)}$$
$$\left\{ \left[p_o + \frac{(1 - T\alpha)}{\beta_T} \right] - \left[p + \frac{(1 - T\alpha)}{\beta_T} \right] e^{-\beta_T(p-p_o)} \right\} \qquad (8.193)$$

$$\Delta h = c_p(T - T_o) - \frac{v_o e^{\alpha(T-T_o)}(1 - T\alpha)}{\beta_T} \left[e^{-\beta_T(p-p_o)} - 1 \right] \qquad (8.194)$$

$$\Delta s = c_p \ln \frac{T}{T_o} + \frac{v_o e^{\alpha(T-T_o)}\alpha}{\beta_T} \left[e^{-\beta_T(p-p_o)} - 1 \right] \qquad (8.195)$$

$$\Delta f = \left[c_p - s(T_o, p_o) \right](T - T_o) - p_o v_o \left[e^{\alpha(T-T_o)} - 1 \right] - c_p T \ln \frac{T}{T_o}$$
$$- v_o e^{\alpha(T-T_o)} \left\{ \left[pe^{-\beta_T(p-p_o)} - p_o \right] + \frac{1}{\beta_T} \left[e^{-\beta_T(p-p_o)} - 1 \right] \right\} \qquad (8.196)$$

$$\Delta \mu = \left[c_p - s(T_o, p_o) \right](T - T_o) - c_p T \ln \frac{T}{T_o}$$

$$-\frac{v_o e^{\alpha(T-T_o)}}{\beta_T}\left[e^{-\beta_T(p-p_o)}-1\right] \tag{8.197}$$

For most solids, the thermal expansion coefficient and the isothermal compressibility are very small, so we can approximate them as $e^{\alpha(T-T_o)} \approx 1+\alpha(T-T_o)$ and $e^{-\beta_T(p-p_o)} \approx 1-\beta_T(p-p_o)+\frac{1}{2}[\beta_T(p-p_o)]^2$. With these two approximations, the finite changes due to simultaneous changes in temperature and pressure can be rewritten as

$$\Delta u = \left(c_p - p_o v_o \alpha\right)(T-T_o)$$
$$+ v_o[1+\alpha(T-T_o)]\left\{\left[p_o + \frac{(1-T\alpha)}{\beta_T}\right]\right.$$
$$\left.-\left[p+\frac{(1-T\alpha)}{\beta_T}\right]\left[1-\beta_T(p-p_o)+\frac{1}{2}[\beta_T(p-p_o)]^2\right]\right\} \tag{8.198}$$

$$\Delta h = c_p(T-T_o) + v_o[1+\alpha(T-T_o)](1-T\alpha)\left[(p-p_o)-\frac{\beta_T}{2}(p-p_o)^2\right] \tag{8.199}$$

$$\Delta s = c_p \ln\left(\frac{T}{T_o}\right) - v_o[1+\alpha(T-T_o)]\left[(p-p_o)-\frac{\beta_T}{2}(p-p_o)^2\right] \tag{8.200}$$

$$\Delta f = \left[c_p - s(T_o, p_o) - p_o v_o \alpha\right](T-T_o) - c_p T \ln(T/T_o)$$
$$+ v_o[1+\alpha(T-T_o)]\beta_T$$
$$\left[p(p-p_o)-\frac{1}{2}(p-p_o)^2 - \frac{\beta_T}{2}p(p-p_o)^2\right] \tag{8.201}$$

$$\Delta\mu = \left[c_p - s(T_o, p_o)\right](T-T_o) - c_p T \ln\left(\frac{T}{T_o}\right)$$
$$+ v_o[1+\alpha(T-T_o)]\left[(p-p_o)-\frac{\beta_T}{2}(p-p_o)^2\right] \tag{8.202}$$

If we make a further assumption that a material is incompressible, i.e., $\beta_T = 0$ and volume is independent of pressure, the finite changes in u, h, s, f, and μ due to simultaneous changes in temperature and pressure can be further rewritten as

$$\Delta u = \left(c_p - p_o v_o \alpha\right)(T-T_o) - v_o[1+\alpha(T-T_o)]T\alpha(p-p_o) \tag{8.203}$$

$$\Delta h = c_p(T-T_o) + v_o[1+\alpha(T-T_o)](1-T\alpha)(p-p_o) \tag{8.204}$$

$$\Delta s = c_p \ln\left(\frac{T}{T_o}\right) - v_o[1+\alpha(T-T_o)](p-p_o) \tag{8.205}$$

$$\Delta f = \left[c_p - s(T_o, p_o) - p_o v_o \alpha\right](T - T_o) - c_p T \ln(T/T_o) \tag{8.206}$$

$$\Delta \mu = \left[c_p - s(T_o, p_o)\right](T - T_o) - c_p T \ln\left(\frac{T}{T_o}\right) + v_o[1 + \alpha(T - T_o)](p - p_o) \tag{8.207}$$

For ideal gases, the finite changes due to simultaneous changes in temperature and pressure are:

$$\Delta u = c_v(T - T_o) \tag{8.208}$$

$$\Delta h = c_p(T - T_o) \tag{8.209}$$

$$\Delta s = c_p \ln \frac{T}{T_o} - R \ln \frac{p}{p_o} \tag{8.210}$$

$$\Delta f = [c_v - s(T_o, p_o)](T - T_o) - c_p T \ln \frac{T}{T_o} + RT \ln \frac{p}{p_o} \tag{8.211}$$

$$\Delta \mu = \left[c_p - s(T_o, p_o)\right](T - T_o) - c_p T \ln \frac{T}{T_o} + RT \ln \frac{p}{p_o} \tag{8.212}$$

8.7 Changes in Thermodynamic Properties for Phase Transitions

A material may undergo a phase transition, such as melting of a solid to become liquid or solidification of a liquid to become solid, or a structural transition from one crystal structure to another, or the development of a spontaneous polarization in a crystal, etc., as the temperature or pressure or both change. We assume that a phase transition takes place at a fixed temperature and pressure. For any given phase transition without specifying pressure, we assume that the pressure is 1 bar.

The chemical potential change in a phase transition determines the spontaneous direction for the phase transition because a system always tends to lower its chemical potential when the temperature (thermal potential) and pressure (mechanical potential) are held constant.

Let us consider a phase transition at temperature T and pressure p,

$$\alpha \overset{T,p}{\to} \beta$$

where α and β represent the parent phase and transformed new product phase, respectively. The chemical potential change $\Delta \mu$, enthalpy change Δh, and entropy change Δs for the transition are defined as

$$\Delta\mu = \mu^{\beta}(T, p) - \mu^{\alpha}(T, p) \tag{8.213}$$

$$\Delta h = h^{\beta}(T, p) - h^{\alpha}(T, p) \tag{8.214}$$

$$\Delta s = s^{\beta}(T, p) - s^{\alpha}(T, p) \tag{8.215}$$

where μ^{β}, h^{β}, and s^{β} are the chemical potential, molar enthalpy, molar entropy of the product phase, and μ^{α}, h^{α}, and s^{α} are the chemical potential, molar enthalpy, and molar entropy of the parent phase, respectively.

The amount of heat absorbed or released Q during a transition is equal to the enthalpy change Δh in magnitude, so the enthalpy change Δh for the transition is also called the heat or latent heat of transition. For example, the enthalpy or heat of fusion or melting is the enthalpy change or the amount of heat required to completely melt one mole of a solid to become liquid. The enthalpy or heat of vaporization is the enthalpy change or heat required to completely vaporize one mole of a liquid while the enthalpy or heat of sublimation is the enthalpy change required to completely vaporize one mole of a solid. The entropy change for the transition is also often simply referred to as entropy of transition.

For a phase transition at a constant temperature and pressure, the change in the chemical potential $\Delta\mu$ is simply related to enthalpy and entropy changes Δh and Δs for the transition,

$$\Delta\mu(T, p) = \Delta h(T, p) - T\Delta s(T, p) \tag{8.216}$$

The chemical energy change $(-\Delta\mu)$, the thermodynamic driving force D, and the amount of entropy produced Δs^{ir} for a phase transition at a constant temperature and pressure are related by

$$D = T\Delta s^{ir} = -\Delta\mu \tag{8.217}$$

If $\Delta\mu$ is negative, the driving force D is positive, and the transition is irreversible, producing Δs^{ir} amount of entropy. The larger the driving force D is, the greater amount of the entropy Δs^{ir} is produced, and thus the higher the degree of irreversibility for the phase transition.

Let us assume that the equilibrium transition temperature and pressure for a phase transition are T_e and p_e, which are defined as the temperature and pressure at which

$$D(T_e, p_e) = T_e\Delta s^{ir}(T_e, p_e) = -\Delta\mu^{\circ}(T_e, p_e) = 0 \tag{8.218}$$

i.e., there is no change in the chemical potential, and thus no driving force for the transition. We label the enthalpy change and entropy change for the transition at the equilibrium temperature T_e and pressure p_e as $\Delta h^{\circ}(T_e, p_e)$ and $\Delta s^{\circ}(T_e, p_e)$, which satisfy the equilibrium condition,

$$\Delta\mu^{\circ}(T_e, p_e) = \Delta h^{\circ}(T_e, p_e) - T_e\Delta s^{\circ}(T_e, p_e) = 0, \tag{8.219}$$

Typical available thermodynamic data for a phase transition of a material include $T_e(p_e = 1\ \text{bar})$, $\Delta h^{\circ}(T_e, p_e = 1\ \text{bar})$, and the constant pressure heat capacity of the parent and product phases, i.e., $c_p^{\alpha}(T, p_e = 1\ \text{bar})$ and $c_p^{\beta}(T, p_e = 1\ \text{bar})$ as a function of temperature. Based on such available data, we can calculate the enthalpy change or heat of transition at another temperature T at 1 bar by breaking down the transition into three imaginary steps with each step being a reversible process (Fig. 8.1). This procedure is based on the fact that chemical potential, enthalpy, and entropy are all properties of a system and thus are state functions. For a state function, its change is independent of process paths starting from the same initial state and ending at the same final state. The first step is a change in the temperature of the parent phase from the transition temperature of interest T to the equilibrium transition temperature T_e. The second step is the phase transition from A to B at equilibrium transition temperature T_e. The last step is the change of temperature of the new phase from T_e to T. Since enthalpy is a state function, we have.

$$\Delta h(T) = \Delta h^{\alpha}(T \rightarrow T_e) + \Delta h^{\circ}(T_e) + \Delta h^{\beta}(T_e \rightarrow T) \tag{8.220}$$

Using the heat capacity of α and β as a function of temperature, we get

$$\Delta h(T) = \int_{T}^{T_e} c_p^{\alpha}dT + \Delta h^{\circ}(T_e) + \int_{T_e}^{T} c_p^{\beta}dT \tag{8.221}$$

or

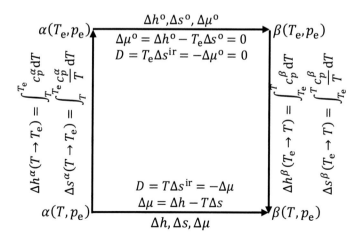

Fig. 8.1 Illustration of an imaginary three-step reversible path to compute the changes in chemical potential, molar enthalpy, and molar entropy for a phase transition taking place at a temperature different from the equilibrium transition temperature

$$\Delta h(T) = \Delta h^\circ(T_e) + \int_{T_e}^{T} \Delta c_p dT \tag{8.222}$$

where

$$\Delta c_p = c_p^\beta - c_p^\alpha$$

Once we have $\Delta h(T)$ as a function of temperature for a phase transition, we can obtain the chemical potential change for the transition as a function of temperature,

$$\left\{ \frac{\partial[\Delta\mu(T)/T]}{\partial T} \right\}_p = -\frac{\Delta h(T)}{T^2} \tag{8.223}$$

By integrating the above equation, we have

$$\frac{\Delta\mu(T)}{T} = \frac{\Delta\mu^\circ(T_e)}{T_e} - \int_{T_e}^{T} \left[\frac{\Delta h(T)}{T^2} \right] dT \tag{8.224}$$

The entropy change for a transition at the equilibrium temperature and pressure can be obtained from the simple fact that

$$\Delta h^\circ(T_e, p_e) = T_e \Delta s^\circ(T_e, p_e) \tag{8.225}$$

Therefore, if we relate the enthalpy change for a phase transition at the equilibrium temperature and pressure to the heat of transition Q_e, the entropy change for a transition can simply be related to the heat of transition and the equilibrium transition temperature,

$$\Delta s^\circ(T_e, p_e) = \frac{\Delta h^\circ(T_e, p_e)}{T_e} = \frac{Q_e}{T_e} \tag{8.226}$$

While a phase transition taking place at the equilibrium temperature and pressure is a reversible process since the parent and new phases are at equilibrium during the transition, a phase transition at another temperature and pressure is irreversible since there will be a finite driving force for the transition. For irreversible processes, we cannot simply use Eq. (8.226) to calculate the entropy change for a transition from the enthalpy change or the heat of transition since there will be entropy produced during an irreversible transition.

Similar to the calculation of enthalpy change for a phase transition at a temperature different from the equilibrium temperature, the entropy change at a transition temperature T different from the equilibrium transition temperature can be calculated using the same three-step reversible process,

$$\Delta s(T, p_e) = \Delta s^{\circ}(T_e, p_e) + \int_{T_e}^{T} \frac{\Delta c_p}{T} dT \tag{8.227}$$

where $\Delta s(T, p_e)$ is the molar entropy change for the transition at temperature T and p_e, and $\Delta s^{\circ}(T_e, p_e)$ is the molar entropy change for the transition at the equilibrium transition temperature T_e and p_e.

Once we have $\Delta s(T, p_e)$ as a function of temperature, we can obtain the chemical potential change for the transition as a function of temperature by using the following equality,

$$\Delta \mu(T, p_e) = \Delta \mu^{\circ}(T_e, p_e) - \int_{T_e}^{T} \Delta s(T, p_e) dT \tag{8.228}$$

Furthermore, if we have both Δh and Δs for a transition as a function of temperature, we can easily compute the chemical potential change as a function of temperature,

$$\Delta \mu(T, p_e) = \Delta h(T, p_e) - T \Delta s(T, p_e) \tag{8.229}$$

Substituting Eqs. (8.222) and (8.227) into Eq. (8.229), we have

$$\Delta \mu(T, p_e) = \Delta h^{\circ}(T_e, p_e) + \int_{T_e}^{T} \Delta c_p dT - T \left(\Delta s^{\circ}(T_e, p_e) + \int_{T_e}^{T} \frac{\Delta c_p}{T} dT \right) \tag{8.230}$$

Rewriting the above equation, we have

$$\Delta \mu(T, p_e) = \Delta h^{\circ}(T_e, p_e) - T \Delta s^{\circ}(T_e, p_e) + \int_{T_e}^{T} \Delta c_p dT - T \int_{T_e}^{T} \frac{\Delta c_p}{T} dT \tag{8.231}$$

If we assume the heat capacity difference between the transformed product phase and original parent phase is a constant independent of temperature, we have

$$\Delta h(T, p_e) = \Delta h^{\circ}(T_e, p_e) + \Delta c_p (T - T_e) \tag{8.232}$$

$$\Delta s(T, p_e) = \Delta s^{\circ}(T_e, p_e) + \Delta c_p \ln(T/T_e) \tag{8.233}$$

$$\Delta \mu(T, p_e) = \Delta h^{\circ} - T \Delta s^{\circ} + \Delta c_p (T - T_e) - \Delta c_p T \ln(T/T_e) \tag{8.234}$$

If we further assume that $\Delta c_p = 0$, i.e., the heat capacity of the parent and product phases is assumed to be the same, or $\Delta h(T, p_e)$ is a constant independent of temperature, we have

$$\Delta\mu(T, p_e) = \Delta h^\circ(T_e, p_e) - T\Delta s^\circ(T_e, p_e) \tag{8.235}$$

If we use the fact that $\Delta\mu^\circ(T_e, p_e) = \Delta h^\circ(T_e, p_e) - T_e\Delta s^\circ(T_e, p_e) = 0$, we can rewrite the above equation as

$$\Delta\mu(T, p_e) = \Delta h^\circ(T_e, p_e) - T\frac{\Delta h^\circ(T_e, p_e)}{T_e} = -\frac{\Delta h^\circ(T_e, p_e)\Delta T}{T_e} \tag{8.236}$$

where $\Delta T = T - T_e$.

We can also readily determine the pressure dependence of $\Delta\mu$ by including the mechanical contribution. Similar to the calculation of enthalpy and entropy changes of a transition at temperatures different from the equilibrium transition temperature, we can compute the chemical potential change at a pressure which is different from the equilibrium pressure for the phase transition. The phase transition at a given pressure p can also be separated into three imaginary steps (see Fig. 8.2). Step 1 involves changing the pressure of the parent phase from pressure p of interest to the equilibrium transition pressure p_e. Step 2 is the phase transition at the equilibrium transition pressure p_e, and the last step is the change of pressure from the equilibrium transition pressure p_e to pressure p for the product phase. Since chemical potential is a state function, the change in the chemical potential for the phase transition at p is the same as the total change in the chemical potential for the three steps, i.e.,

$$\Delta\mu(T, p) = \Delta\mu^\alpha(p \to p_e) + \Delta\mu^{\alpha\to\beta}(T, p_e) + \Delta\mu^\beta(p_e \to p) \tag{8.237}$$

Fig. 8.2 Illustration of an imaginary three-step reversible path to compute the changes in chemical potential, molar enthalpy, and molar entropy for a phase transition taking place at a pressure different from the equilibrium transition pressure

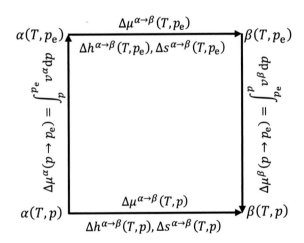

Using the information α and β about the molar volumes of v^α and v^β, as a function of pressure, we have.

$$\Delta\mu(T, p) = \int_{T,p}^{T,p_e} v^\alpha dp + \Delta\mu^{\alpha\to\beta}(T, p_e) + \int_{T,p_e}^{T,p} v^\beta dp \qquad (8.238)$$

or

$$\Delta\mu(T, p) = \Delta\mu^{\alpha\to\beta}(T, p_e) + \int_{T,p_e}^{T,p} \Delta v dp \qquad (8.239)$$

where $\Delta v = v^\beta - v^\alpha$ is the molar volume change for the phase transition.

Combining the temperature and pressure dependences of chemical potential change for a phase transition, we have

$$\Delta\mu(T, p) = \Delta h^\circ - T\Delta s^\circ + \int_{T_e}^{T} \Delta c_p dT - T\int_{T_e}^{T} \frac{\Delta c_p}{T} dT + \int_{T,p_e}^{T,p} \Delta v dp \qquad (8.240)$$

If we assume both Δc_p and Δv are independent of temperature and pressure, we obtain

$$\Delta\mu(T, p) = \Delta h^\circ - T\Delta s^\circ + \Delta c_p\left(T - T_e - T\ln\frac{T}{T_e}\right) + \Delta v(p - p_e) \qquad (8.241)$$

The assumption of constant Δv is equivalent to neglecting the compressibility of the parent and product phases. Equation (8.241) can also be employed to find the new equilibrium temperature under a given applied pressure by setting

$$\Delta\mu(T, p) = 0 \qquad (8.242)$$

For example, if we want to shift the transition temperature by ΔT, the required applied pressure $\Delta p = (p - p_e)$ is

$$\Delta p = \left\{\Delta T\Delta s^\circ - \Delta c_p\left[\Delta T - (T_e + \Delta T)\ln\frac{T_e + \Delta T}{T_e}\right]\right\}/\Delta v \qquad (8.243)$$

If $\Delta T \ll T_e$, we can approximate the solution,

$$\Delta p = \frac{\Delta T}{\Delta v}\left(\frac{\Delta h^\circ + \Delta c_p\Delta T}{T_e}\right) \qquad (8.244)$$

Equation (8.244) is essentially the Clapeyron equation which we will discuss in Chap. 11 on the pressure–temperature equilibrium of a single-component system.

For a phase transition in a crystal, one could consider the effect of an applied stress on the chemical potential change for the transition and thus the shift in transition temperature due to the applied stress,

$$\Delta\mu(T, \sigma_{ij}) = \Delta h^\circ - T\Delta s^\circ + \int_{T_e}^{T} \Delta c_p dT - T\int_{T_e}^{T} \frac{\Delta c_p}{T} dT + \int_{T,p_e}^{T,\sigma_{ij}} v\varepsilon_{ij}^\circ d\sigma_{ij} \quad (8.245)$$

where v is the molar volume of the parent phase, ε_{ij}° is the stress-free transition strain[7] which is a measure of the relative stress-free lattice parameter change from the parent to the product phase, and σ_{ij} is the applied stress. Similar to the effect of pressure on phase transition temperature, one can tune the phase transition temperature of a structural phase transition using an applied stress σ_{ij}.

If the chemical potential change $\Delta\mu$ for a phase transition is known as a function of temperature and pressure, we can obtain the entropy change Δs for the transition,

$$\Delta s(T, p) = -\left[\frac{\partial\Delta\mu(T, p)}{\partial T}\right]_p \quad (8.246)$$

Similarly, the enthalpy change or heat of a transition can also be directly obtained from the knowledge of the chemical potential change for the transition as a function of temperature and pressure,

$$\Delta h(T, p) = -T^2\left[\frac{\partial(\Delta\mu(T, p)/T)}{\partial T}\right]_p = \Delta\mu(T, p) + T\Delta s(T, p) \quad (8.247)$$

Finally, the volume difference can be obtained using the derivative of the chemical potential change with respect to pressure, i.e.,

$$\Delta v(T, p) = \left[\frac{\partial\Delta\mu(T, p)}{\partial p}\right]_T \quad (8.248)$$

The amount of entropy produced during a phase transition is given by

$$\Delta s^{ir}(T, p) = -\frac{\Delta\mu(T, p)}{T} \quad (8.249)$$

---

[7] For example, for a cubic to tetragonal transformation, $\varepsilon_{11}^\circ = \frac{a_t - a_c}{a_c}$, $\varepsilon_{22}^\circ = \frac{a_t - a_c}{a_c}$, $\varepsilon_{33}^\circ = \frac{c_t - a_c}{a_c}$, $\varepsilon_{12}^\circ = \varepsilon_{21}^\circ = \varepsilon_{13}^\circ = \varepsilon_{31}^\circ = \varepsilon_{23}^\circ = \varepsilon_{32}^\circ = 0$, where a_t and c_t are the lattice parameters of the tetragonal phase measured at the stress-free condition, and a_c is the lattice parameter of the cubic phase measured at the stress-free condition.

where $\Delta\mu(T, p)$ is the chemical potential change for the phase transition. If there is a chemical potential reduction for a given phase transition at constant temperature and pressure, there will be entropy produced, and the phase transition is irreversible. The amount of chemical energy dissipated is $-\Delta\mu(T, p)$ which is also the driving force D for the transition.

8.8 Changes in Thermodynamic Properties for Chemical Reactions

Performing thermodynamic calculations for chemical reactions is similar to that for phase transitions. As a matter of fact, it is actually not necessary to distinguish the two in many cases in terms of thermodynamic calculations. The majority of the calculations for both phase transitions and chemical reactions are performed for 1 bar although the pressure dependence can be readily taken into account.

The main difference between phase transitions and chemical reactions include: (1) chemical reactions often involve more than one species in both the reactants and products of a reaction while a phase transition often involves a single component except for phase transitions involving compositional changes in binary and multi-component systems; (2) thermodynamic calculations for chemical reactions usually rely on available thermodynamic properties of the reactants and products at room temperature 298 K and 1 bar and the heat capacity data of all the involved chemical species or components as a function of temperature at 1 bar while thermodynamic calculations for a phase transition often employ data for the thermodynamic properties measured or computed at the equilibrium transition temperature at 1 bar as well as the heat capacity data as a function of temperature at 1 bar; and (3) while the phase transitions taking place at equilibrium temperature and pressure are reversible processes, chemical reactions taking place at room temperature 298 K and ambient pressure 1 bar are almost always irreversible.

Let us consider a chemical reaction at temperature T and pressure p,

$$\nu_A A + \nu_B B \rightleftharpoons \nu_C C + \nu_D D \tag{8.250}$$

where ν_A, ν_B, ν_C, and ν_D are stoichiometric reaction coefficients. It is important that those coefficients are chosen such that the mass is balanced between the reacting and product species because we are only interested in the changes in chemical energy due to the chemical bonding changes resulted from a reaction, i.e., the mass is fixed or constant during a chemical reaction or a set of chemical reactions. Otherwise, the energy associated the rest mass will dominate and overwhelm the energy changes from chemical bonding energy changes. While each reacting and product species has its chemical potential, enthalpy, and entropy, we can also define a chemical potential as well as its associated enthalpy and entropy for the fixed mass existing in the form of reactants and a chemical potential and its associated enthalpy and entropy for the same mass existing in the form of products. However, we will still call the total chemical potential for the reactants as Gibbs free energy of reactants, and the total chemical potential of products as Gibbs free energy of products in order to avoid the confusion between the discussions here and the existing literature.

Let us denote the enthalpy, entropy, and Gibbs free energy of reactants of a reaction as H^R, S^R, and G^R and those of products as H^P, S^P, and G^P, respectively. The changes in enthalpy, entropy and Gibbs free energy for a reaction at a given temperature T and pressure p are given by

$$\Delta G(T, p) = G^P(T, p) - G^R(T, p) \tag{8.251}$$

$$\Delta H(T, p) = H^P(T, p) - H^R(T, p) \tag{8.252}$$

$$\Delta S(T, p) = S^P(T, p) - S^R(T, p) \tag{8.253}$$

where $\Delta H(T, p)$ is also called the heat of reaction since for a reaction at constant pressure, the amount of heat involved has the same magnitude as the enthalpy change for the reaction. According to the first law of thermodynamics, the enthalpy change ΔH for a constant pressure process of a closed system is equal to the heat (thermal energy) Q absorbed or released by the system:

$$\Delta H(T, p) = Q$$

If $Q > 0$, heat is absorbed by the system, the corresponding reaction is called endothermic reaction. If $Q < 0$, heat is released from the system, the corresponding reaction is called an exothermic reaction. Endothermic reactions increase the enthalpy of the system while exothermic reactions decrease the enthalpy of the system.

It should be pointed out that although the enthalpy change for a constant pressure process is equal to the magnitude of heat involved, the concepts of enthalpy and heat are entirely different. Enthalpy is a property of a system and thus is a state function. Heat is associated with a process such as a temperature change or a chemical reaction or a phase transition, and its magnitude is the same as the enthalpy change for a system if the process is carried out under constant pressure.

We can express the enthalpy, entropy, and Gibbs free energy of both the reactants and the products in terms of the molar enthalpy, molar entropy, and chemical potential of each individual species involved in the reaction, i.e.,

$$G^R(T, p) = \nu_A \mu_A(T, p) + \nu_B \mu_B(T, p) \tag{8.254}$$

$$H^R(T, p) = \nu_A h_A(T, p) + \nu_B h_B(T, p) \tag{8.255}$$

$$S^R(T, p) = \nu_A s_A(T, p) + \nu_B s_B(T, p) \tag{8.256}$$

$$G^P(T, p) = v_C \mu_C(T, p) + v_D \mu_D(T, p) \tag{8.257}$$

$$H^P(T, p) = v_C h_C(T, p) + v_D h_D(T, p) \tag{8.258}$$

$$S^P(T, p) = v_C s_C(T, p) + v_D s_D(T, p) \tag{8.259}$$

where for each species i ($i = $ A, B, C, D) involved in the reaction,

$$\mu_i(T, p) = h_i(T, p) - T s_i(T, p) \tag{8.260}$$

The Gibbs free energy change $\Delta G(T, p)$ for a chemical reaction determines the spontaneous direction of the reaction because a system always tends to lower its Gibbs free energy when the temperature and pressure are held constant.

The change in the Gibbs free energy $\Delta G(T, p)$ for a reaction is simply related to the changes in enthalpy $\Delta H(T, p)$ and entropy $\Delta S(T, p)$ at the same temperature and pressure, i.e.,

$$\Delta G(T, p) = \Delta H(T, p) - T \Delta S(T, p) \tag{8.261}$$

For chemical reactions, most of the available thermodynamic data are for room temperature and ambient pressure, i.e., 298 K and 1 bar, and the heat capacity of each chemical species as a function of temperature at 1 bar. Therefore, at room temperature 298 K and ambient pressure 1 bar, the changes in Gibbs free energy, enthalpy, and entropy for the reaction can be calculated using the chemical potential, molar enthalpy, and molar entropy of each individual species from the available data:

$$\Delta G^o_{298\,K,\,1\,bar} = \left[v_C \mu^o_C + v_D \mu^o_D \right] - \left[v_A \mu^o_A + v_B \mu^o_B \right] \tag{8.262}$$

$$\Delta H^o_{298\,K,\,1\,bar} = \left[v_C h^o_C + v_D h^o_D \right] - \left[v_A h^o_A + v_B h^o_B \right] \tag{8.263}$$

$$\Delta S^o_{298\,K,\,1\,bar} = \left[v_C s^o_C + v_D s^o_D \right] - \left[v_A s^o_A + v_B s^o_B \right] \tag{8.264}$$

$$\Delta G^o_{298\,K,\,1\,bar} = \Delta H^o_{298\,K,\,1\,bar} - 298 \Delta S^o_{298\,K,\,1\,bar} \tag{8.265}$$

It should be noted that the chemical reactions at 298 K and 1 bar or at any given temperature T other than the equilibrium temperature and 1 bar are irreversible since

$$\Delta G^o_{298\,K,\,1\,bar} \neq 0 \tag{8.266}$$

Therefore, one cannot obtain the entropy change for a reaction at room temperature using the heat of reaction and the reaction temperature because the entropy change must also include the entropy produced due to the irreversible reaction in addition to the entropy exchange with the surrounding, i.e.,

$$\Delta S^o_{298\,K,1\,bar} \neq \frac{\Delta H^o_{298\,K,1\,bar}}{298\,K}$$

Similar to the case of phase transitions, with the knowledge of Gibbs free energy change or enthalpy change (heat), and entropy change for a reaction at 298 K and 1 bar as well as the heat capacity of each species involved in the reaction, we can compute the Gibbs free energy change, enthalpy change (heat), and entropy change for the reaction at any other temperature.

For reactions taking place at a different temperature other than room temperature, there are two ways to calculate the enthalpy change or the heat of the reaction. One is to calculate the molar enthalpies, molar entropies, and chemical potentials of all reactant and product species at the temperature of reaction, and then take the differences between the total enthalpy, total entropy, and Gibbs free energy of the reactants and those of the products. Another approach is to break the reaction process into three imaginary steps (see Fig. 8.3). The first step is a change in the temperature of the reactants from the temperature T of interest to 298 K. The second step is the imaginary reaction at 298 K, and the third step is the change of temperature for the products from 298 K to T. Since enthalpy, entropy, and Gibbs free energy are state functions, their changes are independent of the reaction paths. For example, the enthalpy change for the reaction at temperature T and 1 bar is given by.

$$\Delta H(T, 1\,bar) = \int_T^{298\,K} C_p^R dT + \Delta H^o_{298\,K,1\,bar} + \int_{298\,K}^T C_p^P dT \qquad (8.267)$$

C_p^R and C_p^P in the above equation are given by

$$C_p^R = \nu_A c_{p,A} + \nu_B c_{p,B} \qquad (8.268)$$

$$C_p^P = \nu_C c_{p,C} + \nu_D c_{p,D} \qquad (8.269)$$

where $c_{p,A}$, $c_{p,B}$, $c_{p,C}$, and $c_{p,D}$ are the molar heat capacities of A, B, C, and D, respectively. Equation (8.267) can be rewritten as

$$\Delta H(T, 1\,bar) = \Delta H^o_{298\,K,1\,bar} + \int_{298\,K}^T \Delta C_p dT \qquad (8.270)$$

where ΔC_p is the difference between the total heat capacity of the products and the total heat capacity of the reactants

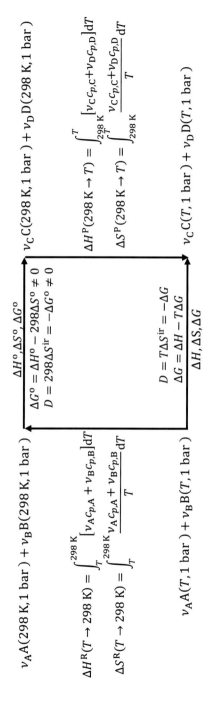

Fig. 8.3 Illustration of an imaginary three-step reversible path to compute the changes in Gibbs free energy, enthalpy, and entropy for a chemical reaction taking place at a temperature different from room temperature 298 K

$$\Delta C_p = C_p^{\text{P}} - C_p^{\text{R}}$$

Once we have $\Delta H(T, 1\,\text{bar})$ as a function of temperature for a reaction, we can obtain the Gibbs free energy change for the reaction as a function of temperature,

$$\left[\frac{\partial(\Delta G(T, p)/T)}{\partial T}\right]_p = -\frac{\Delta H(T, p)}{T^2} \tag{8.271}$$

By integrating the above equation, we have

$$\frac{\Delta G(T, 1\,\text{bar})}{T} = \frac{\Delta G^{\text{o}}_{298\,\text{K},1\,\text{bar}}}{298\,\text{K}} - \int_{298\,\text{K}}^{T} \left(\frac{\Delta H(T, 1\,\text{bar})}{T^2}\right) dT \tag{8.272}$$

If $\Delta H(T, p)$ is independent of temperature, which is only true if the total heat capacity of the reactants and the total heat capacity of the products as a function of temperature are assumed to be the same,

$$\frac{\Delta G(T, 1\,\text{bar})}{T} - \frac{\Delta G^{\text{o}}_{298\,\text{K},1\,\text{bar}}}{298\,K} = \Delta H^{\text{o}}_{298\,\text{K},1\,\text{bar}}\left(\frac{1}{T} - \frac{1}{298\,K}\right) \tag{8.273}$$

The above equation allows one to calculate the Gibbs free energy change at any other temperature, T, based on the Gibbs free energy change at room temperature and the information about enthalpy change or heat involved in the reaction, $\Delta H^{\text{o}}_{298\,\text{K},1\,\text{bar}}$.

The entropy change for a reaction can be calculated by either calculating the entropies of the products and reactants of the chemical reaction at the temperature of interest or by breaking down into three imaginary processes like in the calculation of enthalpy changes for chemical reactions. Both reach the same result,

$$\Delta S(T, 1\,\text{bar}) = \Delta S^{\text{o}}_{298,1\,\text{bar}} + \int_{298}^{T} \frac{\Delta C_p}{T} dT \tag{8.274}$$

where $\Delta S(T, 1\,\text{bar})$ is the entropy change for the reaction at temperature T and 1 bar, and $\Delta S^{\text{o}}_{298\,\text{K},1\,\text{bar}}$ is the entropy change for the reaction at 298 K and 1 bar.

Once we have $\Delta S(T, 1\,\text{bar})$ as a function of temperature for a reaction, we can obtain the Gibbs free energy change for the reaction as a function of temperature by using the following equality,

$$\Delta G(T, 1\,\text{bar}) = \Delta G^{\text{o}}_{298\,\text{K},1\,\text{bar}} - \int_{298\,\text{K}}^{T} \Delta S(T, 1\,\text{bar}) dT \tag{8.275}$$

Finally, if we have both ΔH and ΔS for a reaction as a function of temperature at 1 bar, we can easily obtain the Gibbs free energy change for the reaction as a function of T at 1 bar,

$$\Delta G(T, 1\,\text{bar}) = \Delta H(T, 1\,\text{bar}) - T \Delta S(T, 1\,\text{bar}) \tag{8.276}$$

On the other hand, if the Gibbs free energy change for a reaction is known as a function of temperature and pressure, we can obtain the entropy change for the transition,

$$\Delta S(T, p) = -\left[\frac{\partial \Delta G(T, p)}{\partial T}\right]_p \tag{8.277}$$

Similarly, the enthalpy change or heat of a reaction can also be directly obtained from the knowledge of the Gibbs free energy change for the reaction as a function of temperature and pressure,

$$\Delta H(T, p) = -T^2 \left[\frac{\partial (\Delta G(T, p)/T)}{\partial T}\right]_p = \Delta G(T, p) + T \Delta S(T, p) \tag{8.278}$$

If the difference in compressibility between reactants and products can be ignored, the pressure dependence for the change in the Gibbs free energy for a reaction at a constant temperature can be approximated in a manner similar to the pressure dependence of chemical potential change for phase transitions, i.e.,

$$\Delta G(T, p) = \Delta G(T, p_o) + \int_{1\,\text{bar}}^{p} \Delta V \, dp \tag{8.279}$$

$$\Delta G(T, p) \approx \Delta G(T, 1\,\text{bar}) + \Delta V(p - 1\,\text{bar}) = \Delta G(T, 1\,\text{bar}) + \Delta V \Delta p \tag{8.280}$$

where ΔV is the volume change for the reaction and $\Delta p = p - 1$ bar.

If $\Delta G(T, p) = 0$, i.e., there is no change in the Gibbs free energy, and thus no driving force for the reaction, the reactants and products are at equilibrium. The corresponding temperature and pressure are the equilibrium temperature and pressure, T_e and p_e for the reaction. From

$$\Delta H(T_e, p_e) - T_e \Delta S(T_e, p_e) = 0, \tag{8.281}$$

we have

$$T_e = \frac{\Delta H(T_e, p_e)}{\Delta S(T_e, p_e)} \tag{8.282}$$

The entropy produced ΔS^{ir} or the driving force D consumed in a reaction at a constant temperature and pressure can be directly related to the Gibbs free energy change or dissipation for the reaction,

$$D = -\Delta G = T \Delta S^{ir} \tag{8.283}$$

If a reaction is spontaneous, ΔG is negative with $-\Delta G$ or D as the amount of chemical energy that is converted to thermal energy, $T \Delta S^{ir}$, leading to an increase in entropy ΔS^{ir} at the reaction temperature and pressure.

It should be noted that if a chemical reaction takes place at constant temperature and volume, the entropy produced ΔS^{ir} for a chemical reaction is

$$S^{ir} = \Delta S(T, V) - \frac{\Delta U(T, V)}{T} = -\frac{\Delta F(T, V)}{T} \tag{8.284}$$

where $\Delta F(T, V)$ is the Helmholtz free energy change for a chemical reaction at constant temperature T and constant volume V. Therefore, if there is a Helmholtz free energy reduction for a given chemical reaction at constant temperature and volume, there will be production of entropy, and thus the reaction is irreversible.

If a chemical reaction takes place adiabatically at constant pressure (e.g., 1 bar), the entropy produced ΔS^{ir} for the chemical reaction is

$$\Delta S^{ir} = \Delta S = S(T, 1 \text{ bar}) - S(T_0, 1 \text{ bar}) \tag{8.285}$$

where $S(T, 1 \text{ bar})$ is the entropy of the final state at temperature T and pressure 1 bar, and $S(T_0, 1 \text{ bar})$ is the entropy of the initial state at temperature T_0 and pressure 1 bar.

If a chemical reaction takes place adiabatically at constant volume V, the entropy produced ΔS^{ir} for the chemical reaction is

$$\Delta S^{ir} = \Delta S = S(T, V) - S(T_0, V) \tag{8.286}$$

where $S(T, V)$ is the entropy of the final state at temperature T and volume V, and $S(T_0, V)$ is the entropy of the initial state at temperature T_0 and volume V. The final temperature T is also referred to as the theoretical flame temperature of a reaction or the maximum temperature that can be achieved from a reaction with all the heat or thermal energy produced from the reaction being used to increase the temperature of the products or all the remaining species after the reaction.

8.9 Examples

Example 1 What is the amount of pressure increase Δp as the temperature of one mole of material is increased from T_0 to T at constant molar volume v?

Solution

The rate of increase in pressure with respect to temperature change at constant volume is given by

$$
\left(\frac{\partial p}{\partial T}\right)_v = -\frac{\left(\frac{\partial v}{\partial T}\right)_p}{\left(\frac{\partial v}{\partial p}\right)_T} = \frac{\frac{1}{v}\left(\frac{\partial v}{\partial T}\right)_p}{-\frac{1}{v}\left(\frac{\partial v}{\partial p}\right)_T} = \frac{\alpha}{\beta_T}
$$

where α is volume thermal expansion coefficient, and β_T is the isothermal compressibility. Therefore, the change in pressure due to a change in temperature at constant volume is

$$
\Delta p = \int_{T_o}^{T} dp = \int_{T_o}^{T} \left(\frac{\partial p}{\partial T}\right)_v dT = \int_{T_o}^{T} \frac{\alpha}{\beta_T} dT
$$

If the thermal expansion coefficient α and the isothermal compressibility β_T are assumed to be constant independent of temperature,

$$
\Delta p = \frac{\alpha}{\beta_T}(T - T_o)
$$

For ideal gases,

$$
\Delta p = \int_{T_o}^{T} \frac{\alpha}{\beta_T} dT = \int_{T_o}^{T} \frac{p}{T} dT = \int_{T_o}^{T} \frac{R}{v} dT = \frac{R}{v}(T - T_o)
$$

where R is gas constant.

Example 2 For a diatomic ideal gas undergoing an isentropic (reversible adiabatic) volume change from v_o to v,

(a) Derive the expressions for the changes in temperature ΔT, the molar internal energy Δu, molar enthalpy Δh, molar entropy Δs, molar Helmholtz free energy Δf, and chemical potential $\Delta \mu$ in terms of heat capacity, compressibility, and thermal expansion coefficient, as well as temperature, pressure, volume or molar entropy at the initial state.

(b) Assuming oxygen gas (O_2) behaves as an ideal gas with constant pressure heat capacity of $7R/2$ (R is gas constant) and its molar entropy $s^o_{298\,K,1\,bar}$ of 205 J/(mol K), evaluate the ΔT, Δu, Δh, Δs, Δf, and $\Delta \mu$ of the oxygen gas (O_2) initially at 298 K and 1 bar due to a reversible adiabatic volume expansion by 10 times.

Solution

(a)   For a reversible adiabatic process, the entropy of the system is constant during the process, the rate of change of temperature with respect to volume change at constant entropy is given by

$$\left(\frac{\partial T}{\partial v}\right)_s = -\frac{(\partial s/\partial v)_T}{(\partial s/\partial T)_v} = -\frac{(\partial s/\partial p)_T}{(\partial v/\partial p)_T (c_v/T)} = \frac{(\partial v/\partial T)_p}{(\partial v/\partial p)_T (c_v/T)} = -\frac{T\alpha}{\beta_T c_v}$$

For ideal gases, $T\alpha = 1$ and $\beta_T = 1/p$, hence

$$\left(\frac{\partial T}{\partial v}\right)_s = -\frac{T\alpha}{\beta_T c_v} = -\frac{p}{c_v} = -\frac{RT}{v c_v}$$

To solve the above first-order ordinary differential equation, we rewrite it as

$$\frac{c_v}{T}dT = -\frac{R}{v}dv$$

or

$$d\ln T^{c_v} + d\ln v^R = 0$$

Integrating both sides of the above equation,

$$T^{c_v} v^R = C$$

where C is an integration constant. If we assume that the initial temperature is T_0, and the initial molar volume is v_0, we can determine the constant C,

$$T^{c_v} v^R = C = T_0^{c_v} v_0^R$$

Solving the above equation for T,

$$T = T_0 \left(\frac{v_0}{v}\right)^{\frac{R}{c_v}} = T_0 \left(\frac{v}{v_0}\right)^{-\frac{R}{c_v}}$$

Therefore, the temperature change is given by

$$\Delta T = T - T_0 = T_0 \left[\left(\frac{v}{v_0}\right)^{-\frac{R}{c_v}} - 1\right]$$

To obtain the internal energy change with volume at constant entropy, we first write down the rate of change of molar internal energy with respect to volume at constant entropy,

$$\left(\frac{\partial u}{\partial v}\right)_s = -p$$

The finite change in the molar internal energy as a result of volume change at constant entropy is

$$\Delta u = -\int_{s_0,v_0}^{s_0,v} p\,dv$$

Therefore, the molar internal energy change of one mole of ideal gas due to a change in volume at constant entropy is

$$\Delta u = -\int_{s_0,v_0}^{s_0,v} p\,dv = -\int_{s_0,v_0}^{s_0,v} \frac{RT}{v}\,dv$$

Substituting the expression for T as a function of v into the above expression for the internal energy change, we have

$$\Delta u = -\int_{s_0,v_0}^{s_0,v} \frac{RT}{v}\,dv = -RT_0 v_0^{\frac{R}{c_v}} \int_{s,v_0}^{s,v} v^{-\frac{c_p}{c_v}}\,dv$$

Integrating the above equation, we have

$$\Delta u = -RT_0 v_0^{\frac{c_p}{c_v}-1}\left[\frac{(v)^{1-\frac{c_p}{c_v}}}{1-\frac{c_p}{c_v}} - \frac{(v_0)^{1-\frac{c_p}{c_v}}}{1-\frac{c_p}{c_v}}\right]$$

Simplifying the above equation, we get

$$\Delta u = c_v T_0\left[\left(\frac{v}{v_0}\right)^{-\frac{R}{c_v}} - 1\right] = c_v[T - T_0]$$

For a reversible adiabatic process, the rate of change of molar enthalpy with respect to volume is given by

$$\left(\frac{\partial h}{\partial v}\right)_s = v\left(\frac{\partial p}{\partial v}\right)_s = -B_s = -\frac{1}{\beta_s}$$

or

$$\left(\frac{\partial h}{\partial v}\right)_s = v\left(\frac{\partial p}{\partial v}\right)_s = v\left(\frac{\frac{\partial p}{\partial T}}{\frac{\partial v}{\partial T}}\right)_s = v\frac{\left(\frac{\partial p}{\partial T}\right)_s}{\left(\frac{\partial v}{\partial T}\right)_s} = v\frac{\left(\frac{\partial s}{\partial T}\right)_p \left(\frac{\partial s}{\partial v}\right)_T}{\left(\frac{\partial s}{\partial p}\right)_T \left(\frac{\partial s}{\partial T}\right)_v} = -\frac{c_p}{c_v \beta_T}$$

Therefore, for ideal gases,

$$\left(\frac{\partial h}{\partial v}\right)_s = -\frac{c_p}{c_v \beta_T} = -\frac{c_p}{c_v}p = -\frac{c_p}{c_v}\frac{RT}{v}$$

Using the result for the volume dependence of temperature along an adiabatic reversible path,

$$T = T_0\left(\frac{v_0}{v}\right)^{\frac{R}{c_v}} = T_0\left(\frac{v}{v_0}\right)^{-\frac{R}{c_v}},$$

the molar enthalpy change of an ideal gas due to a change in volume at constant entropy is

$$\Delta h = \int_{s_0,v_0}^{s_0,v}\left(\frac{\partial h}{\partial v}\right)_s dv = -\frac{c_p}{c_v}RT_0 v_0^{\frac{R}{c_v}}\int_{s_0,v_0}^{s_0,v} v^{-\frac{c_p}{c_v}} dv$$

Integrating the above equation, we have

$$\Delta h = c_p T_0\left[\left(\frac{v}{v_0}\right)^{-\frac{R}{c_v}} - 1\right] = c_p[T - T_0]$$

The entropy change along an adiabatic reversible path, Δs, is zero by definition. The change in the molar Helmholtz free energy as a result of volume change at constant entropy is given by

$$\Delta f = \int_{s_0,v_0}^{s_0,v}\left(\frac{\partial f}{\partial v}\right)_s dv$$

To obtain the rate of change of molar Helmholtz free energy with respect to volume at constant entropy, we start from its differential form,

$$df = -sdT - pdv$$

Dividing both sides of the above equation by dv and keeping entropy constant for the derivatives, we have

$$\left(\frac{\partial f}{\partial v}\right)_s = -s\left(\frac{\partial T}{\partial v}\right)_s - p = s\frac{\left(\frac{\partial s}{\partial v}\right)_T}{\left(\frac{\partial s}{\partial T}\right)_v} - p = \frac{sT\alpha}{\beta_T c_v} - p$$

For ideal gases, $T\alpha = 1$ and $\beta_T = 1/p$, and thus

$$\left(\frac{\partial f}{\partial v}\right)_s = \left(\frac{s}{c_v} - 1\right)p = \left(\frac{s}{c_v} - 1\right)\frac{RT}{v}$$

Therefore,

$$\Delta f = \int_{s_0,v_0}^{s_0,v}\left(\frac{\partial f}{\partial v}\right)_s dv = \int_{s_0,v_0}^{s_0,v}\left(\frac{s}{c_v} - 1\right)\frac{RT}{v}dv = \left(\frac{s_0}{c_v} - 1\right)RT_0 v_0^{\frac{R}{c_v}}\int_{s_0,v_0}^{s_0,v} v^{-\frac{c_p}{c_v}}dv$$

Integrating the above equation, we have the final result,

$$\Delta f = (c_v - s_0)T_0\left[\left(\frac{v}{v_0}\right)^{-\frac{R}{c_v}} - 1\right] = (c_v - s_0)[T - T_0]$$

For ideal gases, we can also directly obtain the change in the molar Helmholtz free energy for the isentropic process from the change in molar internal energy,

$$\Delta f = \Delta u - s_0(T - T_0) = c_v(T - T_0) - s_0(T - T_0) = (c_v - s_0)[T - T_0]$$

The chemical potential change as a result of volume change along a reversible adiabatic path is given by

$$\Delta\mu = \int_{s_0,v_0}^{s_0,v}\left(\frac{\partial\mu}{\partial v}\right)_s dv$$

To obtain the rate of change of chemical potential with respect to volume at constant entropy, we use its differential form or the Gibbs–Duhem relation,

$$d\mu = -sdT + vdp$$

Dividing both sides of the above equation by dv and keep entropy constant for the derivatives, we get

$$\left(\frac{\partial\mu}{\partial v}\right)_s = -s\left(\frac{\partial T}{\partial v}\right)_s + v\left(\frac{\partial p}{\partial v}\right)_s = s\frac{\left(\frac{\partial s}{\partial v}\right)_T}{\left(\frac{\partial s}{\partial T}\right)_v} - \frac{1}{\beta_s} = \frac{sT\alpha}{\beta_T c_v} - \frac{c_p}{c_v\beta_T}$$

For ideal gases, $T\alpha = 1$ and $\beta_T = 1/p$, and thus

$$\left(\frac{\partial \mu}{\partial v}\right)_s = \frac{1}{c_v}(s - c_p)p = \frac{R}{c_v}(s - c_p)\frac{T}{v}$$

Therefore,

$$\Delta \mu = \frac{R}{c_v}(s_0 - c_p) \int\limits_{s_0, v_0}^{s_0, v} \frac{T}{v} dv = (c_p - s_0)T_0\left[\left(\frac{v}{v_0}\right)^{-\frac{R}{c_v}} - 1\right] = (c_p - s_0)[T - T_0]$$

For ideal gases, we can also directly obtain the change in the chemical potential for the isentropic process from the change in molar enthalpy,

$$\Delta \mu = \Delta h - s_0(T - T_0) = c_p(T - T_0) - s_0(T - T_0) = (c_p - s_0)[T - T_0]$$

(b)   For oxygen gas (O_2), the temperature change for the isentropic expansion process with 10 times of increase in volume is given by

$$\Delta T = T - T_0 = T_0\left[\left(\frac{v}{v_0}\right)^{-\frac{R}{c_v}} - 1\right] = 298 \times \left(10^{-\frac{2}{5}} - 1\right) = -179.4 \, \text{K}$$

Therefore, the changes in the molar internal energy, molar enthalpy, Helmholtz free energy, and chemical potential are

$$\Delta u = c_v(T - T_0) = \frac{5}{2}R(T - T_0) = \frac{5}{2} \times 8.314 \times (-179.4) = -3729 \, \text{J/mol}$$

$$\Delta h = c_p(T - T_0) = \frac{7}{2}R(T - T_0) = \frac{7}{2} \times 8.314 \times (-179.4) = -5220 \, \text{J/mol}$$

$$\Delta f = (c_v - s_0)(T - T_0) = \left(\frac{5}{2} \times 8.314 - 205\right) \times (-179.4) = 33,048 \, \text{J/mol}$$

$$\Delta \mu = (c_p - s_0)(T - T_0) = \left(\frac{7}{2} \times 8.314 - 205\right) \times (-179.4) = 31,557 \, \text{G}$$

Example 3 For the following processes, write down the general expressions for the changes in the molar internal energy Δu, molar enthalpy Δh, molar entropy Δs, molar Helmholtz free energy Δf, and chemical potential $\Delta \mu$ in terms of heat capacity, compressibility, and thermal expansion coefficient, as well as temperature, pressure, volume, or molar entropy at the initial state:

(a)   Isochoric (constant volume) change in temperature from T_0 to T.
(b)   Isobaric (constant pressure) temperature change from T_0 to T.
(c)   Isothermal (constant temperature) volume change from v_0 to v.
(d)   Isothermal (constant temperature) pressure change from p_0 to p.

Assuming oxygen gas (O_2) behaves as an ideal gas with constant pressure heat capacity of $7R/2$ (R is gas constant) and molar entropy $s^o_{O_2(298\,K,1\,bar)}$ of 205 J/(mol K), evaluate the following:

(e) The $\Delta u, \Delta h, \Delta s, \Delta f$, and $\Delta \mu$ of oxygen gas (O_2) due to simultaneous changes in temperature from 298 K to 600 K and in volume increasing by 10 times.

(f) The $\Delta u, \Delta h, \Delta s, \Delta f$, and $\Delta \mu$ of oxygen gas (O_2) due to simultaneous changes in temperature and pressure from 298 K and 1 bar to 600 K and 10 bar.

Solution

(a) The expressions for the changes in the molar internal energy Δu, molar enthalpy Δh, molar entropy Δs, molar Helmholtz free energy Δf, and chemical potential $\Delta \mu$ due to a change in temperature from T_o to T at constant volume are given in Eqs. (8.26)–(8.30), and thus they are provided below without rederivation:

$$\Delta u = \int_{T_o}^{T} c_v dT$$

$$\Delta h = \int_{T_o}^{T} \left(c_v + v \frac{\alpha}{\beta_T} \right) dT$$

$$\Delta s = \int_{T_o}^{T} \frac{c_v}{T} dT$$

$$\Delta f = \int_{T_o}^{T} [c_v - s(T_o, v)] dT - T \int_{T_o}^{T} \frac{c_v}{T} dT$$

$$\Delta \mu = \int_{T_o}^{T} \left[c_v + v \frac{\alpha}{\beta_T} - s(T_o, v) \right] dT - T \int_{T_o}^{T} \frac{c_v}{T} dT$$

(b) The expressions for the changes in the internal energy Δu, molar enthalpy Δh, molar entropy Δs, molar Helmholtz free energy Δf, and chemical potential $\Delta \mu$ as a result of a change in temperature from T_o to T at constant pressure are given in Eqs. (8.62)–(8.66) and reproduced below:

$$\Delta u = \int_{T_o}^{T} (c_p - pv\alpha) dT$$

$$\Delta h = \int_{T_o}^{T} c_p \mathrm{d}T$$

$$\Delta s = \int_{T_o}^{T} \frac{c_p}{T} \mathrm{d}T$$

$$\Delta f = \int_{T_o}^{T} [c_p - s(T_o, p) - pv\alpha] \mathrm{d}T - T \int_{T_o}^{T} \frac{c_p}{T} \mathrm{d}T$$

$$\Delta \mu = \int_{T_o}^{T} [c_p - s(T_o, p)] \mathrm{d}T - T \int_{T_o}^{T} \frac{c_p}{T} \mathrm{d}T$$

(c)   The expressions for the changes in the internal energy Δu, molar enthalpy Δh, molar entropy Δs, molar Helmholtz free energy Δf, and chemical potential $\Delta \mu$ due to an isothermal volume change from v_o to v are given in Eqs. (8.99)–(8.104) which are rewritten below:

$$\Delta u = \int_{v_o}^{v} \left(T \frac{\alpha}{\beta_T} - p \right) \mathrm{d}v$$

$$\Delta h = \int_{v_o}^{v} \frac{1}{\beta_T} (T\alpha - 1) \mathrm{d}v$$

$$\Delta s = \int_{v_o}^{v} \frac{\alpha}{\beta_T} \mathrm{d}v$$

$$\Delta f = -\int_{v_o}^{v} p \, \mathrm{d}v$$

$$\Delta \mu = -\int_{v_o}^{v} B_T \mathrm{d}v = -\int_{v_o}^{v} \frac{1}{\beta_T} \mathrm{d}v$$

(d)   The expressions for the changes in the internal energy Δu, molar enthalpy Δh, molar entropy Δs, molar Helmholtz free energy Δf, and chemical potential

$\Delta\mu$ as a result of isothermal pressure change from p_o to p are reproduced below from Eqs. (8.130)–(8.134):

$$\Delta u = \int_{p_o}^{p} v(p\beta_T - T\alpha)\mathrm{d}p$$

$$\Delta h = \int_{p_o}^{p} v(1 - T\alpha)\mathrm{d}p$$

$$\Delta s = -\int_{p_o}^{p} v\alpha\mathrm{d}p$$

$$\Delta f = \int_{p_o}^{p} p v\beta_T\mathrm{d}p$$

$$\Delta\mu = \int_{p_o}^{p} v\mathrm{d}p$$

(e) The expressions for the changes Δu, Δh, Δs, Δf and $\Delta\mu$ of oxygen gas (O_2) due to simultaneous changes in temperature and volume are given in Eqs. (8.165)–(8.169) and reproduced below:

$$\Delta u = \int_{T_o,v_o}^{T,v_o} c_v\mathrm{d}T + \int_{T,v_o}^{T,v} \left(\frac{T\alpha}{\beta_T} - p\right)\mathrm{d}v$$

$$\Delta h = \int_{T_o,v_o}^{T,v_o} \left(c_v + v_o\frac{\alpha}{\beta_T}\right)\mathrm{d}T + \int_{T_o,v_o}^{T,v} \frac{1}{\beta_T}(T\alpha - 1)\mathrm{d}v$$

$$\Delta s = \int_{T_o,v_o}^{T,v_o} \frac{c_v}{T}\mathrm{d}T + \int_{T_o,v_o}^{T,v} \frac{\alpha}{\beta_T}\mathrm{d}v$$

$$\Delta f = \int_{T_o,v_o}^{T,v_o} [c_v - s(T_o, v_o)]\mathrm{d}T - T\int_{T_o,v_o}^{T,v_o} \frac{c_v}{T}\mathrm{d}T - \int_{T,v_o}^{T,v} p\mathrm{d}v$$

$$\Delta\mu = \int_{T_0, v_0}^{T, v_0} \left[c_v + \frac{v_0 \alpha}{\beta_T} - s(T_0, v_0) \right] dT - T \int_{T_0, v_0}^{T, v_0} \frac{c_v}{T} dT - \int_{T, v_0}^{T, v} \frac{1}{\beta_T} dv$$

For ideal gases, the above equations are reduced to

$$\Delta u = c_v (T - T_0)$$

$$\Delta h = c_p (T - T_0)$$

$$\Delta s = c_v \ln \frac{T}{T_0} + R \ln \frac{v}{v_0}$$

$$\Delta f = \left[c_v - s^\circ(T_0, v_0) \right](T - T_0) - c_v T \ln \frac{T}{T_0} - RT \ln \frac{v}{v_0}$$

$$\Delta\mu = \left[c_p - s^\circ(T_0, v_0) \right](T - T_0) - c_v T \ln \frac{T}{T_0} - RT \ln \frac{v}{v_0}$$

For oxygen gas subject to simultaneous changes in temperature from 298 K to 600 K and in volume increasing by 10 times,

$$\Delta u = c_v(T - T_0) = \frac{5}{2} R(600 - 298) = 6277 \text{ J/mol}$$

$$\Delta h = c_p(T - T_0) = \frac{7}{2} R(600 - 298) = 8788 \text{ J/mol}$$

$$\Delta s = c_v \ln \frac{T}{T_0} + R \ln \frac{v}{v_0} = R \left(\frac{5}{2} \ln \frac{600}{298} + \ln 10 \right)$$
$$= R(1.750 + 2.303) = 33.70 \text{ J/(mol K)}$$

$$\Delta f = \left[\frac{5}{2} R - 205 \right](600 - 298) - 600 \left[R \left(\frac{5}{2} \ln \frac{600}{298} + \ln 10 \right) \right] = -77,853 \text{ J/mol}$$

$$\Delta\mu = \left[\frac{7}{2} R - 205 \right](600 - 298) - 600 \left[R \left(\frac{5}{2} \ln \frac{600}{298} + \ln 10 \right) \right] = -73,336 \text{ G}$$

(f)   The expressions for the changes Δu, Δh, Δs, Δf, and $\Delta\mu$ of oxygen gas (O_2) due to simultaneous changes in temperature and pressure are given in Eqs. (8.185)–(8.189) and rewritten below:

$$\Delta u = \int_{T_{\mathrm{o}},p_{\mathrm{o}}}^{T,p_{\mathrm{o}}} (c_p - pv\alpha)\mathrm{d}T + \int_{T,p_{\mathrm{o}}}^{T,p} v(p\beta_T - T\alpha)\mathrm{d}p$$

$$\Delta h = \int_{T_{\mathrm{o}},p_{\mathrm{o}}}^{T,p_{\mathrm{o}}} c_p\mathrm{d}T + \int_{T,p_{\mathrm{o}}}^{T,p} v(1 - T\alpha)\mathrm{d}p$$

$$\Delta s = \int_{T_{\mathrm{o}},p_{\mathrm{o}}}^{T,p_{\mathrm{o}}} \frac{c_p}{T}\mathrm{d}T - \int_{T,p_{\mathrm{o}}}^{T,p} v\alpha\mathrm{d}p$$

$$\Delta f = \int_{T_{\mathrm{o}},p_{\mathrm{o}}}^{T,p_{\mathrm{o}}} \left[c_p - s(T_{\mathrm{o}}, p_{\mathrm{o}}) - pv\alpha\right]\mathrm{d}T - T\int_{T_{\mathrm{o}},p_{\mathrm{o}}}^{T,p_{\mathrm{o}}} \frac{c_p}{T}\mathrm{d}T + \int_{T,p_{\mathrm{o}}}^{T,p} pv\beta_T\mathrm{d}p$$

$$\Delta\mu(T, p) = \int_{T_{\mathrm{o}},p_{\mathrm{o}}}^{T,p_{\mathrm{o}}} \left[c_p - s(T_{\mathrm{o}}, p_{\mathrm{o}})\right]\mathrm{d}T - T\int_{T_{\mathrm{o}},p_{\mathrm{o}}}^{T,p_{\mathrm{o}}} \frac{c_p}{T}\mathrm{d}T + \int_{T,p_{\mathrm{o}}}^{T,p} v\mathrm{d}p$$

For ideal gases, the finite changes due to simultaneous changes in temperature and pressure are:

$$\Delta u = c_v(T - T_{\mathrm{o}})$$

$$\Delta h = c_p(T - T_{\mathrm{o}})$$

$$\Delta s = c_p \ln \frac{T}{T_{\mathrm{o}}} - R \ln \frac{p}{p_{\mathrm{o}}}$$

$$\Delta f = [c_v - s(T_{\mathrm{o}}, p_{\mathrm{o}})](T - T_{\mathrm{o}}) - c_p T \ln \frac{T}{T_{\mathrm{o}}} + RT \ln \frac{p}{p_{\mathrm{o}}}$$

$$\Delta\mu = \left[c_p - s(T_{\mathrm{o}}, p_{\mathrm{o}})\right](T - T_{\mathrm{o}}) - c_p T \ln \frac{T}{T_{\mathrm{o}}} + RT \ln \frac{p}{p_{\mathrm{o}}}$$

For oxygen gas subject to simultaneous changes in temperature and pressure from 298 K and 1 bar to 600 K and 10 bar,

$$\Delta u = c_v(T - T_{\mathrm{o}}) = \frac{5}{2}R(600 - 298) = 6277 \text{ J/mol}$$

$$\Delta h = c_p(T - T_{\mathrm{o}}) = \frac{7}{2}R(600 - 298) = 8788 \text{ J/mol}$$

$$\Delta s = c_p \ln \frac{T}{T_o} - R \ln \frac{P}{P_o} = R\left(\frac{7}{2}\ln \frac{600}{298} - \ln 10\right) = R(2.449 - 2.303) = 1.21 \text{ J/(mol}$$

$$\Delta f = \left[\frac{5}{2}R - 205\right](600 - 298) - 600\left[R\left(\frac{7}{2}\ln \frac{600}{298} - \ln 10\right)\right] = -56,366 \text{ J/mol}$$

$$\Delta \mu = \left[\frac{7}{2}R - 205\right](600 - 298) - 600\left[R\left(\frac{7}{2}\ln \frac{600}{298} - \ln 10\right)\right] = -53,855 \text{ G}$$

Example 4 The molar volume of graphite $v_{298 \text{ K}, 1 \text{ bar}}^o$ is 6.0 cm³/mol. The molar constant pressure heat capacity of graphite c_p is 8.5 J/(K mol). The thermal expansion coefficient α of graphite is 3×10^{-5}/K. The isothermal compressibility β_T of graphite is 3×10^{-6}(1/bar). The molar entropy of graphite at room temperature and ambient pressure 1 bar is $s_{298 \text{ K}, 1 \text{ bar}}^o = 5.6$ J/(mol K), and the molar enthalpy of graphite is set at $h_{298 \text{ K}, 1 \text{ bar}}^o = 0$.

(a)   Estimate the changes in molar internal energy, molar enthalpy, molar entropy, molar Helmholtz free energy, and chemical potential when graphite is heated under constant pressure of 1 bar from 298 to 3000 K.

(b)   Estimate the changes in molar internal energy, molar enthalpy, molar entropy, molar Helmholtz free energy, and chemical potential when the pressure of graphite is increased from 1 bar to 40 Gpa at room temperature 298 K.

Solution

(a)   The expressions for the changes in molar internal energy, molar enthalpy, molar entropy, molar Helmholtz free energy, and chemical potential as a result of temperature change under constant pressure of 1 bar by assuming constant heat capacity c_p and thermal expansion coefficient α are given in Eqs. (8.79)–(8.83) and are reproduced here:

$$\Delta u = c_p(T - T_o) - p_o v_o\left[e^{\alpha(T-T_o)} - 1\right]$$

$$\Delta h = c_p(T - T_o)$$

$$\Delta s = c_p \ln(T/T_o)$$

$$\Delta f = [c_p - s(T_o, p_o)](T - T_o) - p_o v_o\left[e^{\alpha(T-T_o)} - 1\right] - c_p T \ln(T/T_o)$$

$$\Delta \mu = [c_p - s(T_o, p_o)](T - T_o) - c_p T \ln(T/T_o)$$

For graphite heated under constant pressure of 1 bar from 298 to 3000 K, we have

$$\Delta u = c_p(T - T_o) - p_o v_o\left[e^{\alpha(T-T_o)} - 1\right]$$

$$= 8.5(3000 - 298) - 1.0 \times 10^5 \times 6 \times 10^{-6}\left[e^{3\times 10^{-5}(3000-298)} - 1\right]$$

$$\approx 22.967 \text{ (kJ/mol)}$$

$$\Delta h = c_p(T - T_o) = 8.5(3000 - 298) = 22.967 \text{ kJ/mol}$$

$$\Delta s = c_p \ln(T/T_o) = 8.5 \ln\frac{3000}{298} \approx 19.63 \text{ J/mol K}$$

$$\Delta f = \left[c_p - s(T_o, p_o)\right](T - T_o) - c_p T \ln\frac{T}{T_o} - p_o v_o\left[e^{\alpha(T-T_o)} - 1\right]$$

$$= [8.5 - 5.6](3000 - 298) - 8.5 \times 3000 \ln\frac{3000}{298} - 1.0 \times 10^5 \times 6$$

$$\times 10^{-6}\left[e^{3\times 10^{-5}(3000-298)} - 1\right] \approx -51.051 \text{ kJ/mol}$$

$$\Delta\mu = \left[c_p - s(T_o, p_o)\right](T - T_o) - c_p T \ln\frac{T}{T_o}$$

$$= [8.5 - 5.6](3000 - 298) - 8.5 \times 3000 \ln\frac{3000}{298} \approx -51.051 \text{ kG}$$

(b) The expressions for the changes in molar internal energy, molar enthalpy, molar entropy, molar Helmholtz free energy, and chemical potential as a result of pressure change at constant temperature assuming constant thermal expansion coefficient and isothermal compressibility are given by in Eqs. (8.140)–(8.144) and are reproduced below:

$$\Delta u = v_o\left\{\left[p_o - \frac{(T_o\alpha - 1)}{\beta_T}\right] - \left[p - \frac{(T_o\alpha - 1)}{\beta_T}\right]e^{-\beta_T(p-p_o)}\right\}$$

$$\Delta h = \frac{(T_o\alpha - 1)v_o}{\beta_T}\left[e^{-\beta_T(p-p_o)} - 1\right]$$

$$\Delta s = \frac{\alpha v_o}{\beta_T}\left[e^{-\beta_T(p-p_o)} - 1\right]$$

$$\Delta f = -v_o\left\{\left(pe^{-\beta_T(p-p_o)} - p_o\right) + \frac{1}{\beta_T}\left[e^{-\beta_T(p-p_o)} - 1\right]\right\}$$

$$\Delta\mu = -\frac{v_o}{\beta_T}\left[e^{-\beta_T(p-p_o)} - 1\right]$$

For graphite subject to a pressure change from 1 bar to 40 Gpa at room temperature:

$$\Delta u = v^o_{298\,\text{K},1\,\text{bar}}\left\{\left[1\,\text{bar} - \frac{(298\alpha - 1)}{\beta_T}\right] - \left[p - \frac{(298\alpha - 1)}{\beta_T}\right]e^{-\beta_T(p-1\,\text{bar})}\right\}$$

$$= 6 \times 10^{-6}\left\{\left[10^5 - \frac{(298 \times 3 \times 10^{-5} - 1)}{3 \times 10^{-6} \times 10^{-5}}\right]\right.$$

$$\left. - \left[40 \times 10^9 - \frac{(298 \times 3 \times 10^{-5} - 1)}{3 \times 10^{-6} \times 10^{-5}}\right]e^{-3\times 10^{-6}\times 10^{-5}\left(40\times 10^9 - 10^5\right)}\right\}$$

$$\approx 6 \times 10^{-6}\left\{3.3 \times 10^{10} - 2.2 \times 10^{10}\right\} \approx 66.0\,\text{kJ/mol}$$

$$\Delta h = \frac{(298\alpha - 1)}{\beta_T}v^o_{298\,\text{K},1\,\text{bar}}\left[e^{-\beta_T(p-1\,\text{bar})} - 1\right]$$

$$\approx -3.30 \times 10^{10} \times 6 \times 10^{-6} \times [-0.699] \approx 138.4\,\text{kJ/mol}$$

$$\Delta s = \frac{\alpha v^o_{298\,\text{K},1\,\text{bar}}}{\beta_T}\left[e^{-\beta_T(p-1\,\text{bar})} - 1\right]$$

$$= \frac{3 \times 10^{-5} \times 6 \times 10^{-6}}{3 \times 10^{-6} \times 10^{-5}}[0.301 - 1] \approx -4.194\,\text{J/mol K}$$

$$\Delta f = v^o_{298\,\text{K},1\,\text{bar}}\left[1\,\text{bar} - pe^{-\beta_T(p-1\,\text{bar})}\right] + \frac{v^o_{298\,\text{K},1\,\text{bar}}}{\beta_T}\left[1 - e^{-\beta_T(p-1\,\text{bar})}\right]$$

$$= 6 \times 10^{-6}\left[10^5 - 4 \times 10^{10} \times 0.301\right] + \frac{6 \times 10^{-6}}{3 \times 10^{-6} \times 10^{-5}}[0.699]$$

$$\approx -0.722 \times 10^5 + 1.398 \times 10^5 = 67.6\,\text{kJ/mol}$$

$$\Delta\mu = \frac{v^o_{298\,\text{K},1\,\text{bar}}}{\beta_T}\left[1 - e^{-\beta_T(p-1\,\text{bar})}\right] = \frac{6 \times 10^{-6}}{3 \times 10^{-6} \times 10^{-5}}[0.699] \approx 139.8\,\text{kG}$$

Example 5 Consider the formation enthalpy of methane gas (CH_4) from graphite (C) and hydrogen gas (H_2). The formation enthalpy at 0 K calculated from quantum mechanical calculations is -66.56 kJ/mol. The constant pressure heat capacities of graphite, hydrogen gas, and methane gas are 8.64 J/(mol K), 28.64 J/(mol K), and 35.52 J/(mol K), respectively. Calculate the formation enthalpy of methane at room temperature 298 K.

Solution
The enthalpy of formation of CH_4 is given by

$$\Delta h^o_{298\,\text{K}} = \Delta h_{0\,\text{K}} + \int_{0\,\text{K}}^{298\,\text{K}} \left(c_{p,CH_4} - c_{p,C} - 2c_{p,H_2}\right)dT$$

$$= \Delta h_{0K} + \left(c_{p,CH_4} - c_{p,C} - 2c_{p,H_2}\right)(298 - 0)$$
$$= -65,560 + (35.52 - 8.64 - 2 \times 28.64) \times 298$$
$$= -75.61 \text{ kJ/mol}$$

Example 6 For the melting of ice to become water at the equilibrium melting temperature of 0 °C under 1 bar, the heat of melting is 6007 J/mol.

(a) What is the enthalpy difference between one mole of water and one mole of ice at 0 °C and 1 bar?

b)   Which one has higher enthalpy, one mole of ice or one mole of water?

c)   What is the entropy difference between one mole of ice and one mole of water at 0 °C and 1 bar?

d)   Which one has higher entropy, one mole of ice or one mole of water at 0 °C and 1 bar?

e)   What is the heat of solidification of water at 0 °C and 1 bar?

f)   What is the chemical potential difference between ice and water at 0 °C and 1 bar?

Solution

(a) According to the first law of thermodynamics, the enthalpy change of a material (the difference in enthalpy between the final state and the initial state of the material) is equal to the heat absorbed or released by the material,

$$\Delta h_m^o = Q_m^o = 6007 \text{ J/mol}$$

(b) Based on the answer in (a), water has higher enthalpy than ice

(c) The entropy difference between one mole of ice and one mole of water is

$$\Delta s_m^o = \frac{\Delta h_m^o}{T_m} = \frac{6007}{273} = 22 \text{ J/(mol K)}$$

(d) Water has higher entropy than ice.

(e) Solidification is the reverse process to melting, therefore,

$$\Delta h_s^o = -\Delta h_m^o = -6007 \text{ J/mol}$$

(f) The chemical potential difference between ice and water at 0 °C and 1 bar is zero.

Example 7 The molar entropy of Al at 298 K and 1 bar , $s^o_{298\,K,\,1\,bar}$, is 28.3 J/mol K. The heat of melting of Al at the equilibrium melting temperature 934 K and pressure 1 bar is 10, 700 J/mol. The heat capacities of solid and liquid Al under 1 bar are $20.67 + 12.38 \times 10^{-3}T$ and 31.76 J/(mol K), respectively. Calculate the following quantities:

(a) The amount of heat required to increase the temperature of one mole of solid aluminum from 300 to 800 K at constant pressure of 1 bar.

(b) The enthalpy change of one mole of Al when it is heated from 298 to 1200 K at 1 bar.

(c) The entropy change of one mole of Al when it is heated from 298 to 1200 K at 1 bar.

(d) The entropy change for the surrounding at 1200 K when one mole of Al is heated from 298 to 1200 K at 1 bar in the system.

(e) The total entropy change for increasing the temperature of one mole of Al from 298 to 1200 K at 1 bar.

(f) The chemical potential change for increasing the temperature of Al from 298 to 1200 K at 1 bar.

Solution

(a) At constant pressure, the amount of heat required to increase the temperature of one mole of solid aluminum from 300 to 800 K at constant pressure of 1 bar is

$$Q = \Delta h = \int_{T_1}^{T_2} c_p dT$$

$$= \int_{300\,K}^{800\,K} (20.67 + 12.38 \times 10^{-3}T) dT$$

$$= \left[20.67T + \frac{1}{2} \times 12.38 \times 10^{-3}T^2 \right]_{300\,K}^{800\,K} = 20497.6 - 6758.1 = 13,739.5\,J$$

(b) The enthalpy change of one mole of Al heated from 298 to 1200 K at 1 bar is given by

$$\Delta h = \int_{298\,K}^{934\,K} c_p^s dT + \Delta H_m^o + \int_{934\,K}^{1200\,K} c_p^l dT$$

$$= \int_{298\,K}^{934\,K} (20.67 + 12.38 \times 10^{-3}T)dT + 10700 + \int_{934\,K}^{1200\,K} 31.76\,dT$$

$$= \left(20.67T + \frac{1}{2} \times 12.38 \times 10^{-3}T^2 \right)_{298\,K}^{934\,K} + 10700 + 31.76 \times (1200 - 934)$$

$$= 37,144.5 \text{ J}$$

(c) The entropy change of one mole of Al when heated from 298 to 1200 K at 1 bar is

$$\Delta s = \int_{298\,K}^{934\,K} \frac{c_p^s}{T}dT + \frac{\Delta h_m^o}{T_m} + \int_{934\,K}^{1200\,K} \frac{c_p^l}{T}dT$$

$$= \int_{298\,K}^{934\,K} \frac{(20.67 + 12.38 \times 10^{-3}T)}{T}dT + \frac{10,700}{934} + \int_{934\,K}^{1200\,K} \frac{31.76}{T}dT$$

$$= 20.67\ln\frac{934}{298} + 12.38 \times 10^{-3}(934 - 298) + 11.46 + 31.76 \times \ln\frac{1200}{934} = 50.90 \text{ J/K}$$

(d) The entropy change for the surrounding at 1200 K when one mole of Al is heated from 298 to 1200 K at 1 bar in the system is

$$\Delta S_{sur} = \frac{-\Delta h}{1200 \text{ K}} = \frac{-37144.5 \text{ J}}{1200 \text{ K}} = -30.95 \text{ J/K}$$

(e) The total entropy change for increasing the temperature of one mole of Al from 298 to 1200 K at 1 bar is

$$\Delta S_{tot} = \Delta s_{Al} + \Delta S_{sur} = 50.90 - 30.95 = 19.95 \text{ J/K}$$

(f) The chemical potential change for increasing the temperature of Al from 298 to 1200 K at 1 bar is

$$\Delta\mu = \mu_{1200\,K} - \mu_{298\,K} = h_{1200\,K} - 1200 s_{1200\,K} - \left(h_{298\,K}^o - 298 s_{298\,K}^o \right)$$

$$= h_{1200\,K} - h_{298\,K}^o - 1200 \left(s_{1200\,K} - s_{298\,K}^o \right) - (1200 - 298)s_{298\,K}^o$$

$$= 37,144.5 - 1200 \times 50.90 - (1200 - 298) \times 28.3$$

$$= -49,462.1 \text{ G}$$

Example 8 The heat capacity of water is $c_p^l = 75\,J/(mol\,K)$, the heat capacity of water vapor is $c_p^v = 36\,J/(mol\,K)$, and the heat of vaporization of water at 100 °C is $\Delta h_v^o = 40.65\,kJ/mol$. Calculate the heat of vaporization at 120 °C or 393 K.

Solution

The heat of vaporization at 393 K is given by

$$\Delta h_{v,393\,K} = \Delta h_{v,373\,K}^o + \int_{373\,K}^{393\,K} (c_p^v - c_p^l)dT$$

$$= \Delta h_{v,373\,K}^o + (c_p^v - c_p^l)(393 - 373)$$

$$= 40,650 + (36 - 75)(393 - 373) = 39,870(\,J/mol)$$

Example 9

(a) Given the data for 1 bar in the following table

Substance	$\Delta h_{298\,K}^o$ (J/mol)	$s_{298\,K}^o$ (J/(K mol))	c_p (J/(K mol))
Ca	0	40.0	26.0
O_2	0	205.1	29.4
Ti	0	30.7	25.06
CaO	-635,000	38.0	42.0
TiO_2	-945,000	50.0	55.0
$CaTiO_3$	-1,660,000	94.0	110.0

What are the enthalpy, entropy, and Gibbs free energy changes for the reaction when 1 mol of CaO reacts with 1 mol of TiO_2 at 298 K to produce 1 mol of $CaTiO_3$ at 298 K?

Solution

For the reaction $CaO + TiO_2 = CaTiO_3$ at 298 K, the enthalpy, entropy, and Gibbs free energy changes are

$$\Delta H_{298\,K}^o = \Delta h_{CaTiO_3}^o - \Delta h_{CaO}^o - \Delta h_{TiO_2}^o$$

$$= -1660,000 + 635,000 + 945,000 = -80,000\,J$$

$$\Delta S_{298\,K}^o = \Delta s_{CaTiO_3}^o - \Delta s_{CaO}^o - \Delta s_{TiO_2}^o$$

$$= 94 - 38 - 50 = 6\ J/K$$

$$\Delta G_{298\,K}^o = \Delta H_{298\,K}^o - 298\Delta S_{298\,K}^o = -80,000 - 298 \times 6 = -81,788\,J$$

Example 10 The molar entropies of silicon (Si), oxygen gas (O_2), and silicon dioxide (SiO_2) at 298 K and 1 bar are 18.70, 205.0, and 41.84 J/(K mol), respectively. The enthalpy of formation of SiO_2 at 298 and 1 bar is -911 kJ/mol. The constant pressure heat capacities of Si, O_2, and SiO_2 1 bar are approximately 19.6, 29.1, and 40.8 J/(K mol), respectively. Calculate:

(a) The entropy change for the oxidation reaction $Si + O_2 = SiO_2$ at room temperature 298 K and 1 bar.
(b) The entropy change for the oxidation reaction $Si + O_2 = SiO_2$ at 500 K and 1 bar, assuming that the heat capacities of Si, O_2, and SiO_2 are independent of temperature.
(c) The enthalpy change for the oxidation reaction $Si + O_2 = SiO_2$ at room temperature 298 K and 1 bar.
(d) The enthalpy change for the oxidation reaction $Si + O_2 = SiO_2$ at 500 K and 1 bar, assuming that the heat capacities of Si, O_2, and SiO_2 are independent of temperature.
(e) The Gibbs free energy change for the oxidation reaction $Si + O_2 = SiO_2$ at room temperature 298 K and 1 bar.
(f) The Gibbs free energy change for the oxidation reaction $Si + O_2 = SiO_2$ at 500 K and 1 bar, assuming that the heat capacities of Si, O_2, and SiO_2 are independent of temperature.

Solution

(a) The entropy change for the oxidation reaction $Si + O_2 = SiO_2$ at room temperature 298 K and 1 bar,

$$\Delta S^o_{298\,K} = s^o_{SiO_2} - s^o_{Si} - s^o_{O_2} = 41.84 - 18.70 - 205.0 = -181.86\,\text{J/K}$$

(b) The entropy change for the oxidation reaction $Si + O_2 = SiO_2$ at 500 K and 1 bar,

$$\Delta S_{500\,K} = \Delta S^o_{298\,K} + \int_{298\,K}^{500\,K} \frac{\left(c_{p,SiO_2} - c_{p,Si} - c_{p,O_2}\right)}{T}\,dT$$

$$= -181.86 + (40.8 - 19.6 - 29.1) \times \ln\frac{500}{298} = -185.95\,\text{J/K}$$

(c) The enthalpy change for the oxidation reaction $Si + O_2 = SiO_2$ at room temperature 298 K and 1 bar,

$$\Delta H_{298\,K}^{o} = h_{SiO_2}^{o} - h_{Si}^{o} - h_{O_2}^{o} = -911\,kJ$$

(d) The enthalpy change for the oxidation reaction $Si + O_2 = SiO_2$ at 500 K and 1 bar,

$$\Delta H_{500\,K} = \Delta H_{298\,K}^{o} + \int_{298\,K}^{500\,K} \left(c_{p,SiO_2} - c_{p,Si} - c_{p,O_2}\right) dT$$

$$= \Delta H_{298\,K}^{o} + \int_{298\,K}^{500\,K} (40.8 - 19.6 - 29.1) dT = -912,596\,J$$

(e) The Gibbs free energy change for the oxidation reaction $Si + O_2 = SiO_2$ at room temperature 298 K and 1 bar,

$$\Delta G_{298\,K}^{o} = \Delta H_{298\,K}^{o} - 298\Delta S_{298\,K}^{o} = -911,000 - 298 \times (-181.86)$$

$$= -856,806\,J$$

(f) The Gibbs free energy change for the oxidation reaction $Si + O_2 = SiO_2$ at 500 K and 1 bar,

$$\Delta G_{500\,K} = \Delta H_{500\,K} - 500\Delta S_{500\,K}$$

$$= -912,595.8 - 500 \times (-185.95) = -819,621\,J$$

8.10 Exercises

1. If the thermal expansion coefficient of a material is positive, does the entropy of the material (i) increase or (ii) decrease as the pressure increases while keeping temperature constant?

2. For a monatomic ideal gas undergoing an isentropic (reversible adiabatic) pressure change from p_o to p,

    (a) Derive the expressions for the changes in temperature ΔT, molar internal energy Δu, molar enthalpy Δh, molar entropy Δs, molar Helmholtz free energy Δf, and chemical potential $\Delta \mu$ in terms of heat capacity, compressibility, and thermal expansion coefficient, as well as temperature, pressure, volume or molar entropy at the initial state.

(b) Assuming Argon gas behaves as an ideal gas with constant pressure heat capacity of $5R/2$ (R is gas constant) and molar entropy $s^o_{298\,K,1\,bar}$ of $155\,J/(mol\ K)$, evaluate the ΔT, Δu, Δh, Δs, Δf, and $\Delta \mu$ of Argon gas initially at 298 K and 1 bar due to a reversible adiabatic compression to reach a final pressure of 1001 bar.

3. For 1 mol of monatomic ideal gas initially at 298 K and 1 bar, calculate the work W done on the gas, the heat Q transferred to the surrounding, and the internal energy change ΔU of the ideal gas for the following two processes:

(a) Compress the gas to 5 bar reversibly and adiabatically.
(b) Compress the gas to 5 bar reversibly and isothermally.

4. Translate the following statements into thermodynamic expressions, then express them in terms of heat capacity, compressibility, thermal expansion coefficient, and other necessary thermodynamic properties of one mole of a material, and evaluate them for the case of monatomic ideal gases:

(a) The volume change as a result of temperature change at constant pressure.
(b) The pressure change as a result of volume change at constant temperature.
(c) The volume change as a result of pressure change at constant temperature.

5. The molar volume of graphite $v^o_{298\,K,1\,bar}$ is $6.0\,cm^3/mol$, and the molar entropy: $s^o_{298\,K,1\,bar} = 5.6\,J/(K\ mol)$. The molar heat capacity of graphite c_p is $8.5\,J/(K\ mol)$. The thermal expansion coefficient of graphite α is $3 \times 10^{-5}/K$. The isothermal compressibility of graphite β_T is $3 \times 10^{-6}/bar$. Obtain expressions for:

(a) The molar internal energy change of graphite when its temperature is changed from 298 K to T and pressure is changed from 1 bar to p.
(b) The molar enthalpy change of graphite when its temperature is changed from 298 K to T and pressure is changed from 1 bar to p.
(c) The molar entropy change of graphite when its temperature is changed from 298 K to T and pressure is changed from 1 bar to p.
(d) The molar Helmholtz free energy change of graphite when its temperature is changed from 298 K to T and pressure is changed from 1 bar to p.
(e) The chemical potential change of graphite when its temperature is changed from 298 K to T and pressure is changed from 1 bar to p.

6. Given data for graphite (g) and diamond (d):
Molar enthalpy: $h^{o,g}_{298\,K,1\,bar} = 0$, $h^{o,d}_{298\,K,1\,bar} = 1900\,J/mol$.
Molar volume: $v^{o,g}_{298\,K,1\,bar} = 6.0\,cm^3/mol$, $v^{o,d}_{298\,K,1\,bar} = 3.4\,cm^3/mol$.
Molar heat capacity: $c^g_p = 8.5\,J/(K\ mol)$, $c^d_p = 6.0\,J/(K\ mol)$.
Isothermal compressibility: $\beta^g_T = 3 \times 10^{-6}/bar$, $\beta^d_T = 2 \times 10^{-7}/bar$.
Thermal expansion coefficient: $\alpha^g = 3 \times 10^{-5}/K$, $\alpha^d = 10^{-5}/K$.
Molar entropy: $s^{o,g}_{298\,K,1\,bar} = 5.6\,J/(K\ mol)$, $s^{o,d}_{298\,K,1\,bar} = 2.4\,J/(K\ mol)$.

Derive expressions for:

(a) The difference in the chemical potential between diamond and graphite as a function of temperature at 1 bar (assuming constant c_p).

(b) The difference in the chemical potential between diamond and graphite as a function of pressure at 298 K (assuming constant α and β_T).

(c) The chemical potential difference between diamond and graphite as a function of temperature and pressure (assuming constant c_p, α, and β_T).

7. The enthalpy change during solidification of water to become ice at 0 °C is negative, and the entropy change for the system is also negative: true or false?

8. When a liquid freezes to become a solid at a constant temperature, it releases heat to the surrounding, true or false?

9. The total entropy change of the system and its surrounding for the solidification of water to become ice at equilibrium temperature 0 °C and ambient pressure of 1 bar is zero: true or false?

10. The molar enthalpy of diamond at room temperature and 1 bar with respect to graphite is $h_{298\,K,1\,bar}^{o,d} = 1900$ J/mol. If one mole of diamond were transformed to graphite at room temperature and 1 bar, heat would be released or absorbed? What is the amount of heat released or absorbed?

11. For an oxidation reaction of a pure solid metal to form a pure solid oxide at 300 K, is the entropy change for the reaction positive or negative?

12. Assuming the constant pressure heat capacities of ice and water are 38 and 75 J/(mol K), respectively, and the heat of melting of ice at 273 K is 6007 J/mol,

(a) Calculate the molar entropy difference between water at 298 K and ice at 268 K.

(b) For the melting of 1 mol of ice initially at −5 °C to become water at 25 °C, what is the corresponding change of entropy in the surrounding at 25 °C?

(c) What is the total entropy change (for the system plus the surrounding) when 1 mol of ice initially at −5 °C melts to become water at 25 °C with the surrounding at 25 °C?

(d) Based on your calculation results of (c), please determine whether the melting process is reversible or irreversible.

13. Calcium carbonate ($CaCO_3$) has two polymorphs, calcite and aragonite. The chemical potential change for the aragonite to calcite transition as a function of temperature at 1 bar is given by

$$\Delta\mu = 210 - 4.2T \text{ (G)}$$

(a) Determine the equilibrium transition temperature under 1 bar.

(b) Determine the heat of transition under 1 bar.

(c) Determine the entropy of transition under 1 bar.

14. The heat of melting of Ag at equilibrium melting temperature 1234 K and 1 bar is 11,090 J/mol, The constant pressure heat capacity of solid Ag is $20.67 + 12.38 \times 10^{-3}T$ J/(mol K), and the constant pressure heat capacity of liquid Ag is 30.50 J/(mol K)

   (a) Calculate the amount of heat required to heat 1 mol of solid Ag from 298 K to become liquid Ag at 1500 K under 1 bar.
   (b) What is the molar entropy difference between liquid Ag at 1500 K and solid Ag at 298 K under 1 bar.

15. Solid Ba has two polymorphic forms α and β. The molar entropy of α phase is $s^{0,\alpha}_{298\,\text{K},1\,\text{bar}} = 62.4$ J/(mol K) at 298 K. The α phase is stable at temperatures up to 648 K at which the α phase is transformed to the β phase with a molar heat of transition of $\Delta h^{\circ}_e = 630$ J/mol. The β phase is stable until its melting temperature which is higher than 1000 K. The molar constant pressure heat capacity of α phase is $c^{\alpha}_p = -438.2 + 1587.0 \times 10^{-3}T$ $+128.2 \times 10^5 T^{-2}$(J/(mol K)), and the molar constant pressure heat capacity of β phase is $c^{\beta}_p = -5.69 + 80.33 \times 10^{-3}T$(J/(mol K)). Calculate the chemical potential difference between β and α as a function of temperature from 298 to 1000 K.

16. The heat of melting for Cu is 13,260 J/mol and the equilibrium melting temperature of Cu is 1358 K. The molar entropy of Cu at 298 K and 1 bar is 33.2 J/(mol K), and the molar enthalpy of Cu at 298 K and 1 bar is set zero. Assuming the constant pressure heat capacity of solid Cu and liquid Cu is the same, 24.0 J/(mol K).

   (a) What are the differences in molar enthalpy, molar entropy, and chemical potential between liquid Cu and solid Cu at the equilibrium melting temperature 1358 K?
   (b) What is the heat of solidification from liquid Cu to solid Cu at the equilibrium melting temperature 1358 K?
   (c) What are the changes in molar enthalpy, molar entropy, and chemical potential for the solidification of liquid Cu at 1000 K?
   (d) What are the differences in molar enthalpy, entropy, and chemical potential between one mole of liquid Cu at 1400 K and one mole of solid Cu at 1300 K?

17. The heat of melting for Si is 50,200 J/mol, and the equilibrium melting temperature of Si is 1687 K. Assuming that the constant pressure heat capacities of solid Si and liquid Si are 32 J/(mol K) and 42 J/(mol K), respectively,

   (a) What are the differences in enthalpy, entropy, and chemical potential between one mole of liquid Si at 1687 K and one mole of solid Si at 1687 K?

(b) What are changes in the molar enthalpy, molar entropy, and chemical potential when liquid Si at 1500 K is solidified to become solid Si at 1500 K?

18. The chemical potentials of Si in diamond (d) structure μ_{Si}^d and liquid (l) Si μ_{Si}^l as a function of temperature at 1 bar are given below using the enthalpy of Si in diamond structure at room temperature under 1 bar as the reference (from SGTE element database):

$$\mu_{Si}^d = -8163 + 137.2T - 22.83T \ln T - 1.913 \times 10^{-3}T^2 - 0.003552$$
$$\times 10^{-6}T^3 + 176700T^{-1} \quad (G) \quad (298.15\,K < T < 1687\,K)$$

$$\mu_{Si}^d = -9458 + 167.3T - 27.20T \ln T - 420.4 \times 10^{28}T^{-9}$$
$$(G) \quad (1687K < T < 3600\,K)$$

$$\mu_{Si}^l = 42530 + 107.1T - 22.83T \ln T - 1.913 \times 10^{-3}T^2 - 0.003552$$
$$\times 10^{-6}T^3 + 176700T^{-1} + 209.3 \times 10^{-23}T^7 \quad (G) \quad (298.15\,K < T < 1687\,K)$$

$$\mu_{Si}^l = 40370 + 137.7T - 27.20T \ln T \quad (G) \quad (1687\,K < T < 3600\,K)$$

Derive the expressions for the enthalpy, entropy and heat capacity differences between liquid Si and Si in diamond structure as a function of temperature.

19. The enthalpies of $Zr(\alpha)$, O_2, and $ZrO_2(\alpha)$ as well as the molar heat capacities of $ZrO_2(\alpha)$, $ZrO_2(\beta)$, O_2, $Zr(\alpha)$, and $Zr(\beta)$ are given below in the table:

Substance	Δh_{298}^o (J/mol)	c_p (J/(K mol))
$ZrO_2(\alpha)$	$-1, 100, 800$	70.0
$ZrO_2(\beta)$		75.0
O_2	0	30.0
$Zr(\alpha)$	0	25.0
$Zr(\beta)$		26.0

For the transition $Zr(\alpha) \rightarrow Zr(\beta)$, the heat of transition at the equilibrium transition temperature 1136 K is 3900 J/mol, and for the transition of $ZrO_2(\alpha) \rightarrow ZrO_2(\beta)$, the heat of transition at equilibrium transition temperature 1478 K is 5900 J/mol. Find the expressions for:

(a) The enthalpy of Zr as a function of temperature.
(b) The heat of transition for $Zr(\alpha) \rightarrow Zr(\beta)$ as a function of temperature.
(c) The heat of transition for $ZrO_2(\alpha) \rightarrow ZrO_2(\beta)$ as a function of temperature.
(d) The heat of reaction $Zr + O_2 \rightarrow ZrO_2$ as a function of temperature.

20. Given the data:
    The heat capacity of liquid Pb, $c_{p,Pb}^l = 32.4 - 3.1 \times 10^{-3}T$ (J/(mol K)).
    The heat capacity of solid Pb, $c_{p,Pb}^s = 23.56 + 9.75 \times 10^{-3}T$ (J/(mol K)).
    The heat capacity of solid PbO, $c_{p,PbO}^s = 36.9 + 26.8 \times 10^{-3}T$ (J/(mol K)).
    The heat capacity of O_2, $c_{p,O_2}^g = 29.96 + 4.18 \times 10^{-3}T - 1.67 \times 10^5 T^{-2}$ (J/(mol K)).
    The heat of melting (fusion) of Pb at $T_{m,Pb} = 600$ K, $\Delta h_{m,Pb}^o = 4810$ J/mol.
    The heat of melting (fusion) of PbO at $T_{m,PbO} = 1158$ K, $\Delta h_{m,PbO}^o = 27,480$ J/mol.
    The enthalpy of PbO $h_{PbO}^o = -217$ kJ/mol at 298 K, and the enthalpies of Pb and O_2 at 298 K are set zero.

    (a) What is the heat of oxidation reaction of solid Pb at 500 K?
    (b) What is the heat of oxidation reaction of liquid Pb at 700 K?

21. Given that the amounts of heat of formation for BaO, TiO_2, and $BaTiO_3$ at 1 bar and 298 K are -548.1, -944, and -1647 kJ/mol, respectively, and assuming that the constant pressure molar heat capacities of BaO, TiO_2, and $BaTiO_3$ are 55.0, 75.0, and 120.0 J/(K mol), respectively.

    (a) What is the heat of reaction when 1 mol of BaO reacts with 1 mol of TiO_2 at 298 K to produce 1 mol of $BaTiO_3$ at 298 K at 1 bar?
    (b) Calculate the heat of reaction as a function of temperature for BaO + $TiO_2 = BaTiO_3$.

Substance	Δh_{298}^o (J/mol)	s_{298}^o (J/(K mol))	c_p (J/(Kmol))
Si_3N_4	$-744, 800$	113.0	70.0
$SiO_2(\alpha\text{-quartz})$	$-910, 900$	41.5	45.0
O_2	0	205.1	30.0
N_2	0	191.5	28.0

22. For the reaction, $Si_3N_4 + 3O_2 = 3SiO_2(\alpha - \text{quartz}) + 2N_2$, please use the data from the table above and answer the following questions:

    (a) What are the heat, entropy, and Gibbs free energy of reaction at 298 K?
    (b) What are the heat, entropy, and Gibbs free energy of reaction at 800 K?
    (c) What are the heat, entropy, and Gibbs free energy of reaction as a function of temperature?

Chapter 9
Construction of Fundamental Equation of Thermodynamics

The differential and integrated forms of the fundamental equation of thermodynamics of a system are discussed in Chap. 3. Once the integrated form of the fundamental equation of thermodynamics for a material system is established, all the thermodynamic properties for the system can be obtained from the first and second derivatives of energy or entropy representations of the fundamental equation of thermodynamics with respect to their natural thermodynamic variables. Therefore, a complete thermodynamic knowledge of a material requires the construction of its fundamental equation of thermodynamics.

Statistical thermodynamics, briefly discussed in Chap. 4, can, in principle, be employed to derive the fundamental equation of thermodynamics for a material system through the establishment of a partition function in combination with quantum mechanical calculations of the energetics of the system. However, the direct analytical derivation of a fundamental equation of thermodynamics is only possible for a few rather idealized model systems which can be considered as consisting of independent particles such as ideal atomic/molecular gases, electron gas, a group of phonon modes, photon gas, etc. Therefore, for the majority of practical materials, the construction of the fundamental equation of thermodynamics of a material system will have to rely on utilizing thermodynamic data obtained from experiments and computation under specific thermodynamic conditions. Unfortunately, the data available to construct a complete fundamental equation of thermodynamics is rarely sufficient for any practical materials system of interest despite the advances in experimental measurements and materials computation. As a result, it is currently only possible to construct approximate or incomplete fundamental equation of thermodynamics for most of the practical materials, and this is true for all existing thermodynamic databases which contain incomplete, and at best, represent the approximate fundamental equations of thermodynamics.

Chapter 8 establishes the expressions for computing the changes for the different forms of energy and entropy functions as the temperature, pressure, or volume vary as well as for phase transitions and chemical reactions at a given temperature and pressure based on experimentally measurable or theoretically computable properties

© Springer Nature Singapore Pte Ltd. 2022
L.-Q. Chen, *Thermodynamic Equilibrium and Stability of Materials*,
https://doi.org/10.1007/978-981-13-8691-6_9

such as heat capacity, mechanical bulk modulus or compressibility, and thermal expansion coefficient. This chapter outlines the procedures to utilize essentially the same set of thermodynamic properties from the second derivatives of energy functions to construct approximate fundamental equations of thermodynamics for single-component systems. One may consider this as an approach to encoding known thermodynamic properties into a single equation as an approximation to the fundamental equation of thermodynamics from which one can recover the thermodynamic properties. The focus is on simple systems with a brief discussion on the possible extensions to systems with stress, electric, and surface effects.

9.1 Choice of Fundamental Equation of Thermodynamics

There are different energy function forms that can be employed to represent the same fundamental equation of thermodynamics. Materials databases of fundamental equations of thermodynamics should be independent of the size of a materials system, and thus, fundamental equations of thermodynamics to be employed should be expressed in molar quantities, e.g., molar entropy (s) and molar volume (v).

For a simple closed system with one mole of chemical substance, one can easily show that there are only four possible functional forms for the fundamental equation of thermodynamics expressed in molar properties by choosing one variable from each of the two conjugate variable pairs (T, s), (p, v) as the two natural variables. The four combinations of two natural variables for a simple closed system with one mole of chemical substance are

$$(s, v), (T, v), (s, p), (T, p) \tag{9.1}$$

Since in the majority of applications in materials science and engineering, the most convenient sets of thermodynamic variables to control in an experiment or to vary in a computation are T, p or v, the two most commonly employed sets of natural variables among the four choices are (T, v) and (T, p). Therefore, the most commonly employed fundamental equations of thermodynamics in molar quantities are

$$f(T, v) = u - Ts = -pv + \mu \tag{9.2}$$

$$g(T, p) = u - Ts + pv = \mu \tag{9.3}$$

where f is molar Helmholtz free energy, g is molar Gibbs free energy, and u is molar internal energy. The differential forms for Eqs. (9.2) and (9.3) are

$$df = -sdT - pdv \tag{9.4}$$

$$dg = d\mu = -sdT + vdp \tag{9.5}$$

As shown in Eqs. (9.3) and (9.5), the molar Gibbs free energy g and chemical potential μ are exactly the same thermodynamic quantity. Since chemical potential μ is a basic variable for a system while molar Gibbs free energy g is an auxiliary function representing the combination of basic variables: $u - Ts + pv$, we will employ chemical potential μ rather than molar Gibbs free energy g to represent the fundamental equation of thermodynamics although the existing literature mostly refers it as molar Gibbs free energy g.

Based on the above discussion, we will primarily focus on the construction of the molar Helmholtz free energy f of a material as a function of temperature T and molar volume v, and the chemical potential μ as a function of temperature T and pressure p.

9.2 Molar Helmholtz Free Energy as a Function of Temperature and Molar Volume

Utilizing the formula derived in Sect. 8.5 for the change of molar Helmholtz free energy with respect to changes in temperature and molar volume in terms of constant volume heat capacity and isothermal bulk modulus or isothermal compressibility as well as the thermodynamic properties of a reference state, we can readily write down the molar Helmholtz free energy as a function of temperature and molar volume,

$$f(T, v) = f^{\circ}(T_0, v_0) - \int_{T_0, v_0}^{T, v_0} sdT - \int_{T, v_0}^{T, v} pdv \tag{9.6}$$

where

$$f^{\circ}(T_0, v_0) = u^{\circ}(T_0, v_0) - T_0 s^{\circ}(T_0, v_0) \tag{9.7}$$

$$-\int_{T_0, v_0}^{T, v_0} sdT = -s^{\circ}(T_0, v_0)(T - T_0) + \int_{T_0, v_0}^{T, v_0} c_v dT - T \int_{T_0, v_0}^{T, v_0} c_v d(\ln T) \tag{9.8}$$

In Eqs. (9.7) and (9.8), T_0, v_0, u°, s°, and f° are the temperature, molar volume, molar internal energy, entropy, and molar Helmholtz free energy of a reference state, respectively, and c_v is the molar constant volume heat capacity.

The molar Helmholtz free energy of a material as a function of temperature and molar volume can also be obtained from the molar internal energy and molar entropy as a function of temperature and volume,

$$f(T, v) = u(T, v) - Ts(T, v) \tag{9.9}$$

where

$$u(T, v) = u^\circ(T_0, v_0) + \int\limits_{T_0, v_0}^{T, v_0} c_v dT + T \int\limits_{T, v_0}^{T, v} \alpha B_T dv - \int\limits_{T, v_0}^{T, v} p dv \tag{9.10}$$

and

$$s(T, v) = s^\circ(T_0, v_0) + \int\limits_{T_0, v_0}^{T, v_0} \frac{c_v}{T} dT + \int\limits_{T, v_0}^{T, v} \alpha B_T dv \tag{9.11}$$

where α is the volume thermal expansion coefficient, and B_T the isothermal bulk modulus which is the inverse of isothermal compressibility, i.e., $B_T = 1/\beta_T$.

If c_v, B_T, and α are assumed to be constant independent of temperature and volume, the molar internal energy, molar entropy, and molar Helmholtz free energy are given by

$$u(T, v) = u^\circ(T_0, v_0) + c_v(T - T_0)$$
$$+ [B_T(T\alpha - 1) - p_0](v - v_0) + B_T v \ln \frac{v}{v_0} \tag{9.12}$$

$$s(T, v) = s^\circ(T_0, v_0) + c_v \ln \frac{T}{T_0} + \alpha B_T(v - v_0) \tag{9.13}$$

$$f(T, v) = f^\circ(T_0, v_0) + \left[c_v - s^\circ(T_0, v_0)\right](T - T_0) - c_v T \ln \frac{T}{T_0}$$
$$- (B_T + p_0)(v - v_0) + B_T v \ln \frac{v}{v_0} \tag{9.14}$$

where p_0 is the pressure of the initial state. Therefore, with the properties, c_v and B_T or β_T as a function of temperature and molar volume, we can construct the fundamental equation of thermodynamics in terms of molar Helmholtz free energy by defining a reference state.

9.3 Chemical Potential as a Function of Temperature and Pressure

Using the expressions derived in Sect. 8.6 for the temperature and pressure dependences of chemical potential change $\Delta\mu$ as a result of simultaneous temperature and pressure changes, we can easily write down the chemical potential of a homogeneous single-component system or a multicomponent system with a fixed composition as a function of temperature and pressure,

$$\mu(T, p) = \mu^\circ(T_o, p_o) + \int_{T_o, p_o}^{T, p_o} [c_p - s^\circ(T_o, p_o)]\mathrm{d}T - T\int_{T_o, p_o}^{T, p_o} \frac{c_p}{T}\mathrm{d}T + \int_{T, p_o}^{T, p} v\mathrm{d}p$$

$$(9.15)$$

where c_p is the molar constant pressure heat capacity, and $s^\circ(T_o, p_o)$ and $\mu^\circ(T_o, p_o)$ are the molar entropy and chemical potential of a reference state at temperature T_o and pressure p_o. The molar entropy $s^\circ(T_o, p_o)$, chemical potential $\mu(T_o, p_o)$, and the molar enthalpy $h^\circ(T_o, p_o)$ of the reference state are related by

$$\mu^\circ(T_o, p_o) = h^\circ(T_o, p_o) - T_o s^\circ(T_o, p_o) \tag{9.16}$$

The chemical potential μ as a function of temperature and pressure can also be directly obtained from the temperature and pressure dependencies of molar enthalpy h and molar entropy s,

$$\mu(T, p) = h(T, p) - T s(T, p) \tag{9.17}$$

Using the results from Sect. 8.6 for the temperature and pressure dependences of Δh and Δs, the molar enthalpy and entropy as a function of temperature and pressure are

$$h(T, p) = h^\circ(T_o, p_o) + \int_{T_o, p_o}^{T, p_o} c_p \mathrm{d}T + \int_{T, p_o}^{T, p} v(1 - T\alpha)\mathrm{d}p \tag{9.18}$$

$$s(T, p) = s^\circ(T_o, p_o) + \int_{T_o, p_o}^{T, p_o} \frac{c_p}{T}\mathrm{d}T - \int_{T, p_o}^{T, p} v\alpha \mathrm{d}p \tag{9.19}$$

If the thermal expansion coefficient α is a constant independent of temperature, and if the isothermal compressibility β_T is constant independent of pressure, the volume dependence on temperature and pressure can be expressed as

$$v(T, p) = v_0(T_0, p_0)e^{\alpha(T-T_0)-\beta_T(p-p_0)} \tag{9.20}$$

The chemical potential as a function of temperature and pressure assuming constant heat capacity, constant thermal expansion coefficient, and constant compressibility with T_0 and p_0 as the initial state is given by

$$\mu(T, p) = \mu^\circ(T_0, p_0) + \left[c_p - s^\circ(T_0, p_0)\right](T - T_0) - c_p T \ln \frac{T}{T_0}$$
$$- \frac{v_0(T_0, p_0)e^{\alpha(T-T_0)}}{\beta_T}\left[e^{-\beta_T(p-p_0)} - 1\right] \tag{9.21}$$

Therefore, with constant pressure heat capacity c_p, volume thermal expansion coefficient α, and isothermal compressibility β_T as a function of temperature and pressure, together with the thermodynamic data at a reference state, we can construct the fundamental equation of thermodynamics in terms of chemical potential.

For most solids, the thermal expansion coefficient and the isothermal compressibility are very small, so we can make the approximations:

$$e^{\alpha(T-T_0)} \approx 1 + \alpha(T - T_0) \tag{9.22}$$

and

$$e^{-\beta_T(p-p_0)} \approx 1 - \beta_T(p - p_0) + \frac{1}{2}[\beta_T(p - p_0)]^2 \tag{9.23}$$

With these two approximations, the fundamental equation of thermodynamics in terms of chemical potential as a function of temperature and pressure can be approximated as

$$\mu(T, p) = \mu^\circ(T_0, p_0) + \left[c_p - s^\circ(T_0, p_0)\right](T - T_0) - c_p T \ln\left(\frac{T}{T_0}\right)$$
$$+ v_0[1 + \alpha(T - T_0)]\left[(p - p_0) - \frac{1}{2}\beta_T(p - p_0)^2\right] \tag{9.24}$$

9.4   Chemical Potential as a Function of Temperature Involving Phase Changes

If, within the temperature range from a reference state T_0 and p_0 to a temperature of interest T and pressure p_0, there is a first-order phase change from a phase, called α, to another phase, called β, at the transition temperature T_e with $T_0 < T_e < T$, there are abrupt changes in the enthalpy and entropy associated with the phase transition of the material; i.e., there are different branches for the fundamental equation of thermodynamics. Let us label the enthalpy and entropy of transition at the equilibrium transition temperature T_e under pressure p_o as Δh_e^o and Δs_e^o. The enthalpy, entropy, and chemical potential of the α phase as a function of temperature T at p_o are given by

$$h^\alpha(T, p_o) = h^{0,\alpha}(T_o, p_o) + \int_{T_o, p_o}^{T, p_o} c_p^\alpha dT \tag{9.25}$$

$$s^\alpha(T, p_o) = s^{0,\alpha}(T_o, p_o) + \int_{T_o, p_o}^{T, p_o} \frac{c_p^\alpha}{T} dT \tag{9.26}$$

$$\mu^\alpha(T, p_o) = h^{0,\alpha}(T_o, p_o) - T s^{0,\alpha}(T_o, p_o) + \int_{T_o, p_o}^{T, p_o} c_p^\alpha dT - T \int_{T_o, p_o}^{T, p_o} \frac{c_p^\alpha}{T} dT \tag{9.27}$$

The enthalpy and entropy of the β phase are given by

$$h^\beta(T, p_o) = h^\alpha(T_e, p_o) + \Delta h_e^o(T_e, p_o) + \int_{T_e, p_o}^{T, p_o} c_p^\beta dT \tag{9.28}$$

$$s^\beta(T, p_o) = s^\alpha(T_e, p_o) + \Delta s_e^o(T_e, p_o) + \int_{T_e, p_o}^{T, p_o} \frac{c_p^\beta}{T} dT \tag{9.29}$$

where

$$\Delta s_e^o(T_e, p_o) = \frac{\Delta h_e^o(T_e, p_o)}{T_e} \tag{9.30}$$

The chemical potential of the β phase $\mu^\beta(T, p_o)$ as a function of temperature can easily be obtained from $h^\beta(T, p_o)$ and $s^\beta(T, p_o)$ since

$$\mu^\beta(T, p_o) = h^\beta(T, p_o) - T s^\beta(T, p_o) \tag{9.31}$$

Substituting Eqs. (9.28) and (9.29) into Eq. (9.31), we have the chemical potential of the β phase as a function of temperature and pressure,

$$\mu^\beta(T, p_o) = h^\alpha(T_e, p_o) - T s^\alpha(T_e, p_o) + \Delta h_e^o(T_e, p_o) - T \Delta s_e^o(T_e, p_o)$$

$$+ \int\limits_{T_e, p_o}^{T, p_o} c_p^\beta dT - T \int\limits_{T_e, p_o}^{T, p_o} \frac{c_p^\beta}{T} dT \tag{9.32}$$

Using the properties and property relations at the phase transition temperature, we can rewrite the above equation as

$$\mu^\beta(T, p_o) = \mu^\beta(T_e, p_o) + \int\limits_{T_e, p_o}^{T, p_o} [c_p^\beta - s^\beta(T_e, p_o)] dT - T \int\limits_{T_e, p_o}^{T, p_o} \frac{c_p^\beta}{T} dT \tag{9.33}$$

where

$$\mu^\beta(T_e, p_o) = \mu^\alpha(T_e, p_o) = h^{o,\alpha}(T_o, p_o) - T_e s^{o,\alpha}(T_o, p_o) + \int\limits_{T_o, p_o}^{T_e, p_o} \frac{(T - T_e)c_p^\alpha}{T} dT$$

$$\tag{9.34}$$

9.5   Reference States

In order to construct the fundamental equations of thermodynamics based on the molar Helmholtz free energy as a function of temperature and molar volume or chemical potential as a function of temperature and pressure, we have to choose a reference state at which we can define the values of the molar Helmholtz free energy or chemical potential and those of associated thermodynamic properties such as molar internal energy, enthalpy, and entropy. Since the thermodynamic data for molar Helmholtz free energy is mostly from computation, and those for chemical potential are from experiments, the practices for choosing the reference states for molar Helmholtz free energy and chemical potential are generally different in literature.

9.5.1   Reference State for Molar Helmholtz Free Energy at 0 K

The reference state for molar Helmholtz free energy has almost always been associated with the state at 0 K at which the molar Helmholtz free energy is equal to the molar internal energy,

$$f(0\,\text{K}, v) = u(0\,\text{K}, v) \tag{9.35}$$

As it is mentioned in Chap. 1, there is the rest mass energy associated with a substance. Therefore, in principle, the absolute energy of a chemical substance can be defined. However, the molar rest mass energy is many orders of magnitude larger than the changes in the molar internal energy of a system due to changes in temperature or volume or changes due to phase transitions and chemical reactions. As a result, it is the usual practice to define the internal energy of a given unit mass of a solid at 0 K relative to the energy of individual atoms. For a solid, the cohesive energy, which is defined as the energy difference between the atoms arranged in a crystalline state and isolated individual atoms is employed to represent the internal energy and stability of a solid at 0 K. As a matter of fact, the cohesive energy of a majority of solids at 0 K can now be computed using quantum mechanical electron density functional theory calculations.

The molar Helmholtz free energy as a function of temperature and volume using 0 K as the reference state is then given by

$$
f(T, v) = u(0\,\mathrm{K}, v) - Ts(0\,\mathrm{K}, v_0) + \int_{0\,\mathrm{K}, v_0}^{T, v_0} c_v \mathrm{d}T - T \int_{0\,\mathrm{K}, v_0}^{T, v_0} c_v \mathrm{d}(\ln T)
$$

$$
- p_0(v - v_0) - \int_{T, v_0}^{T, v} B_T \mathrm{d}v + v \int_{T, v_0}^{T, v} B_T \mathrm{d}(\ln v) \tag{9.36}
$$

where $s(0\,\mathrm{K}, v_0)$ is the entropy of a substance at 0 K, and v_0 is the molar volume at 0 K typically obtained under vacuum.

The entropy of a solid at 0 K is specified by the third law of thermodynamics, according to which the entropy of a homogeneous, stable substance (pure elemental substance or ordered compounds of several elements) at 0 K is zero, i.e.,

$$
\lim_{T \to 0\,\mathrm{K}} s(T, v_0) \to 0 \tag{9.37}
$$

This is derived from the Nernst postulate:

$$
\Delta s \to 0 \text{ in an isothermal process as } T \to 0,
$$

i.e., the entropy change for any process is approaching to zero when T goes to zero.

One could infer from the Nernst postulate that as $T \to 0$,

$$
c_v(T \to 0) = T\left(\frac{\partial s}{\partial T}\right)_v (T \to 0) \to 0 \tag{9.38}
$$

since $(\partial s / \partial T)_v$ is finite as $T \to 0$. Similarly,

$$c_p(T \to 0) = T\left(\frac{\partial s}{\partial T}\right)_p (T \to 0) \to 0 \tag{9.39}$$

Since the thermal expansion coefficient of a material is related to the rate of entropy change with respect to pressure change at constant temperature by a Maxwell relation, one can infer that the thermal expansion coefficient is also 0 at 0 K, i.e.,

$$\alpha(T \to 0) = \frac{1}{v}\left(\frac{\partial v}{\partial T}\right)_p (T \to 0) = -\frac{1}{v}\left(\frac{\partial s}{\partial p}\right)_T (T \to 0) \to 0 \tag{9.40}$$

Therefore, the molar Helmholtz free energy is given by

$$f(T, v) = u(v) + \int_{0\,\mathrm{K}, v_0}^{T, v_0} c_v dT - T \int_{0\,\mathrm{K}, v_0}^{T, v_0} c_v d(\ln T) - p_0(v - v_0)$$

$$- \int_{T, v_0}^{T, v} B_T dv + v \int_{T, v_0}^{T, v} B_T d(\ln v) \tag{9.41}$$

9.5.2 Reference State for Chemical Potential at 298 K and 1 bar

Although most calculations of molar Helmholtz free energy as a function of temperature and volume use 0 K and the molar volume corresponding to the minimum cohesive energy at 0 K as the reference state, existing calculations and databases for chemical potentials (often called molar Gibbs free energy or simply Gibbs energy in the existing literature) as a function of temperature and pressure employ room temperature 298 K and pressure 1 bar as the reference state.

As it is briefly discussed above, according to the third law of thermodynamics, the entropy of a material at 0 K is zero, and hence, the entropy of a material at 298 K and 1 bar must be positive. Many of the elemental solids and compounds at 298 K and 1 bar are available in thermodynamic databases and online sources.

In addition to the value of entropy at the reference state, we also need either the value of chemical potential or the molar enthalpy at the reference state of 298 K and 1 bar. One option is to define the chemical potential of an elemental, stable substance at 298 K and 1 bar as the reference with its value set to zero while the chemical potential of a compound is the formation of chemical potential from its elemental constituents at 298 K and 1 bar.[1] However, the majority of existing thermodynamic databases employ the convention by setting the enthalpy of a thermodynamically

---

[1] G. Job and F. Herrmann, "Chemical potential—a quantity in search of recognition." Institute of Physics Publishing, Eur. J. Physics 27, 353 (2006).

stable, pure elemental substance to zero at room temperature 298 K and ambient pressure 1 bar, i.e.,

$$h^\circ_{298\,\mathrm{K},1\,\mathrm{bar}} = 0 \tag{9.42}$$

The enthalpy of a compound at 298 K and 1 bar is then equal to the heat of formation of the compound from the corresponding stable elemental substances at 298 K and 1 bar.

The enthalpy or heat of formation is the enthalpy change or heat released when one mole of compound forms from its constituent elements. For example, for the formation of one mole of solid oxide SiO_2 from the solid Si and O_2 gas at 298 K and 1 bar,

$$Si(s) + O_2(g) = SiO_2(s)$$

The heat of formation, Q, is about -860 kJ/mol. Using the above convention, we have

$$h^\circ_{SiO_2} = -860\,\mathrm{kJ/mol}$$
$$h^\circ_{Si} = 0$$
$$h^\circ_{O_2} = 0$$
$$\Delta h^\circ_{SiO_2} = Q = h^\circ_{SiO_2} - h^\circ_{Si} - h^\circ_{O_2}$$
$$= -860\,\mathrm{kJ/mol}.$$

Similarly, the enthalpy of formation for methane (CH_4) from hydrogen gas (H_2) and graphite (C), $\Delta h^\circ_{CH_4}$, is about -74.8 kJ/mol at 298 K and 1 bar, and hence, using the convention, we have

$$h^\circ_{CH_4} = -74.8\,\mathrm{kJ/mol}$$
$$h^\circ_{H_2} = 0$$
$$h^\circ_C = 0$$

Therefore, with 298 K and 1 bar as the reference, we can construct the fundamental equation of thermodynamics of a system in terms of its chemical potential as a function of temperature and pressure,

$$\mu(T, p) = \mu^\circ(298\,\mathrm{K},\,1\,\mathrm{bar}) + \int_{298\,\mathrm{K},1\,\mathrm{bar}}^{T,1\,\mathrm{bar}} \left[c_p - s^\circ(298\,\mathrm{K},\,1\,\mathrm{bar})\right]dT$$

$$- T \int_{298\,\mathrm{K},1\,\mathrm{bar}}^{T,1\,\mathrm{bar}} \frac{c_p}{T}dT + \int_{T,1\,\mathrm{bar}}^{T,p} v\,dp \tag{9.43}$$

9.6   Fundamental Equations with Multiphysics Effects

The fundamental equation of thermodynamics in terms of molar Helmholtz free energy as a function of temperature and molar volume or the chemical potential as a function of temperature and pressure discussed above is only applicable to simple systems. We can incorporate the contributions from both internal and external fields as well as surface effects in the fundamental equation of thermodynamics. Examples include the presence of gravitational potential gz or the gravitational potential energy per unit mass m, electric potential ϕ or the electrostatic potential energy per unit charge q, surface/interfacial potential or specific surface/interfacial energy γ, i.e., the surface/interfacial energy per unit area A, as well as stress, electric, and magnetic fields. The general differential form for the fundamental equation of thermodynamics in terms of internal energy including these additional contributions is

$$dU = TdS - pdV + V_{o}\left(\sigma_{ij}^{d}d\varepsilon_{ij}^{d} + E_{i}dD_{i} + H_{i}dB_{i}\right)$$
$$+ gzdm + \phi dq + \gamma dA + \mu_{i}dN_{i} \tag{9.44}$$

in which all the coefficients in front of the differentials of the above equation are either potentials $T, p, gz, \phi, \gamma, \mu_{1}, \mu_{2}, \ldots, \mu_{n}$ or stress (σ_{ij}), electric (E_{i}) and magnetic (H_{i}) fields. In Eq. (9.44), σ_{ij}^{d} and ε_{ij}^{d} are the deviatoric components of stress and strain, respectively, V_{o} is the volume of the reference state before the mechanical deformation, D_{i} is the electric displacement, and B_{i} is the magnetic induction or flux density.

We can also rewrite Eq. (9.44) in terms of Helmholtz free energy or Gibbs free energy,

$$dF = -SdT - pdV + V_{o}\left(\sigma_{ij}^{d}d\varepsilon_{ij}^{d} + E_{i}dD_{i} + H_{i}dB_{i}\right)$$
$$+ gzdm + \phi dq + \gamma dA + \mu_{i}dN_{i} \tag{9.45}$$

$$dG = -SdT + Vdp - V_{o}\varepsilon_{ij}^{d}d\sigma_{ij}^{d} + V_{o}(E_{i}dD_{i} + H_{i}dB_{i})$$
$$+ gzdm + \phi dq + \gamma dA + \mu_{i}dN_{i} \tag{9.46}$$

For applications to inhomogeneous systems, we often express the fundamental equation in terms of volume densities. For example, we can rewrite Eq. (9.44) as

$$du_{v} = Tds_{v} - pd\varepsilon + \sigma_{ij}^{d}d\varepsilon_{ij}^{d} + E_{i}dD_{i} + H_{i}dB_{i} + gzd\rho_{m} + \phi d\rho_{q} + \gamma da_{v} + \mu_{i}dc_{i} \tag{9.47}$$

where $u_{v}, s_{v}, \rho_{m}, \rho_{q}, a_{v}$ are volume densities of internal energy, entropy, mass, charge, and surface area, c_{i} is the volume concentration of species i, and $d\varepsilon$ is dV/V_{o}.

The Helmholtz free energy density f_v at constant temperature and the Gibbs free energy density g_v at constant temperature and pressure can be written as

$$df_v = -p d\varepsilon + \sigma_{ij}^d d\varepsilon_{ij}^d + E_i dD_i + H_i dB_i + gz d\rho_m + \phi d\rho_q + \gamma da_v + \mu_i dc_i \tag{9.48}$$

$$dg_v = -\varepsilon_{ij}^d d\sigma_{ij}^d + E_i dD_i + H_i dB_i + gz d\rho_m + \phi d\rho_q + \gamma da_v + \mu_i dc_i \tag{9.49}$$

Therefore, in principle, similar to simple systems, we can construct fundamental equations of thermodynamics for inhomogeneous systems in terms of Helmholtz free energy density or molar Gibbs free energy density,

$$f_v = f_v\left(T, \varepsilon, \varepsilon_{ij}^d, D_i, B_i, \rho_m, \rho_q, a_v, c_i\right) \tag{9.50}$$

$$g_v = g_v\left(T, p, \sigma_{ij}^d, D_i, B_i, \rho_m, \rho_q, a_v, c_i\right) \tag{9.51}$$

There is also a common practice to incorporate some of these contributions by combining their potentials with chemical potential into a total potential. Below we discuss a number of examples.

9.6.1 Chemogravitational Potential

Let us first look at a simple case of only considering gravitational potential and chemical potential at constant temperature and volume. The differential form for the Helmholtz free energy is given by

$$dF = gz dm + \mu dN \tag{9.52}$$

where g is the gravitational acceleration, m is mass, and z is height of the system. Let us rewrite above equation as follows

$$dF = \left(gz\frac{dm}{dN} + \mu\right)dN = (gzM + \mu)dN \tag{9.53}$$

where M is the molar mass.

In terms of Helmholtz free energy density, Eq. (9.53) becomes

$$df_v = \left(gz\frac{dm}{dN} + \mu\right)dc = (gzM + \mu)dc \tag{9.54}$$

Therefore, in the presence of gravitation potential, the total potential of a material is the chemogravitational potential given by

$$\mu^{cg} = gzM + \mu \tag{9.55}$$

where μ is the chemical potential.

9.6.2 Electrochemical Potential

The total potential for a charged species i, e.g., ions, electrons, holes, is the sum of chemical potential and electric potential. Here we write the differential form for the Helmholtz free energy at constant temperature and volume,

$$dF = \phi dq + \mu_i dN_i \tag{9.56}$$

where ϕ is electric potential, q is charge, μ_i is chemical potential of species i, and N_i is number of moles of species i.

We can rewrite the above equation as

$$dF = \left(\phi \frac{dq}{dN_i} + \mu_i\right)dN_i = (\phi z_i e N_o + \mu_i)dN_i = (\phi z_i \mathcal{F} + \mu_i)dN_i \tag{9.57}$$

where z_i is the valence of species i, e is the elementary charge, N_o is the Avogadro's constant, and $e N_o$ is the Faraday's constant $\mathcal{F}$. Therefore, the electrochemical potential for a charged species i is

$$\tilde{\mu}_i = \mu_i + z_i \mathcal{F} \phi \tag{9.58}$$

For example, the electrochemical potential of Li^+ is

$$\tilde{\mu}_{Li^+} = \mu_{Li^+} + \mathcal{F} \phi$$

9.6.3 Chemical Potential Including Surface Effects

If the size of a material becomes nanoscale, the contribution of surface energy to the thermodynamics cannot be ignored. Atoms sitting at the surface have a different atomic bonding environment and normally have higher energy than those inside the bulk. The surface energy is this extra energy associated with a surface, and the extra energy per unit surface area is the specific surface energy. Therefore, a material always tends to minimize its surface area in order to minimize the surface energy and thus the total energy of the material. Considering the variation in the Gibbs free energy of a particle with the amount of substance at constant temperature and pressure while taking into account the surface energy contribution of the particle, we have

$$dG = \gamma dA + \mu dN \tag{9.59}$$

where γ is the specific surface energy, and A is surface area.

We can rewrite the surface energy term as follows,

$$dG = \gamma \left(\frac{\partial A}{\partial V}\right)\left(\frac{\partial V}{\partial N}\right) dN + \mu dN = \gamma v \left(\frac{\partial A}{\partial V}\right) dN + \mu dN \tag{9.60}$$

Therefore,

$$dG = \left[\gamma v \left(\frac{\partial A}{\partial V}\right) + \mu\right] dN \tag{9.61}$$

and

$$\mu_r = \gamma v \left(\frac{\partial A}{\partial V}\right) + \mu \tag{9.62}$$

where v is the molar volume. For example, for a spherical solid particle with radius r,

$$\mu_r = \gamma v \left(\frac{\partial A/\partial r}{\partial V/\partial r}\right)_{T,p} + \mu \tag{9.63}$$

For a spherical particle, we can easily show that

$$\frac{\partial A}{\partial r} = 8\pi r \tag{9.64}$$

and

$$\frac{\partial V}{\partial r} = 4\pi r^2 \tag{9.65}$$

Thus, we have

$$\mu_r = \frac{2\gamma v}{r} + \mu \tag{9.66}$$

, i.e., the chemical potential of a spherical particle including the surface effect is

$$\mu_r = \frac{2\gamma v}{r} + \mu_\infty \tag{9.67}$$

where μ_r is the chemical potential of a particle with radius r, and μ_∞ is the chemical potential of a solid with infinite size.

Therefore, chemical potentials of atoms in particles with different sizes would be different at the same temperature and pressure. Atoms in a small particle have higher chemical potentials and are thus thermodynamically less stable than those in a larger particle. This is the very reason for particle coarsening; i.e., large particles grow at the expense of smaller ones if the atoms from the small particles have transport paths to reach large particles.

The chemical potential difference between a finite-size particle and an infinite-size particle can be translated to the pressure difference,

$$d\mu = vdp \tag{9.68}$$

Ignoring the pressure dependence of molar volume, integrating the above equation from μ_∞ to μ_r for the chemical potential and pressure from p_∞ to p_r, we have

$$\mu_r - \mu_\infty = v(p_r - p_\infty) \tag{9.69}$$

Using the result from Eq. (9.67), we can write the pressure inside a particle with radius r as a function of r,

$$p_r = \frac{2\gamma}{r} + p_\infty \tag{9.70}$$

i.e., the pressure inside a finite-size particle is higher than outside at equilibrium.

9.7   Examples

Example 1 At ambient pressure of 1 bar, the molar entropy of graphite $s^o_{298\,K,1\,bar}$ is $5.6\,J/(K\,mol)$. The molar volume of graphite $v^o_{298\,K,1\,bar}$ is $6.0\,cm^3/mol$. The molar constant pressure heat capacity c_p is $8.5\,J/(K\,mol)$. The isothermal compressibility of graphite β_T is $3 \times 10^{-6}\,1/bar$. The constant pressure volume thermal expansion coefficient α is $3 \times 10^{-5}\,1/K$. Determine the chemical potential of graphite as a function of temperature and pressure.

Solution

$$\mu(T, p) = h^o_{298\,K,1\,bar} - s^o_{298\,K,1\,bar}T + c_p(T - 298\,K)$$
$$- c_p T \ln\left(\frac{T}{298\,K}\right) + \frac{v^o_{298\,K,1\,bar}e^{\alpha(T-298\,K)}}{\beta_T}\left[1 - e^{-\beta_T(p-1\,bar)}\right]$$

$$\mu(T, p) = -5.6T + 8.5 \left[(T - 298 \text{ K}) - T \ln \left(\frac{T}{298 \text{ K}} \right) \right]$$
$$+ 2.0 \times 10^{-6} e^{3 \times 10^{-5}(T - 298 \text{ K})} \left[1 - e^{-3 \times 10^{-6}(p - 1 \text{ bar})} \right]$$

Example 2 At ambient pressure of 1 bar, the entropy of liquid (l) water at 298 K $s_{298\text{ K}}^{o,l}$, is 70 J/(mol K). The heat of formation of water from hydrogen and oxygen gases at 298 K and $\Delta h_{298\text{ K}}^{o,l}$ is -286 kJ/mol. The heat capacity of solid (s) ice c_p^s is 38 J/(mol K). The heat capacity of water c_p^l is 75 J/(mol K). The heat capacity of water vapor c_p^v is 36 J/(mol K). The heat of melting Δh_m^o at 0°C is 6007 J/mol. The heat of vaporization of water Δh_v^o at 100°C is 40.65 kJ/mol. Obtain the chemical potential of ice at 263 K and chemical potential of water vapor at 393 K.

Solution
The enthalpy of ice at 263 K is given by

$$h_{263\text{ K}}^s = h_{298\text{ K}}^{o,l} + \int_{298\text{ K}}^{273\text{ K}} c_p^l dT + \Delta h_s^o + \int_{273\text{ K}}^{263\text{ K}} c_p^s dT$$
$$= h_{298\text{ K}}^{o,l} + c_p^l (273 - 298) - \Delta h_m^o + c_p^s (263 - 273)$$
$$= -286,000 - 75 \times 25 - 6007 - 38 \times 10 = -294,262 \text{ J/mol}$$

The entropy of ice at 263 K is

$$s_{263\text{ K}}^s = s_{298\text{ K}}^{o,l} + \int_{298\text{ K}}^{273\text{ K}} \frac{c_p^l}{T} dT + \Delta s_s^o + \int_{273\text{ K}}^{263\text{ K}} \frac{c_p^s}{T} dT$$
$$= s_{298\text{ K}}^{o,l} + c_p^l \ln \left(\frac{273}{298} \right) - \frac{\Delta h_m^o}{273} + c_p^s \ln \left(\frac{263}{273} \right)$$
$$= 70 + 75 \times (-0.0876) - \frac{6007}{273} + 38 \times (-0.0373) = 40 \text{ J/(K mol)}$$

Therefore, the chemical potential of ice at 263 K is

$$\mu_{263\text{ K}}^s = h_{298\text{ K}}^{o,l} + \int_{298\text{ K}}^{273\text{ K}} c_p^l dT + \Delta h_s^o + \int_{273\text{ K}}^{263\text{ K}} c_p^s dT$$
$$- 263 \left[s_{298\text{ K}}^{o,l} + \int_{298\text{ K}}^{273\text{ K}} \frac{c_p^l}{T} dT + \Delta s_s^o + \int_{273\text{ K}}^{263\text{ K}} \frac{c_p^s}{T} dT \right]$$
$$= h_{298\text{ K}}^{o,l} + c_p^l (273 - 298) - \Delta h_m^o + c_p^s (263 - 273)$$

$$-263\left[s_{298\,K}^{o,l}+c_p^l\ln\left(\frac{273}{298}\right)-\frac{\Delta h_m^o}{273\,K}+c_p^s\ln\left(\frac{263}{273}\right)\right]$$

$$=-286,000-75\times25-6007-38\times10$$

$$-263\left[70+75\times(-0.0876)-\frac{6007}{273}+38\times(-0.0373)=40\,J/(K\,mol)\right]$$

$$=-304,782\,G$$

The enthalpy of water vapor at 393 K is

$$h_{393\,K}^v=h_{298\,K}^{o,l}+\int_{298\,K}^{373\,K}c_p^l dT+\Delta h_v^o+\int_{373\,K}^{393\,K}c_p^v dT$$

$$=h_{298\,K}^{o,l}+c_p^l(373-298)+\Delta h_v^o+c_p^v(393-373)$$

$$=-286,000+75\times75+40,650+36\times20=-239,005\,J/mol$$

The entropy of water vapor at 393 K is

$$s_{393\,K}^v=s_{298\,K}^{o,l}+\int_{298\,K}^{373\,K}\frac{c_p^l}{T}dT+\Delta s_v^o+\int_{373\,K}^{393\,K}\frac{c_p^v}{T}dT$$

$$=s_{298\,K}^{o,l}+c_p^l\ln\left(\frac{373}{298}\right)+\frac{\Delta h_v^o}{373\,K}+c_p^v\ln\left(\frac{393}{373}\right)$$

$$=70+75\times(0.2245)+\frac{40,650}{373}+36\times(0.0522)=198\,J/(K\,mol)$$

Therefore, the chemical potential of water vapor at 393 K is

$$\mu_{393\,K}^v=h_{298\,K}^{o,l}+\int_{298\,K}^{373\,K}c_p^l dT+\Delta h_v^o+\int_{373\,K}^{393\,K}c_p^v dT$$

$$-393\left[s_{298\,K}^{o,l}+\int_{298\,K}^{373\,K}\frac{c_p^l}{T}dT+\Delta s_v^o+\int_{373\,K}^{393\,K}\frac{c_p^v}{T}dT\right]$$

$$=h_{298\,K}^{o,l}+c_p^l(373-298)+\Delta h_v^o+c_p^v(393-373)$$

$$-393\left[s_{298\,K}^{o,l}+c_p^l\ln\left(\frac{373}{298}\right)+\frac{\Delta h_v^o}{373\,K}+c_p^v\ln\left(\frac{393}{373}\right)\right]$$

$$=-286,000+75\times75+40,650+36\times20$$

$$-393\left[70+75\times(0.2245)+\frac{40,650}{373}+36\times(0.0522)\right]$$

$$=-316,819\,G$$

9.8 Exercises

1. If we use the convention that the enthalpy of oxygen gas at room temperature 298 K and 1 bar is zero, the enthalpy of oxygen gas at 0 °C is less than zero: true or false?

2. The entropy of one mole of water cannot be negative: true or false?

3. The enthalpy of oxygen gas at room temperature 298 K and 1 bar is defined as zero by convention while the entropy of oxygen gas at room temperature is greater than zero: true or false?

4. Based on the convention determining the enthalpy of a system and given that the heat of formation of NiO at 1 bar and 298 K is $-244.6\,kJ/mol$, what are the enthalpies of Ni, O_2, and NiO at 1 bar and 298 K?

5. Based on the convention determining the enthalpy of a system, rank the enthalpies of the following systems in an increasing order, all for 1 mol and under 1 bar: O_2 at 298 K, Al at 500 K, and Al_2O_3 at 298 K.

6. Rank the enthalpies of the following systems in an increasing order, all for 1 mol and under 1 bar: water at 0°C, ice at 0 K, water vapor at 100°C, ice at 0°C, and water at 100°C.

7. Rank the molar entropies of solid, liquid, and vapor Cu from the lowest to the highest.

8. Rank the entropies of the following systems in an increasing order for all at one mole and 1 bar: one mole of pure aluminum solid in fcc structure at 0 K, one mole of water at 298 K, and one mole of oxygen gas at 298 K.

9. Rank the magnitudes for the entropies of Al (solid with fcc structure at 0 K, 1 bar), Al (solid with fcc structure at 298 K, 1 bar), Al (solid with fcc structure at 934 K), and Al (liquid at 934 K, 1 bar) from lowest to highest.

10. Rank the magnitudes for the entropies of C (graphite at 298 K, 1 bar), O_2 (298 K, 1 bar), and O_2 (298 K, 100 bar) from lowest to highest.

11. Rank the entropies of the following systems in an increasing order (all at 1 mol and 1 bar): water at 0°C, ice at 0 K, water vapor at 100°C, ice at 0°C, and water at 100°C.

12. Rank the entropies of the following materials in an increasing order (all at one mole and 1 bar): liquid Ag at 2435 K, vapor Ag at 2435 K, solid Ag at 0 K, solid Ag at 298 K, solid Ag at 1234 K, and liquid Ag at 1234 K.

13. The Helmholtz free energy of a pure substance at constant volume always decreases with temperature, true or false?

14. The Helmholtz free energy of a pure substance at constant temperature always decreases with volume, true or false?

15. At constant temperature, the rate of change of the Helmholtz free energy of a system with respect to volume is always positive, true or false?

16. At a given temperature, when you squeeze a piece of solid, does the Helmholtz free energy of the solid (i) increase or (ii) decrease?

17. At 0 K, the chemical potential of a system is the same as its molar enthalpy, true or false?

18. Given the data in the table for 1 bar

Substance	$h^{o}_{298\,K}$ (J/mol)	$s^{o}_{298\,K}$ (J/(K mol))	c_P (J/(K mol))
Ba		62.4	28.07
O_2		205.1	29.4
Ti		30.7	25.06
BaO	$-582,000$	70.0	50.0
TiO_2	$-945,000$	50.0	55.0
$BaTiO_3$	$-1,647,000$	107.9	110.0

Calculate the chemical potential as a function of temperature at 1 bar for Ba, Ti, O_2, BaO, TiO_2, and $BaTiO_3$.

19. For solid Ba, it has two polymorphic forms α and β. The α phase is stable at temperatures up to 648 K. At 648 K, the α phase is transformed to the β phase. The β phase is stable until its melting temperature which is higher than 1000 K. The molar constant pressure heat capacity of α phase: $c^{\alpha}_p = -438.2 + 1587.0 \times 10^{-3}T + 128.2 \times 10^5 T^{-2}$ J/(mol K). The molar constant pressure heat capacity of β phase, $c^{\beta}_p = -5.69 + 80.33 \times 10^{-3}T$ J/(mol K). The molar entropy of α phase, $s^{o,\alpha}_{298\,K} = 62.4$ J/(mol K). The equilibrium phase transition temperature from α to β is $T^{\alpha \to \beta}_e = 648$ K. The molar heat of transition for the α to β transition is $\Delta h^{\alpha \to \beta}_e = 630$ J/mol. Calculate the chemical potential μ_{Ba} of Ba as a function of temperature from 298 to 1000 K.

20. Given data: molar enthalpy of diamond $h^{o,d}_{298\,K,\,1\,bar} = 1900$ J/mol; molar entropy of graphite $s^{o,g}_{298\,K,\,1\,bar} = 5.6$ J/(K mol); molar entropy of diamond $s^{o,d}_{298\,K,\,1\,bar} = 2.4$ J/(K mol); molar volume of graphite $v^{o,g}_{298\,K,\,1\,bar} = 6.0$ cm^3/mol; molar volume of diamond $v^{o,d}_{298\,K,\,1\,bar} = 3.4$ cm^3/mol; molar constant pressure heat capacity of graphite $c^g_p = 8.5$ J/(K mol); molar constant pressure heat capacity of diamond $c^d_p = 6.0$ J/(K mol); molar isothermal compressibility of graphite $\beta^g_T = 3 \times 10^{-6}$/bar; molar isothermal compressibility of diamond $\beta^d_T = 2 \times 10^{-7}$/bar volume thermal expansion coefficient of graphite $\alpha^g = 3 \times 10^{-5}$/K, and volume thermal expansion coefficient of diamond $\alpha^d = 10^{-5}$/K. Please determine:

(a)   The chemical potential of graphite as a function of temperature at 1 bar.

(b)   The chemical potential of diamond as a function of temperature at 1 bar.

(c)   The chemical potential of graphite as a function of pressure at 298 K.

(d)   The chemical potential of diamond as a function of pressure at 298 K.

(e)   The chemical potential of graphite as a function of temperature and pressure.

(f)   The chemical potential of diamond as a function of temperature and pressure.

(g)   The chemical potential difference between graphite and diamond as a function of temperature and pressure (assuming that the compressibility $\beta = 0$ for both diamond and graphite).

21.   Consider a cube-shaped finite-size solid particle, find the chemical potential dependence on the edge length (a) of the particle, assuming that the chemical potential of the corresponding infinite-sized particle is μ_∞, molar volume is v, and the surface energy is γ.

22.   Using the chemogravitational potential and the equilibrium condition that the total potential must be uniform, show that the pressure for gas molecules of molar mass M at an altitude z above sea level is given by

$$p(z) = p(0)\exp\left(-\frac{Mgz}{RT}\right)$$

where $p(0)$ is the pressure at the sea level, R is gas constant, and g is the gravitational constant.

Chapter 10
Chemical Potentials of Gases, Electrons, Crystals, and Defects

The most convenient fundamental equation of thermodynamics in applications of thermodynamics in materials science and engineering is in terms of chemical potential as a function of temperature and pressure. In theoretical calculations, one often employs the equation of state in terms of chemical potential as a function of temperature and molar volume in addition to molar Helmholtz free energy as a function of temperature and volume. This chapter will discuss the thermodynamics of a number of well-known classical systems, for which the fundamental equations and thus chemical potentials can be derived from statistical mechanics. Examples include the ideal gases, valence electrons in solids, crystals with their lattice vibrations approximated by the Einstein or Debye model, as well as non-interacting lattice point defects. In this chapter, their fundamental equations and/or chemical potentials are presented without derivation, and all the important thermodynamic properties for these model systems are then obtained from their fundamental equations of thermodynamics.

10.1 Chemical Potential of Ideal Gases

The chemical potential μ (in unit of energy per atom) for a single component ideal gas as a function of temperature T and pressure p is given by

$$\mu(T, p) = -k_B T \left\{ \ln \left[\frac{k_B T}{p} \left(\frac{2\pi m k_B T}{h^2} \right)^{3/2} \right] \right\} \tag{10.1}$$

where m is atomic mass, k_B is the Boltzmann constant, and h is the Planck constant.

© Springer Nature Singapore Pte Ltd. 2022
L.-Q. Chen, *Thermodynamic Equilibrium and Stability of Materials*,
https://doi.org/10.1007/978-981-13-8691-6_10

In the above equation, the term

$$\left(\frac{2\pi mk_BT}{h^2}\right)^{3/2} = N^g \tag{10.2}$$

can be viewed as the density of states for an ideal gas, and its inverse as the volume for each state, i.e.,

$$v^g = \frac{1}{N^g} = \left(\frac{h^2}{2\pi mk_BT}\right)^{3/2} \tag{10.3}$$

The term

$$\frac{k_BT}{p} = \frac{V}{N} = v = \frac{1}{n} \tag{10.4}$$

is the volume per atom for the gas.

In the application of thermodynamics in materials science and engineering, we often employ the unit of G ($= J/mol$) rather than on a per atom basis for the chemical potential. On a per mole basis, the chemical potential of an ideal gas is given by

$$\mu(T, p) = -RT\left\{\ln\left[\frac{k_BT}{p}\left(\frac{2\pi mk_BT}{h^2}\right)^{3/2}\right]\right\} \tag{10.5}$$

where R is the gas constant.

10.1.1 Standard State, Activity, and Fugacity

For gases, the convention is to define the standard state for pure gases with their pressure p^o being at 1 bar ($= 10^5$ pa). The chemical potential of an ideal gas at $p^o = 1$ bar is

$$\mu^o(T, p^o) = -RT\left\{\ln\left[\frac{k_BT}{p^o}\left(\frac{2\pi mk_BT}{h^2}\right)^{3/2}\right]\right\} \tag{10.6}$$

By defining a standard state for the gases, the dependence of chemical potential on temperature and pressure can be rewritten as

$$\mu(T, p) = \mu^o(T, p^o) + RT \ln\frac{p}{p^o} \tag{10.7}$$

One can define a unit-less quantity, called activity,

$$a = \frac{p}{p^o} \tag{10.8}$$

With the pure gas at 1 bar as the standard state, the chemical potential of a gas can also be rewritten as

$$\mu(T, p) = \mu^o(T, 1\,\text{bar}) + RT \ln \frac{p}{1\,\text{bar}}$$
$$= \mu^o(T, 1\,\text{bar}) + RT \ln a \tag{10.9}$$

with

$$a = \frac{p}{1\text{bar}} \tag{10.10}$$

Therefore, the activity a for an ideal gas has the same magnitude as the pressure of the gas but without the pressure unit bar. For example, for an ideal gas at 0.05 bar, its activity is 0.05.

For real gases, the ideal gas law is no longer valid. However, to keep the same functional form for the chemical potential of a real gas as that for an ideal gas (Eq. 10.7), a quantity called the fugacity, f, is defined through

$$d\mu = RT d \ln f \tag{10.11}$$

Fugacity has the same unit as pressure. At a given temperature, the fugacity is a function of pressure, and it can be approximated by the corresponding pressure as the gas pressure approaches zero, i.e.,

$$\lim_{p \to 0} f \to p \tag{10.12}$$

With the introduction of fugacity, the chemical potential of a gas can be expressed as

$$\mu(T, f) = \mu^o(T, f^o) + RT \ln\left(\frac{f}{f^o}\right) \tag{10.13}$$

where f^o is the fugacity at the standard state at pressure of 1 bar.

In terms of activity, the chemical potential of a gas can be expressed as

$$\mu(T, f) = \mu^o(T, f^o) + RT \ln a \tag{10.14}$$

where

$$a = \frac{f}{f^{\circ}} \tag{10.15}$$

Therefore, the activity a involves the ratio of fugacity at any state to the fugacity at the standard state and thus has no unit.

If one considers a vapor phase in equilibrium with a solid or liquid, the most convenient choice for the standard state for the vapor phase is the equilibrium vapor pressure $p^{v,o}$ above the condensed phase, i.e.,

$$\mu^{v}(T, p) = \mu^{v,o}(T, p^{v,o}) + RT \ln\left(\frac{p}{p^{v,o}}\right) \tag{10.16}$$

The standard state for the corresponding solid or liquid is a pure solid or liquid under the equilibrium vapor pressure,

$$\mu^{l}(T, p) = \mu^{l,o}(T, p^{v,o}) \tag{10.17}$$

Therefore, at equilibrium, the pressure in the vapor phase is the equilibrium vapor pressure, and

$$\mu^{l}(T, p) = \mu^{l,o}(T, p^{v,o}) = \mu^{v}(T, p) = \mu^{v,o}(T, p^{v,o}) \tag{10.18}$$

It should be noted that for most other applications, the standard state for a solid or liquid is pure solid or liquid under 1 bar. The difference between results obtained under the equilibrium vapor pressure or 1 bar for a solid or liquid can, in general, be neglected.

10.1.2 Chemogravitational Potential of Ideal Gases

The fundamental equation of thermodynamics for ideal gases in a gravitational field in the entropic representation is given by

$$S = Nk_{B}\left\{\frac{5}{2} + \ln\left[\frac{V}{N}\left(\frac{4\pi m(U - Nmgz)}{3h^2 N}\right)^{3/2}\right]\right\} \tag{10.19}$$

where S is entropy, N is the number of gas atoms, U is internal energy, V is volume, m is atomic mass, k_B is the Boltzmann constant, h is the Planck constant, g is gravitational acceleration, and z is the height.

The chemical potential of an ideal gas in the presence of the gravitational field, or the chemogravitational potential, can be readily obtained from entropy through one of the equations of state,

$$\mu(T, p, z) = -T\left(\frac{\partial S}{\partial N}\right)_{U,V} = -k_B T\left\{\ln\left[\frac{k_B T}{p}\left(\frac{2\pi m k_B T}{h^2}\right)^{3/2}\right]\right\} + mgz$$

$$(10.20)$$

At equilibrium, the chemogravitational potential of gas atoms should be uniform, i.e.,

$$\mu(T, p, z) = \text{constant} \qquad (10.21)$$

Assuming that the temperature T is independent of z while the pressure p is a function of z, we have

$$\mu[T, p(z), z] = \mu[T, p(0), 0] \qquad (10.22)$$

Therefore,

$$-k_B T\left\{\ln\left[\frac{k_B T}{p(z)}\left(\frac{2\pi m k_B T}{h^2}\right)^{3/2}\right]\right\} + mgz = -k_B T\left\{\ln\left[\frac{k_B T}{p(0)}\left(\frac{2\pi m k_B T}{h^2}\right)^{3/2}\right]\right\}$$

$$(10.23)$$

or

$$k_B T \ln p(z) + mgz = k_B T \ln p(0) + 0 \qquad (10.24)$$

which can be rewritten as

$$k_B T \ln \frac{p(z)}{p(0)} = -mgz \qquad (10.25)$$

Therefore, the pressure distribution of ideal gases in the presence of a gravitational field is given by

$$p(z) = p(0)e^{-\frac{mgz}{k_B T}} \qquad (10.26)$$

We can also obtain the density distribution $\rho(z)[= p(z)/k_B T]$ as a function of z,

$$\rho(z) = \rho(0)e^{-\frac{mgz}{k_B T}} \qquad (10.27)$$

10.2   Chemical Potentials of Electrons and Holes

The chemical potential of electrons, holes, atomic vacancies, and interstitials in solids can be defined in the same way as the chemical potential of a chemical species. In the case of electrons, the chemical potential is usually expressed in energy per electron rather than energy per mole of electrons, and the energy per electron is conventionally given in the unit of electron-volt (eV/electron). Therefore, the chemical potential of electrons is defined as the change in the free energy of an electron system when one electron is added or removed from the electron system. A more easily understandable definition of chemical potential for electrons will simply be the chemical energy possessed per electron.

10.2.1   Electrons in Metals

For metals, the electrons fill up the quantum states of electron energy bands without an energy gap between the filled states and the unfilled states at 0 K. At finite temperatures, some of the electrons near the top of the filled states are excited to empty states of higher energy within the same energy band. The valence electrons of a metal can be considered as a gas of electrons which obey the Pauli exclusion principle. The fundamental equation of thermodynamics for the valence electrons at relatively low temperatures in terms of Helmholtz free energy can be approximated as[1]

$$F(T, V, N) = \frac{3}{5} N \mu_o \left[1 - \frac{5\pi^2}{12} \left(\frac{k_B T}{\mu_o} \right)^2 \right], \quad \mu_o = \frac{h^2}{2m} \left(\frac{3N}{8\pi V} \right)^{2/3} \tag{10.28}$$

where N is number of valence electrons in a volume V at temperature T, h is the Planck constant, m the effective electron mass, and μ_o is the chemical potential of electrons at 0 K, which is usually called the Fermi energy E_f.

Based on the differential form of fundamental equation of thermodynamics in terms of Helmholtz free energy, we can easily obtain other thermodynamic properties including the chemical potential as a function of temperature and volume. For example, the entropy of the electron gas as a function of temperature, volume, and the number of electrons is

$$S(T, V, N) = -\left(\frac{\partial F}{\partial T} \right)_{V,N} = \frac{N \mu_o}{T} \left[\frac{\pi^2}{2} \left(\frac{k_B T}{\mu_o} \right)^2 \right] \tag{10.29}$$

The pressure of the electron gas system as a function of temperature and volume per electron is given by

---

[1] Donald A. McQuarrie, Statistical Mechanics, Harper & Row, New York, 1976.

$$p(T, V/N) = -\left(\frac{\partial F}{\partial V}\right)_{T,N} = \frac{2}{5}\frac{N\mu_o}{V}\left[1 + \frac{5\pi^2}{12}\left(\frac{k_B T}{\mu_o}\right)^2\right] \tag{10.30}$$

Similarly, we can obtain the chemical potential or Fermi level as a function of temperature and volume per electron,

$$\mu(T, V/N) = \left(\frac{\partial F}{\partial N}\right)_{T,V} = \mu_o\left[1 - \frac{\pi^2}{12}\left(\frac{k_B T}{\mu_o}\right)^2\right] \tag{10.31}$$

We can readily obtain the rest of the thermodynamic properties. For example, the internal energy of the electron system as a function of temperature, volume, and the number of electrons is given by

$$U(T, V, N) = \frac{3}{5}N\mu_o\left[1 + \frac{5\pi^2}{12}\left(\frac{k_B T}{\mu_o}\right)^2\right] \tag{10.32}$$

The thermal energy of the electron gas as a function of temperature, volume, and the number of electrons is

$$TS(T, V, N) = N\mu_o\left[\frac{\pi^2}{2}\left(\frac{k_B T}{\mu_o}\right)^2\right] \tag{10.33}$$

We can also obtain the grand potential energy or mechanical energy of the electron system as a function of temperature, volume, and the number of electrons,

$$\Xi(T, V, N) = -p(T, V/N)V = -\frac{2}{5}N\mu_o\left[1 + \frac{5\pi^2}{12}\left(\frac{k_B T}{\mu_o}\right)^2\right] \tag{10.34}$$

The Gibbs free energy or chemical energy of the electron system as a function of temperature, volume, and the number of electrons is simply given by

$$G(T, V, N) = N\mu(T, N/V) = N\mu_o\left[1 - \frac{\pi^2}{12}\left(\frac{k_B T}{\mu_o}\right)^2\right] \tag{10.35}$$

It is easy to show that the sum of thermal, mechanical, and chemical energy is the internal energy of the electron system.

The enthalpy of the electron system as a function of temperature and volume, and the number of electrons is

$$H(T, V, N) = U + pV = TS + N\mu = N\mu_o\left[1 + \frac{5\pi^2}{12}\left(\frac{k_B T}{\mu_o}\right)^2\right] \tag{10.36}$$

From the internal energy or entropy as a function of temperature, we can obtain the constant volume heat capacity of the electron system

$$C_V(T, V, N) = \left(\frac{\partial U}{\partial T}\right)_{V,N} = T\left(\frac{\partial S}{\partial T}\right)_{V,N} = \frac{\pi^2}{2}Nk_B^2\frac{T}{\mu_o} = \frac{\pi^2}{2}Nk_B\left(\frac{T}{T_f}\right)$$

(10.37)

where T_f is called the Fermi temperature defined as

$$T_f = \frac{\mu_o}{k_B}$$

(10.38)

The isothermal bulk modulus for the electron gas can be obtained from the pressure as a function of temperature and volume per electron as

$$B_T(T, V/N) = -V\left(\frac{\partial p}{\partial V}\right)_{T,N} = \frac{2}{3}\frac{N\mu_o}{V}\left[1 - \frac{\pi^2}{3}\left(\frac{k_B T}{\mu_o}\right)^2\right]$$

(10.39)

It is easy to show that the thermodynamic properties of electron system at 0 K are given by

$$U(0, V, N) = F(0, V, N) = \frac{3}{5}N\mu_o$$

(10.40)

$$H(0, V, N) = G(0, V, N) = N\mu_o$$

(10.41)

$$p(0, V/N) = \frac{2}{5}\frac{N}{V}\mu_o$$

(10.42)

$$S(0, V, N) = 0$$

(10.43)

$$C_V(0, V, N) = 0$$

(10.44)

$$B_T(0, V/N) = \frac{2}{3}\frac{N}{V}\mu_o$$

(10.45)

10.2.2   Electrons and Holes in Semiconductors and Insulators

For semiconductors and insulators, the valence electrons fill up the lower energy electron energy bands completely, and there is an energy band gap between the valence band entirely filled by electrons at 0 K, and another band above the valence band separated by a band gap is the conduction band with no electrons at 0 K. At a finite temperature, a small number of electrons are thermally excited from the valence band to the conduction band leaving electron holes in the valence band. For the same material, the number of electrons excited from the valence band to the conduction band increases with an increase in temperature. At the same temperature, the amount of conduction band electrons and valence band holes is higher for materials with a small band gap between the top of the valence band and the bottom of the conduction band. These electrons in the conduction band and the electron holes in the valence band are responsible for the electric conductivity of a semiconductor or an insulator, similar to lattice vacancies and interstitials that are responsible for the atomic diffusion in solids.

The fundamental equation of thermodynamics for the electrons in the conduction band in the entropic representation can be approximated as

$$S_e(U_e, V, N_e) = N_e k_B \left\{ \frac{5}{2} + \ln\left[\frac{V}{N_e} 2\left(\frac{4\pi m_e(U_e - N_e E_c)}{3h^2 N_e} \right)^{3/2} \right] \right\} \tag{10.46}$$

where S_e is the entropy of the electron system, N_e is the number of electrons in the conduction band, k_B is the Boltzmann constant, m_e is the effective mass of an electron, U_e is the internal energy of the electrons, E_c is the conduction band bottom energy, and h is the Planck constant.

The effective electron density of states N_c at the conduction band edge E_c is defined as

$$N_c = 2\left(\frac{2\pi m_e k_B T}{h^2} \right)^{3/2} \tag{10.47}$$

The entropy of the conduction band electrons can then be rewritten as

$$S_e = N_e k_B \left\{ \frac{5}{2} + \ln\left[\frac{V}{N_e} N_c \right] \right\} \tag{10.48}$$

We can write down the three equations of state for the conduction band electrons

$$\frac{1}{T} = \left(\frac{\partial S_e}{\partial U_e} \right)_{V,N} = \frac{3}{2} \frac{N_e k_B}{(U_e - N_e E_c)} \tag{10.49}$$

$$\frac{p}{T} = \left(\frac{\partial S_e}{\partial V} \right)_{U_e,N} = \frac{N_e k_B}{V} \tag{10.50}$$

$$-\frac{\mu_e}{T} = \left(\frac{\partial S_e}{\partial N_e} \right)_{U_e,V} = k_B \ln\left(\frac{V}{N_e} N_c \right) - \frac{3}{2} \frac{N_e k_B E_c}{(U_e - N_e E_c)} \tag{10.51}$$

where p is pressure, and μ_e is the chemical potential of electrons (or Gibbs free energy per electron).

From Eq. (10.49), it is easy to show that the internal energy per electron u_e of the electrons in the conduction band is given by

$$u_e = E_c + \frac{3}{2}k_B T \tag{10.52}$$

The enthalpy per electron h_e for the electrons in the conduction band is

$$h_e = u_e + pv_e = E_c + \frac{5}{2}k_B T \tag{10.53}$$

where v_e is the volume per electron given by V/N_e.

According to the equation of state (Eq. 10.50), $pV = N_e k_B T$, or $p = n_e k_B T$, i.e., the conduction band electrons follow the ideal gas law if their interactions are ignored.

From Eq. (10.51), the chemical potential of electrons μ_e as a function of temperature and pressure is given by

$$\mu_e = E_c - k_B T \ln\left(\frac{k_B T}{p}N_c\right) \tag{10.54}$$

We can rewrite the chemical potential in terms of temperature and electron density

$$\mu_e = E_c - k_B T \ln\left(\frac{V}{N_e}N_c\right) = E_c + k_B T \ln\left(\frac{n_e}{N_c}\right) \tag{10.55}$$

where n_e is the electron density given by N_e/V. We can regard E_c, the electron energy at the bottom of the conduction band, as the chemical potential at the standard state of electrons for a semiconductor or an insulator. The quantity n_e/N_c in Eq. (10.55) can be regarded as the fraction of the states in the conduction band which are occupied by electrons. The term $k_B \ln(n_e/N_c)(= -s$, where s is the entropy per electron) is the contribution of electron configuration entropy (mixing of occupied and unoccupied states among N_c conduction band density of states) to chemical potential. In semiconductor physics, the chemical potential of electrons is called the Fermi level E_f, i.e.,

$$E_f = \mu_e = E_c + k_B T \ln\left(\frac{n_e}{N_c}\right) \tag{10.56}$$

Similarly, for electron holes, the internal energy per electron hole in the valence band is

$$u_h = -E_v + \frac{3}{2}k_B T \tag{10.57}$$

where E_v is the valence band top edge energy. Please note that an electron hole has a charge opposite to that of an electron, and the energy band diagram plots electron energies. Therefore, the energies associated with the band diagram for holes are negative, and hence, the negative sign appears before the valence band edge energy E_v in the above equation.

The enthalpy per electron hole in the valence band is

$$h_h = u_h + p v_h = -E_v + \frac{5}{2}k_B T \tag{10.58}$$

where v_h is the volume occupied by each electron hole, i.e., $V/N_h = v_h$. The entropy per electron hole in the valence band, including both the thermal entropy and configurational entropy, is given by

$$s_h = k_B \left\{ \frac{5}{2} + \ln\left[\frac{2}{n_h} \left(\frac{2\pi m_h k_B T}{h^2} \right)^{3/2} \right] \right\} \tag{10.59}$$

where m_h is the effective mass of an electron hole. The effective electron density of states N_v at the valence band is given by

$$N_v = 2 \left(\frac{2\pi m_h k_B T}{h^2} \right)^{3/2} \tag{10.60}$$

The entropy per electron hole can be written as

$$s_h = k_B \left[\frac{5}{2} + \ln\left(\frac{N_v}{n_h} \right) \right] = k_B \left[\frac{5}{2} - \ln\left(\frac{n_h}{N_v} \right) \right] \tag{10.61}$$

The chemical potential for the electron holes in the valence band is then given by

$$\mu_h = h_h - T s_h = -E_v + k_B T \ln\left(\frac{n_h}{N_v} \right) \tag{10.62}$$

where $-E_v$ can be considered as the standard state chemical potential of holes. Therefore, the Fermi level for holes E_f can be written as

$$-E_f = \mu_h = -E_v + k_B T \ln\left(\frac{n_h}{N_v} \right) \tag{10.63}$$

or

Fig. 10.1 Excitation of
electron from the valence
band to the conduction band

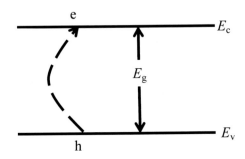

$$E_f = E_v - k_B T \ln\left(\frac{n_h}{N_v}\right) \tag{10.64}$$

The thermal creation of an electron–hole pair can be treated similar to chemical reactions,

$$\text{Null} = e' + h^{\cdot}$$

At equilibrium

$$\mu_e + \mu_h = 0$$

If we substitute the chemical potentials of electrons and holes using Eqs. (10.55) and (10.62), we have

$$n_e n_h = N_v N_c \exp\left(-\frac{E_g}{k_B T}\right) \tag{10.65}$$

where $E_g = E_c - E_v$ is the electron energy band gap (see Fig. 10.1).

10.2.3   Electrochemical Potential of Electrons

The electrochemical potential of electrons when electrons are subject to an electrical potential ϕ is

$$\tilde{\mu}_e = \mu_e - e\phi = E_c - e\phi + k_B T \ln\left(\frac{n_e}{N_c}\right) \tag{10.66}$$

The corresponding electrochemical potential of holes is

$$\tilde{\mu}_h = \mu_h + e\phi = -E_v + e\phi + k_B T \ln\left(\frac{n_h}{N_v}\right) \tag{10.67}$$

The electrochemical potentials of electrons and holes are related to the quasi-Fermi levels, E_{fe} and E_{fh}, when the electrons and holes are not at thermodynamic equilibrium,

$$\tilde{\mu}_e = E_{fe} = E_c - e\phi + k_B T \ln\left(\frac{n_e}{N_c}\right) \tag{10.68}$$

$$\tilde{\mu}_h = -E_{fh} = -E_v + e\phi + k_B T \ln\left(\frac{n_h}{N_v}\right) \tag{10.69}$$

Since quasi-Fermi levels for electrons can be considered as the electron electrochemical potential under nonequilibrium conditions, the gradient of E_{fn} is the driving force for electron transport.

At equilibrium, the chemical potential of the electron-hole pair is zero, i.e.,

$$\mu_{eh} = \tilde{\mu}_e + \tilde{\mu}_h = E_{fe} - E_{fh} = \mu_e + \mu_h = 0 \tag{10.70}$$

which indicates that the local electric potential does not affect the product of equilibrium n_e and n_h carrier concentrations, and the Fermi levels of electrons and holes are equal at equilibrium. However, it should be kept in mind that if the electron–hole system is not at equilibrium, the Fermi levels of electrons and holes are not the same, which is the case in photovoltaic devices. As a matter of fact, $\mu_{eh} = E_{fe} - E_{fh}$ is the available chemical energy of the electron–hole system created by light for photovoltaic devices.

At equilibrium, E_{fe} should be uniform. Let us write $E_{fe} = E_{fh} = E_f$, we have

$$n_e = N_c \exp\left(-\frac{E_c - E_f - e\phi}{k_B T}\right) \tag{10.71}$$

$$n_h = N_v \exp\left(-\frac{E_f - E_v + e\phi}{k_B T}\right) \tag{10.72}$$

For example, the electrons in a n-type silicon have a higher chemical potential than those in a p-type silicon. In a p-n junction diode at equilibrium, the chemical potential of electrons varies in space with a lower chemical potential in the p-type to the higher chemical potential in the n-type. However, the electrochemical potential of electrons, i.e., the Fermi level, is uniform within the entire p-n junction.

10.2.4 Chemical Potentials of Electrons in Trapped Electronic States of Dopants or Impurities

Impurity atoms or deliberate doping of a crystal with solute atoms create electronic energy states within the band gap of an insulator or semiconductor. Let us first discuss the relatively simple case of one electron trapping state d with the electron energy level of E_d located within a band gap and with the density of states N_d which is the number density of donors. $N_{d^\times}$ is the total number of neutral atoms or the number of electrons occupying the trapped states.

The excitation of an electron from a trapped state at E_d to a conduction band state at E_c can be expressed as

$$e'(\text{trapped state d}) \rightarrow e'(\text{conduction band}), E_c - E_d \qquad (10.73)$$

where $E_c - E_d$ is the ionization energy, which can be viewed as the standard chemical potential or free energy change for the above electron excitation reaction. When an electron is localized at a doped atom, or an electron occupies the trapped state created by the doped atoms, the doped atom is neutral or un-ionized. Therefore, the total number of electrons in the trapped state is the same as the total number of neutral dopant atoms at 0 K. When an electron is delocalized or excited from the trapped state to the conduction band at a finite temperature, an atom is ionized with an effective charge of $+e$. The total number of empty trapped states is the same as the number of ionized dopant atoms (because some of the electrons are excited to the conduction band). Therefore, the configurational entropy of electrons in the trapped state can be obtained from the number of configurations distributing $N_{d^\times}$ electrons among the N_d sites, i.e.,

$$S = k_B ln \frac{2^{N_{d^\times}}(N_d)!}{(N_{d^\times})!(N_{d^+})!} = k_B ln \frac{2^{N_{d^\times}}(N_d)!}{(N_{d^\times})![(N_d - N_{d^\times})]!} \qquad (10.74)$$

where k_B is Boltzmann constant, and N_d^+ is the total number of singly ionized donors with a positive charge or the number of electrons excited from the donor state to the conduction band. The degeneracy factor is 2 because of the spin-up or spin-down possibilities for electron occupancy. Applying the Stirling's approximation for large N, i.e., $\ln N! = N \ln N - N$,

$$S = k_B N_{d^\times} \ln 2 + k_B N_d \ln N_d - k_B N_{d^\times} \ln N_{d^\times}$$
$$- k_B (N_d - N_{d^\times}) \ln(N_d - N_{d^\times}) \qquad (10.75)$$

The entropy per electron at the trapped state is then

$$\frac{dS}{dN_{d^\times}} = k_B \ln 2 - k_B \ln N_{d^\times} + k_B \ln(N_d - N_{d^\times}) = k_B \ln \frac{2(N_d - N_{d^\times})}{N_{d^\times}} \qquad (10.76)$$

Therefore, the electron chemical potential is

$$\mu_e = E_d - TS = E_d + k_B T \ln\left(\frac{N_{d^\times}}{2(N_d - N_{d^\times})}\right) \tag{10.77}$$

where $N_d = N_{d^+} + N_{d^\times}$, and E_d and μ_e are the energy and chemical potential of electrons at the donor level d, respectively. E_d in the above equation can be viewed as the standard state chemical potential of electrons at the trapped states.

If the above reaction is at equilibrium, the chemical potential of electrons is the Fermi level E_f,

$$\mu_e = E_d - TS = E_d + k_B T \ln\left(\frac{N_{d^\times}}{2(N_d - N_{d^\times})}\right) = E_f \tag{10.78}$$

Solving for the above equation for the ratio of ionized dopants to the neutral dopants, we have

$$\frac{N_{d^+}}{N_{d^\times}} = \frac{1}{2}\exp\left(-\frac{E_f - E_d}{k_B T}\right) \tag{10.79}$$

Or solving for the number of ionized dopants in terms of the total dopant concentration

$$N_{d^+} = \frac{N_d}{1 + 2\exp\left(\frac{E_f - E_d}{k_B T}\right)} \tag{10.80}$$

Similar expressions can be derived for the case of electrons occupying acceptor states within a band gap.

10.3 Chemical Potentials of Einstein and Debye Crystals

One of the classical models to include the contribution of lattice vibration to the thermodynamics of a crystalline solid is the Einstein model which considers the lattice vibrations as $3N$ harmonic oscillators with the same frequency ω for a crystal of N atoms. In this model, the total energy of the system is the sum of static potential energy of the N number of atoms in the crystal $U(0\,\text{K}, V)$ and the energy of the $3N$ harmonic oscillators. It should be noted that the number of atoms in a crystal and the number of phonons should be distinguished. The number of physical particles N is fixed, but the average number of phonons depends on the temperature. Therefore, the number of phonons in the Einstein solid is not equal to N nor equal to $3N$.

The canonical partition function Z_C for an Einstein model of $3N$ vibrational modes with the same vibrational frequency, ω, is given by

$$Z_C = \left(\frac{e^{-\frac{\hbar\omega}{2k_BT}}}{1 - e^{-\frac{\hbar\omega}{k_BT}}} \right)^{3N} e^{-\frac{Nu(0,v)}{k_BT}} \tag{10.81}$$

where k_B is Boltzmann constant, N is the number of atoms in a crystal, and $u(0, v)$ is the static potential energy per atom of the crystal, and v is the volume per atom or simply called atomic volume. Therefore, the fundamental equation of thermodynamics of the Einstein crystal in terms of Helmholtz free energy is given by

$$F(T, V, N) = -k_BT \ln Z_C = -k_BT \ln \left[\left(\frac{e^{-\frac{\hbar\omega}{2k_BT}}}{1 - e^{-\frac{\hbar\omega}{k_BT}}} \right)^{3N} e^{-\frac{Nu(0,v)}{k_BT}} \right] \tag{10.82}$$

which can be rewritten as

$$F(T, V, N) = Nu(0, v) + 3Nk_BT \ln \left[2\sinh\left(\frac{\hbar\omega}{2k_BT} \right) \right] \tag{10.83}$$

The chemical potential of an Einstein crystal is given by

$$\mu(T, v) = \left(\frac{\partial F}{\partial N} \right)_{T,V} = u(0, v) + 3k_BT \ln \left[2\sinh\left(\frac{\hbar\omega}{2k_BT} \right) \right] \tag{10.84}$$

Please note this is the chemical potential of an Einstein crystal, whereas the chemical potential of non-conserved phonons is zero.

We can also easily obtain the entropy and internal energy of the Einstein crystal as a function of temperature, volume, and number of atoms,

$$S = -\left(\frac{\partial F}{\partial T} \right)_{V,N} = -3Nk_B \ln\left(1 - e^{-\frac{\hbar\omega}{k_BT}}\right) + \frac{3N\hbar\omega}{T} \frac{e^{-\frac{\hbar\omega}{k_BT}}}{1 - e^{-\frac{\hbar\omega}{k_BT}}} \tag{10.85}$$

$$U = F + TS = Nu(0, v) + \frac{3}{2}N\hbar\omega + \frac{3N\hbar\omega}{e^{\frac{\hbar\omega}{k_BT}} - 1} \tag{10.86}$$

Therefore, the constant volume heat capacity of an Einstein crystal is given by

$$C_V = \left(\frac{\partial U}{\partial T} \right)_{V,N} = T\left(\frac{\partial S}{\partial T} \right)_{V,N} = 3N \frac{(\hbar\omega)^2}{k_BT^2} \frac{e^{\frac{\hbar\omega}{k_BT}}}{\left(e^{\frac{\hbar\omega}{k_BT}} - 1 \right)^2} \tag{10.87}$$

At low temperatures, the heat capacity can be approximated as

$$C_V = \left(\frac{\partial U}{\partial T}\right)_{V,N} = 3Nk_B \left(\frac{\hbar\omega}{k_B T}\right)^2 e^{-\frac{\hbar\omega}{k_B T}} \tag{10.88}$$

At high temperature, the constant volume heat capacity is almost a constant given by

$$C_V = \left(\frac{\partial U}{\partial T}\right)_{V,N} = 3Nk_B \tag{10.89}$$

A more accurate description of lattice vibrations is given by the Debye model in which the density of the vibrational modes is assumed to have a parabolic dependence on frequency with an upper limit on the lattice vibration frequency called the Debye frequency determined by the $3N$ number of vibrational modes in a crystal of volume V, in contrast to the Einstein model in which there is only a single vibration frequency.

In the Debye model, it can be shown[2] that the Helmholtz free energy of a Debye crystal as a function of temperature, volume, and number of atoms is given by

$$F = Nu(0, v) - \int_0^{\omega_D} \left[\ln\left(\frac{e^{-\frac{\hbar\omega}{2k_B T}}}{1 - e^{-\frac{\hbar\omega}{k_B T}}}\right)\right]\frac{9N}{\omega_D^3}\omega^2 d\omega \tag{10.90}$$

where ω_D is the Debye frequency. Performing integration by parts, we can rewrite the above equation as

$$F = Nu(0, v) + \frac{9Nh\omega_D}{8} + 3Nk_B T \ln(1 - e^{-x})$$
$$- \frac{3Nh}{\omega_D^3}\left(\frac{k_B T}{h}\right)^4 \int_0^x \frac{x^3}{(e^x - 1)}dx \tag{10.91}$$

where

$$x = \frac{\hbar\omega}{k_B T} \tag{10.92}$$

At low temperatures,

$$\ln(1 - e^{-x}) \to 0, \quad \int_0^x \frac{x^3}{(e^x - 1)}dx = \frac{\pi^4}{15} \tag{10.93}$$

Therefore, the low temperature approximation for the Helmholtz free energy is given by

---

[2] Derivation of partition function and thus the fundamental equation of thermodynamics of a Debye crystal can be found most of the solid state physics textbooks.

$$F = Nu(0, v) + \frac{9Nh\omega_D}{8} - \frac{Nh}{5\omega_D^3}\left(\frac{\pi k_B T}{h}\right)^4 \tag{10.94}$$

From the Helmholtz free energy, we can obtain other thermodynamic properties. For example, the entropy of the Debye crystal at low temperature is given by

$$S = -\left(\frac{\partial F}{\partial T}\right)_{N,V} = \frac{4\pi^4 Nk_B}{5}\left(\frac{k_B T}{h\omega_D}\right)^3 \tag{10.95}$$

The corresponding low temperature constant volume heat capacity is then given by

$$C_V = T\left(\frac{\partial S}{\partial T}\right)_{N,V} = \frac{12\pi^4 Nk_B}{5}\left(\frac{k_B T}{h\omega_D}\right)^3 = \frac{12\pi^4 Nk_B}{5}\left(\frac{T}{T_D}\right)^3 \tag{10.96}$$

where T_D is the Debye temperature,

$$T_D = \frac{h\omega_D}{k_B} \tag{10.97}$$

At high temperatures,

$$F \approx Nu(0, v) + \frac{9Nh\omega_D}{8} + 3Nk_B T \ln\left(\frac{h\omega_D}{k_B T}\right) - Nk_B T \tag{10.98}$$

Therefore,

$$S = -\left(\frac{\partial F}{\partial T}\right)_{N,V} = -3Nk_B \ln\left(\frac{h\omega_D}{k_B T}\right) + 4Nk_B \tag{10.99}$$

The corresponding high temperature constant volume heat capacity is given by

$$C_V = T\left(\frac{\partial S}{\partial T}\right)_{N,V} = 3Nk_B \tag{10.100}$$

For one mole of atoms, i.e., the molar constant volume heat capacity at high temperatures is given by

$$c_v = 3N_o k_B = 3R \tag{10.101}$$

10.4 Chemical Potentials of Atomic Defects

A crystal at finite temperatures always contains a certain amount of point defects at thermodynamic equilibrium despite the fact that it costs energy for point defects to form. The driving force for the formation of defects at finite temperature is the increase in the entropy of a solid. Defects are critical in determining materials properties as they are mainly responsible for the movement of atoms in a crystal, additional electron energy states within band gaps, etc., and they can interact with other point defects, dislocations, and grain boundaries, as well as external stimuli. Therefore, it is important that we have a good understanding of the thermodynamics of atomic defects.

Following Maier's analogy[3] between atomic defects and electronic defects using the band diagram, atoms become movable only when defects are formed, and electrons become movable only when they go to the conduction band (so conduction band electrons are like defects) of a semiconductor or an insulator. One could think of conduction electrons as interstitial atoms, and electron holes as atomic vacancies. Even impurity effects are similar for ionic and electronic motions, i.e., the formation energy or energy states of impurities are analogous to the electronic states, e.g., band gap states, created due to the impurities. One can also define energy gap levels for ionic defects such as vacancies and interstitials. The chemical potentials and concentrations of atomic and electronic defects almost have a complete analogy to each other.

10.4.1 Chemical Potentials of Vacancies in Elemental Crystals

The most common type of defects in a crystal are atomic vacancies, i.e., lattice sites with atoms missing. To perform the thermodynamic analysis of vacancies, one can imagine that a vacancy is created when an atom moves from inside a single crystal to the surface of the crystal (Fig. 10.2). If we ignore the difference between an atom sitting on the surface of a crystal and an atom inside the crystal, the net result of this process is the creation of a vacancy inside the crystal while the total number of atoms remains the same. During this process, the total number of crystalline lattice sites is increased by one, and therefore, without considering atomic relaxation, the volume of the crystal is increased by one atomic volume.

Because the creation of vacancies breaks the atomic bonds, the energy of the crystal increases. We define the formation energy of a vacancy as the energy increase in the system due to the creation of the vacancy while keeping the total number of atoms n in the crystal the same,

---

[3] Joachim Maier, Physical Chemistry of Ionic Materials: Ions and Electrons in Solids, 2005, John Wiley & Sons, Ltd.

Fig. 10.2 Creation of a vacancy by moving an atom from inside to the crystal surface

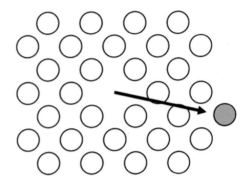

$$\Delta u_v = U_{n+v} - U_n \tag{10.102}$$

If the creation of a vacancy is carried out at constant pressure, which is usually the case, the formation enthalpy of a vacancy, Δh_v, is defined as the enthalpy change by removing an atom inside a crystal and placing it on the surface.

$$\Delta h_v = H_{n+v} - H_n = \Delta u_v + p\Delta v_v \tag{10.103}$$

where p is pressure, and Δv_v is the vacancy formation volume.

There is also an entropy change for the system associated with the creation of a vacancy. The formation entropy of a vacancy, ΔS, is the entropy difference between a system with n number of atoms and a vacancy and a perfect crystal with n atoms. It consists of two contributions: vibrational entropy Δs_v of formation of a vacancy and the configuration entropy ΔS_c of a crystal consisting of atoms and vacancies. The vibrational entropy of a vacancy can be estimated as

$$\Delta s_v = -k_B \sum_i \ln \frac{\omega_i'}{\omega_i^0} \tag{10.104}$$

where k_B is the Boltzmann constant, ω_i' are the vibrational frequencies of atoms in a crystal with a vacancy, ω_i^0 are the vibrational frequencies of atoms in a perfect crystal.

To calculate ΔS_c, consider a crystal with $n + n_v$ lattice sites and n_v vacancies,

$$\Delta S_c = k_B \ln \Omega = k_B \ln \frac{(n + n_v)!}{n! n_v!} \tag{10.105}$$

where Ω is the number of possible configurations with $n + n_v$ total lattice sites that are occupied by n atoms and n_v vacancies.

The total free energy of a crystal with n atoms and n_v vacancies is then

$$G = G^\circ + n_v \Delta h_v - T\left(n_v \Delta s_v + k_B \ln \frac{(n+n_v)!}{n! n_v!}\right) \tag{10.106}$$

where G° is the free energy of a perfect crystal. Using the Stirling approximation, we have

$$G = G^\circ + n_v(\Delta h_v - T \Delta s_v)$$
$$- k_B T[(n+n_v)\ln(n+n_v) - n \ln n - n_v \ln n_v]$$

or

$$G = G^\circ + n_v \Delta g_v - k_B T[(n+n_v)\ln(n+n_v) - n \ln n - n_v \ln n_v]$$

where $\Delta g_v = \Delta h_v - T \Delta s_v$.

The chemical potential of vacancies is then

$$\mu_v = \left(\frac{\partial G}{\partial n_v}\right)_{T,p,n}$$

$$\mu_v = \Delta g_v + k_B T \ln\left(\frac{n_v}{n+n_v}\right)$$

$$\mu_v = \Delta g_v + k_B T \ln x_v$$

At equilibrium,

$$\mu_v = \Delta g_v + k_B T \ln\left(\frac{n_v^\circ}{n+n_v^\circ}\right) = 0$$

Solving for the above equation for the number of vacancies, we have

$$\frac{n_v^\circ}{n+n_v^\circ} = x_v^\circ = \exp\left(-\frac{\Delta g_v}{k_B T}\right) \tag{10.107}$$

If we identify Δg_v as the chemical potential of vacancies at their standard state, i.e., $\mu_v^\circ = \Delta g_v$, at equilibrium,

$$\mu_v = \mu_v^\circ + k_B T \ln x_v^\circ = 0 \tag{10.108}$$

Under non-equilibrium conditions, e.g., a crystal under irradiation (Fig. 10.3),

$$\mu_v = \mu_v^\circ + k_B T \ln x_v = k_B T \ln\left(\frac{x_v}{x_v^\circ}\right) \tag{10.109}$$

Fig. 10.3 Chemical potential of vacancy as a function of vacancy concentration

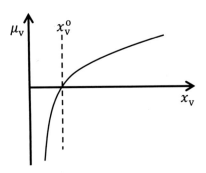

Another way to look at the chemical potential of vacancies is to consider vacancy creation as a reaction,

$$A_A \rightleftharpoons A_s + V_A \qquad (10.110)$$

where A_A represents an A atom occupying an A lattice site inside a crystal, A_s represents an A atom occupying a surface site, and V_A represents a vacant site on the A lattice. If we ignore the difference between A_A and A_s, we have

$$\text{Null} \rightleftharpoons V_A \qquad (10.111)$$

Therefore, at equilibrium,

$$\mu_{\text{Null}} = 0 = \mu_v \qquad (10.112)$$

Assuming the ideal entropy of mixing between the vacancies and the atoms, we have

$$\mu_v = \mu_v^o + k_B T \ln x_v \qquad (10.113)$$

In the above discussion, the creation of a vacancy leads to an addition of a lattice site to the crystal while the total number of atoms is fixed.

If the total number of lattice sites is conserved during the creation of a vacancy, then the reaction of vacancy creation is expressed as

$$A_A \rightleftharpoons V_A + A_{\text{res}} \qquad (10.114)$$

or

$$\text{Null} \rightleftharpoons (V_A - A_A) + A_{\text{res}} \qquad (10.115)$$

where A_{res} represents an A atom in a pure A atom reservoir, e.g., a perfect A crystal at the same temperature and pressure with the standard chemical potential of μ_A^o. In the notation by Schottky, $(V_A - A_A)$ is called a building element. A building element does not change the number of lattice sites, $(V_A - A_A)$ means removing an A atom from the lattice site and replacing it with a vacancy, $+A_{res}$ means placing the removed A atom into the A atom reservoir. At equilibrium,

$$\mu_{(V_A - A_A)} + \mu_A^o = 0 \tag{10.116}$$

If the total number of lattice sites N is fixed,

$$\mu_{(V_A - A_A)} = \mu_{(V_A - A_A)}^o + k_B T \ln \frac{n_v}{N - n_v} \tag{10.117}$$

It should be pointed out that μ_v^o and $\mu_{(V_A - A_A)}^o$ are different: μ_v^o is the formation free energy of a vacancy, defined as the difference between the free energy (including the vibrational entropy contribution) of a crystal with n atoms plus one vacancy and the energy of a perfect crystal with n atoms, while $\mu_{(V_A - A_A)}^o$ is the difference between the energy of a crystal with $n - 1$ atoms and a vacancy and the energy of a perfect crystal with n atoms.

Solving Eq. (10.117) for n_v, we have

$$n_v = \frac{N}{1 + \exp\left(\frac{\mu_{(V_A - A_A)}^o - \mu_{(V_A - A_A)}}{k_B T}\right)} \tag{10.118}$$

which is the same as the Fermi–Dirac distribution for electron holes: the chemical potential of vacancies $\mu_{(V_A - A_A)}$ corresponds to the chemical potential of electron holes which is the negative of Fermi level of electrons E_f, and the vacancy formation free energy $\mu_{(V_A - A_A)}^o$ corresponds to the negative of valence band edge energy E_v.

At equilibrium,

$$\mu_{(V_A - A_A)}^o + k_B T \ln \frac{n_v}{N - n_v} + \mu_A^o = 0 \tag{10.119}$$

If the total number of lattice sites N is fixed,

$$n_v^o = \frac{N}{1 + \exp\left(\frac{\mu_{(V_A - A_A)}^o + \mu_A^o}{k_B T}\right)} \tag{10.120}$$

For very small vacancy concentrations,

$$x_v^o = \frac{n_v^o}{N} = \exp\left(-\frac{\mu_{(V_A - A_A)}^o + \mu_A^o}{k_B T}\right) \tag{10.121}$$

Therefore, in the dilute approximation,

$$\mu_v^o = \Delta g_v = \mu_{(V_A - A_A)}^o + \mu_A^o \tag{10.122}$$

or

$$\mu_{(V_A - A_A)}^o = \mu_v^o - \mu_A^o \tag{10.123}$$

10.4.2 Chemical Potentials of Interstitials

The chemical potential of interstitials can be obtained similar to that of vacancies. Consider the reaction of a surface atom A_s goes to an interstitial site inside a crystal to become A_i,

$$A_s = A_i \tag{10.124}$$

With this reaction, the number of lattice sites is reduced by one. The chemical potential of an atom on the surface is assumed to be the same as that in the bulk crystal, i.e.,

$$\mu_{A_s} = \mu_A^o \tag{10.125}$$

Here the standard chemical potential is the formation chemical potential of an interstitial that is the sum of formation energy (or enthalpy) and the vibrational entropy contribution.

At equilibrium, $\mu_{A_s} = \mu_{A_i}$ or $\mu_{A_i} - \mu_{A_s} = 0$, and thus

$$\mu_{A_i} = \mu_{A_i}^o + k_B T \ln\left(\frac{n_i^o}{N - n_i^o}\right) \tag{10.126}$$

$$\mu_{A_s} = \mu_A^o = \mu_{A_i} \tag{10.127}$$

where n_i^o is the equilibrium number of interstitials, and N is the total number of interstitial sites.

Therefore,

$$-\left(\mu_{A_i}^o - \mu_A^o\right) = -\Delta\mu_{A_i}^o = k_B T \ln\left(\frac{n_i^o}{N - n_i^o}\right) \approx k_B T \ln\left(\frac{n_i^o}{N}\right) = k_B T \ln x_i^o$$

$$x_i^o = \exp\left(-\frac{\Delta\mu_{A_i}^o}{k_B T}\right) \tag{10.128}$$

where $\Delta\mu^o_{A_i}$ is the standard chemical potential difference between an interstitial atom and a lattice atom, or usually referred to as the formation energy or enthalpy of an interstitial.

10.4.3 Chemical Potential of Frenkel Defects

A Frenkel defect is referred to a pair of defects, one is a lattice vacancy, and the other is an interstitial atom. The creation of a Frenkel defect involves moving a lattice atom A from a regular lattice site to an interstitial site, and at the same time, a lattice vacancy is created, i.e.,

$$A_A + V_i = A_i + V_A \quad \text{or} \quad \text{Null} = (A_i - V_i) + (V_A - A_A) \qquad (10.129)$$

At equilibrium

$$\mu^o_A + \mu_{V_i} = \mu_{A_i} + \mu_{V_A} = \mu_F \quad \text{or} \quad 0 = \mu_{(A_i-V_i)} + \mu_{(V_A-A_A)}$$

where μ_F is the chemical potential of a Frenkel defect.

The individual chemical potentials in the ideal solution approximation can be expressed as

$$\mu_{A_i} = \mu^o_{A_i} + k_B T \ln\left(\frac{n_i}{N_i}\right)$$

$$\mu_{V_A} = \mu^o_{V_A} + k_B T \ln\left(\frac{n_v}{N_A}\right)$$

$$\mu_A = \mu^o_A + k_B T \ln\left(\frac{N_A - n_v}{N_A}\right)$$

$$\mu_{V_i} = k_B T \ln\left(\frac{N_i - n_i}{N_i}\right)$$

where n_i is the number of interstitial atoms, n_v is the number of vacancies, N_i is the number of interstitial sites, and N_A is the total number of lattice sites. The formation energy of a Frenkel defect is then given by,

$$\Delta\mu^o_F = \mu^o_{A_i} + \mu^o_v - \mu^o_A = \Delta\mu^o_{A_i} + \mu^o_v$$

where $\left(\mu^o_{A_i} - \mu^o_A\right) = \Delta\mu^o_{A_i}$ is the formation energy of an interstitial (Fig. 10.4).

In the dilute approximation,

Fig. 10.4 Energy diagram illustrating the creation of a Frenkel defect

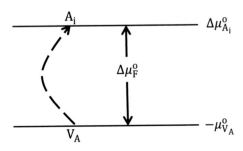

$$\mu_{A_i} = \mu^o_{A_i} + k_B T \ln(x_i)$$

$$\mu_{V_A} = \mu^o_{V_A} + k_B T \ln(x_v)$$

$$\mu_F = \mu^o_F + k_B T \ln(x_i x_v)$$

where x_i is the site fraction of interstitials, and x_v is the site fraction of vacancies,

$$\mu_F = \mu^o_A = \mu^o_F + k_B T \ln(x_i x_v)$$

or

$$k_B T \ln(x_i x_v) = -\left(\mu^o_F - \mu^o_A\right) = -\Delta\mu^o_F$$

$$x_i x_v = \exp\left(-\frac{\Delta\mu^o_F}{k_B T}\right) \tag{10.130}$$

If $x_i = x_v$, the equilibrium interstitial and vacancy concentrations are given by

$$x_i = x_v = \exp\left(-\frac{\Delta\mu^o_F}{2k_B T}\right) \tag{10.131}$$

10.4.4   Chemical Potential of Schottky Defects

The creation of a Schottky defect can be expressed as

$$A^{AB}_A + B^{AB}_B \rightleftharpoons V^{AB}_A + V^{AB}_B + AB$$

where A^{AB}_A represents an A atom occupying on the A sublatice of compound AB, B^{AB}_B represents a B atom occupying on the B sublatice of compound AB, V^{AB}_A is a vacancy on the A sublattice of compound AB, and V_B^{AB} is a vacancy on the B sublattice of compound AB.

At equilibrium,

$$\mu_A^{AB} + \mu_B^{AB} = \mu_{V_A^{AB}} + \mu_{V_B^{AB}} + \mu_{AB}^o$$

In the dilute solution approximation,

$$\mu_A^{AB} = \mu_A^{AB,o} + k_B T \ln\left(\frac{N_A - n_{V_A}^{AB}}{N_A}\right)$$

$$\mu_B^{AB} = \mu_B^{AB,o} + k_B T \ln\left(\frac{N_B - n_{V_B}^{AB}}{N_B}\right)$$

$$\mu_{V_A^{AB}} = \mu_{V_A^{AB}}^o + k_B T \ln\left(\frac{n_{V_A^{AB}}}{N_A}\right) = \mu_{V_A^{AB}}^o + k_B T \ln\left(x_{V_A^{AB}}\right)$$

$$\mu_{V_B^{AB}} = \mu_{V_B^{AB}}^o + k_B T \ln\left(\frac{n_{V_B^{AB}}}{N_B}\right) = \mu_{V_B^{AB}}^o + k_B T \ln\left(x_{V_B^{AB}}\right)$$

Since the chemical potential of a compound is the sum of the chemical potentials for each component, i.e.,

$$\mu_A^{AB,o} + \mu_B^{AB,o} = \mu_{AB}^o,$$

we have

$$x_{V_A^{AB}} x_{V_B^{AB}} = \exp\left[-\frac{\left(\mu_{V_A^{AB}}^o + \mu_{V_B^{AB}}^o\right)}{k_B T}\right] = \exp\left[-\frac{\mu_S^o}{k_B T}\right] \qquad (10.132)$$

where μ_S^o is the chemical potential of a Schottky defect.
If $x_{V_A^{AB}} = x_{V_B^{AB}}$,

$$x_{V_A^{AB}} = x_{V_B^{AB}} = \exp\left[-\frac{\mu_S^o}{2k_B T}\right] \qquad (10.133)$$

In Schottky notation using building elements,

$$0 \rightleftharpoons \left(V_A^{AB} - A_A^{AB}\right) + \left(V_B^{AB} - B_B^{AB}\right) + (AB)_{res}$$

where $(AB)_{res}$ represents a molecule of AB compound from a chemical reservoir of compound AB. Dopant-controlled defect concentration and chemical potentials involving chemical Schottky and Frenkel defects in ionic crystals are similar to the trapped states of electrons in semiconductors.

10.4.5   Chemical Potentials of Neutral Dopants

The chemical potential of neutral dopants, $\mu_{d^\times}$, can be expressed as

$$\mu_{d^\times} = \mu_d^o + \Delta\mu_d^o + k_B T \ln\left(\frac{N_{d^\times}}{N_L}\right) \tag{10.134}$$

where μ_d^o is chemical potential of pure dopant d at temperature T and ambient pressure, and N_L is the total number of host lattice sites. The formation energy $\Delta\mu_d^o$ of dopant d in a host includes the formation energy, formation entropy from the lattice vibration contribution, and formation volume (which can be neglected under ambient pressure). It should be noted that $\Delta\mu_d^o$ does not include the configurational entropy contribution. We can rewrite Eq. (10.134) as

$$\mu_{d^\times} - \mu_d^o = \Delta\mu_d^o + k_B T \ln x_{d^\times} \tag{10.135}$$

where $x_{d^\times} = N_{d^\times}/N_L$. The solubility limit $x_{d^\times}^o$ of d in host is given by

$$0 = \Delta\mu_d^o + k_B T \ln x_{d^\times}^o \tag{10.136}$$

If the composition of d in host is below the solubility limit, $\mu_{d^\times} - \mu_d^o = \Delta\mu_d^o + k_B T \ln x_{d^\times} < 0$.

10.5   Examples

Example 1 Estimate the amount of oxygen at a given height z, e.g., $10,000$ m, relative to the ground level at $z = 0$. Assume constant temperature 300 K and use the mass of 32 g/mol for O_2.

Solution
The ratio of the oxygen level at $10,000$ m in the sky to that at the ground level, assuming the temperature is uniform at room temperature, is

$$\frac{\rho(z)}{\rho(0)} = e^{-\frac{mgz}{k_B T}} = e^{-\frac{32\times 10^{-3}\times 9.81\times 10000}{8.314\times 300}} = 0.284$$

Example 2 What is the electrical potential drop across a semiconductor p-n junction with the n-type of semiconductor doped with N_d number of electron donors per unit volume, and the p-type of semiconductor is doped with N_a number of electron acceptors per unit volume? One can assume that the number of conduction band electrons in the n-type semiconductor is approximately equal to the number of donors, and the number of valence band electron holes in the p-type semiconductor is approximately equal to the number of electron acceptors in the p-type semiconductor.

Solution
For an n-type semiconductor, the chemical potential μ_e of electrons can be approximated as

$$\mu_e = E_c + k_B T \ln\left(\frac{n}{N_c}\right) \approx E_c + k_B T \ln\left(\frac{N_d}{N_c}\right)$$

For a p-type semiconductor, the chemical potential μ_h of holes can be approximated as

$$\mu_h = -E_v + k_B T \ln\left(\frac{p}{N_v}\right) \approx -E_v + k_B T \ln\left(\frac{N_a}{N_v}\right)$$

The electrons in the n-type of semiconductor have a higher chemical potential than those in the p-type semiconductor, and thus, when an n-type semiconductor and a p-type of semiconductor are brought into contact, electrons will move from the n-type semiconductor to the p-type semiconductor. As a result, the side of the n-type semiconductor at the junction will be positively charged due to the loss of electrons whereas the side of the p-type semiconductor will be negatively charged due to the gain of electrons, creating an electric potential difference and thus electric field across the p-n junction. At equilibrium, the electrochemical potential of electrons should be uniform throughout the entire crystal containing the p-n junction, and there is no longer any net flow of electrons within the entire system. Let us assume that the electric potential in the n-type semiconductor side is ϕ^n, and that in the p-type semiconductor side is ϕ^p. The electrochemical potentials of electrons and holes in the p-n junction are given by

$$\tilde{\mu}_e = E_c - e\phi^n + k_B T \ln\left(\frac{N_d}{N_c}\right)$$

and

$$\tilde{\mu}_h = -E_v + e\phi^p + k_B T \ln\left(\frac{N_a}{N_v}\right)$$

At equilibrium

$$0 = \tilde{\mu}_e + \tilde{\mu}_h$$

Therefore,

$$0 = E_c - e\phi^n + k_B T \ln\left(\frac{N_d}{N_c}\right) - E_v + e\phi^p + k_B T \ln\left(\frac{N_a}{N_v}\right)$$

Solving for the electrical potential difference from the above equation, we have

$$\Delta\phi = \phi^n - \phi^p = \frac{E_g}{e} + \frac{k_B T}{e} \ln\left(\frac{N_d N_a}{N_c N_v}\right) = \frac{k_B T}{e} \ln\left(\frac{N_d N_a}{n_i^2}\right)$$

where E_g is the band gap, and n_i is the intrinsic concentration of electrons and holes without any dopants given by

$$n_i^2 = N_c N_v \exp\left(-\frac{E_g}{k_B T}\right)$$

Example 3 The chemical potential of a Debye crystal A at low temperatures can be approximated by.

$$\mu_A^{DC} = u(0, v) + \frac{9h\omega_D}{8} - \frac{h}{5\omega_D^3}\left(\frac{\pi k_B T}{h}\right)^4$$

where $u(0, v)$ is the static energy per A atom at $0\,K$, v is molar volume, h is the Planck constant, k_B is the Boltzmann constant, and ω_D is the Debye frequency. It is also shown that the chemical potential of an ideal gas as a function of temperature T and pressure p is given by

$$\mu_A^g = -k_B T \ln\left[\frac{(2\pi m)^{3/2}(k_B T)^{5/2}}{ph^3}\right]$$

where m is atomic mass. Find the vapor pressure which is at equilibrium with the Debye crystal.

Solution

At equilibrium, $\mu_A^{DC} = \mu_A^g$

$$u(0, v) + \frac{9h\omega_D}{8} - \frac{h}{5\omega_D^3}\left(\frac{\pi k_B T}{h}\right)^4 = -k_B T \ln\left[\frac{(2\pi m)^{3/2}(k_B T)^{5/2}}{ph^3}\right]$$

which can be rewritten as

$$\frac{u(0, v)}{k_B T} + \frac{9h\omega_D}{8k_B T} - \frac{\pi^4}{5}\left(\frac{k_B T}{h\omega_D}\right)^3 = -\ln\left[\frac{(2\pi m)^{3/2}(k_B T)^{5/2}}{ph^3}\right]$$

Solving the above equation for pressure, we have the equilibrium vapor pressure above a Debye crystal,

$$p = \frac{(2\pi m)^{3/2}(k_B T)^{5/2}}{h^3}\exp\left[\frac{u(0, v)}{k_B T} + \frac{9h\omega_D}{8k_B T} - \frac{\pi^4}{5}\left(\frac{k_B T}{h\omega_D}\right)^3\right]$$

10.6 Exercises

1. What is the value of the chemical potential of a vacancy at equilibrium in a pure elemental metal?

2. What is the relationship between the chemical potential of electrons and holes at equilibrium in a semiconductor?

3. Approximately what is the internal energy of electrons in the conduction band of a semiconductor or an insulator?

4. What is the standard chemical potential of electrons in the conduction band of a semiconductor or an insulator ?

5. What is the standard chemical potential of electrons in the valence band of a semiconductor or an insulator?

6. What is the approximate internal energy of electron holes in the valence band of a semiconductor or an insulator?

7. What is the standard chemical potential of electron holes in the valence band of a semiconductor or an insulator?

8. What is the relationship between the chemical potential of electrons in the conduction band and the chemical potential of electron holes in the valence band at equilibrium in a semiconductor?

9. If the equilibrium vacancy mole fraction in a single crystal near its melting temperature is 10^{-5}, estimate the standard chemical potential of vacancies, or often called the formation free energy of a vacancy.

10. In a pure semiconductor, the concentrations of electrons and holes are the same. Doping a semiconductor with electronic donors increases the electron concentration and decrease the hole concentration.

    a. Does the entropy of electrons increase or decrease with donor doping?

b.   Does the entropy of holes increase or decrease with donor doping?

11.   In a donor-doped semiconductor, what is the relationship between the chemical potential of electrons at the conduction band and the chemical potential of electrons occupying the donor states within the band gap?

12.   The chemical potential of electrons is given by,

$$\mu_e = E_c + k_B T \ln\left(\frac{n_e}{N_c}\right)$$

Assuming E_c is a constant independent of temperature, please.

(a)   Show that the expression for entropy per electron from chemical potential of electrons is

$$s_e = -\left(\frac{\partial \mu_e}{\partial T}\right)_p = k_B\left[\frac{5}{2} - \ln\left(\frac{n_e}{N_c}\right)\right]$$

where p is pressure.

(b)   Show that the constant pressure heat capacity per electron is $(5/2)k_B$, i.e.,

$$c_{p,e} = -T\left(\frac{\partial^2 \mu_e}{\partial T^2}\right)_p = \frac{5}{2}k_B$$

13.   At room temperature, the equilibrium mole fraction of vacancies in a crystal is 10^{-12}. Now the crystal is subjected to irradiation which creates a great number of vacancies leading to a vacancy mole fraction of 0.01. What is the chemical potential of the vacancies in the irradiated crystal?

14.   Consider the equilibrium between an Einstein crystal A and its vapor. The chemical potential of an ideal monatomic gas A is

$$\mu_A^g = -k_B T \ln\left[\frac{(2\pi m)^{3/2}(k_B T)^{5/2}}{ph^3}\right]$$

The chemical potential of A in an Einstein crystal at low temperatures can be approximated as

$$\mu_A^{EC} = u(0, v) + 3k_B T \ln\left[2\sinh\left(\frac{\hbar\omega}{2k_B T}\right)\right]$$

Show that the equilibrium vapor pressure above an Einstein crystal is given by

$$p = \left[k_B T \frac{(2\pi m k_B T)^{3/2}}{h^3}\right]\left[2\sinh\left(\frac{\hbar\omega}{2k_B T}\right)\right]^3 e^{\frac{u(0, v)}{k_B T}}$$

Chapter 11
Phase Equilibria of Single-Component Materials

A phase in thermodynamics is defined as a substance that exhibits a certain type of atomic and electronic structural state, e.g., gas, liquid, amorphous solid, crystalline solid with a specific lattice symmetry and electronic structure, over certain ranges of chemical composition, temperature, pressure (stress), electric field, and magnetic field. Therefore, when we refer to a phase, we have to specify its chemical composition as well as its physical state: gas, liquid, or solid with a specific crystal structure, lattice symmetry, and electronic state (metallic, semiconducting, insulating, superconducting, magnetic, etc.).

Most materials in practical applications contain more than one phase and are thus heterogeneous. The chemical composition for the whole system is the overall composition, and the composition of a phase is called the phase composition. A phase boundary in thermodynamics can mean either a physical boundary describing an interface separating two phases in space in a multiphase mixture or a boundary describing the temperatures, pressures, or compositions along which two phases are at thermodynamic equilibrium, i.e., a phase boundary on a phase diagram.

The thermodynamics of a material or a phase in a material is completely described by its fundamental equation of thermodynamics or by its chemical potential as a function of temperature and pressure for a material which can be considered as a simple system. The stability of a phase and the relative stability of different phases under a particular thermodynamic condition can then be determined from the fundamental equations of thermodynamics for each phase.

An originally stable phase may become unstable due to the changes in thermodynamic conditions and thus transform to a thermodynamically more stable phase under the new thermodynamic condition. The transition or transformation or change from one phase to another is called a phase transition, a phase transformation, or a phase change. Examples include the transitions from one physical state to another at a fixed chemical composition, e.g., melting, solidification, vaporization, etc., from one crystal structure to another, e.g., cubic to tetragonal transition, crystallization of an amorphous phase, etc., from one electronic state to another, e.g., superconducting

© Springer Nature Singapore Pte Ltd. 2022
L.-Q. Chen, *Thermodynamic Equilibrium and Stability of Materials*,
https://doi.org/10.1007/978-981-13-8691-6_11

transition, metal to insulator transition, magnetic transition, etc. Another type of transitions involves a spatial rearrangement of chemical species such as compositional phase separation or atomic rearrangement of atomic species on a fixed crystalline lattice such as order–disorder transition.

The graphical representations of stable equilibrium phases under different thermodynamic conditions are called phase diagrams. There are several options to represent a phase diagram for a single-component simple system, including stable equilibrium states represented on a temperature versus molar volume diagram, a pressure versus molar volume diagram, a temperature versus molar entropy diagram, a pressure versus molar entropy, and a pressure versus temperature diagram. This chapter will mainly discuss the stability of a material with respect to temperature and pressure at a fixed chemical composition and thus focus on the pressure–temperature diagrams or simply called p–T diagrams which represent the stable equilibrium phases at different temperatures and pressures.

11.1   General Temperature and Pressure Dependencies of Chemical Potential

From the discussions in previous chapters, we understand that the Gibbs free energy per mole or molar Gibbs free energy of a species or a group of species is the chemical potential of the species or the group of species. Therefore, for a homogeneous single-component system or a homogeneous multicomponent system with a fixed composition, its thermodynamic properties are most naturally and conveniently defined using the chemical potential for essentially all practical purposes. The change in chemical potential $d\mu$ of a material with respect to temperature change dT and pressure change dp is given by the differential form of the fundamental equation of thermodynamics,

$$d\mu = -sdT + vdp \tag{11.1}$$

Since the three intensive variables, T, p, μ, are related by the above relation, only two of them can be independently varied at the same time for a homogeneous system. For example, once the temperature and pressure are fixed, the chemical potential of a homogeneous system at equilibrium is fixed. Therefore, Eq. (11.1) is essentially the Gibbs–Duhem relation that relates all intensive potential variables of a homogeneous thermodynamic system. Gibbs–Duhem relation for binary and multicomponent systems will be further discussed in Chap. 12 by including the chemical potentials for each component.

Let us discuss the temperature and pressure dependences of chemical potential by examining its first and second derivatives with respect to temperature and pressure. From Eq. (11.1), one can immediately see that the chemical potential μ of a homogeneous system decreases with its temperature T (Fig. 11.1) since the rate of change of chemical potential with temperature at constant pressure is the negative

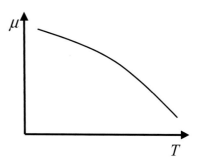

Fig. 11.1 Schematic dependence of chemical potential on temperature at constant pressure

of molar entropy, s, of the homogeneous system, i.e.,

$$\left(\frac{\partial \mu}{\partial T}\right)_p = -s \tag{11.2}$$

The curvature of the curve describing the chemical potential as a function of temperature can be determined by examining the second derivative of chemical potential with respect to temperature at constant pressure,

$$\left(\frac{\partial^2 \mu}{\partial T^2}\right)_p = -\frac{c_p}{T} \tag{11.3}$$

Since the heat capacity c_p of a stable material is always positive, and T is always positive, the second derivative of chemical potential with respect to temperature at constant pressure is always negative. Therefore, at constant pressure, the curvature of the curve describing the temperature dependence of chemical potential is always negative, with the curvature given by $-c_p/T$.

At constant temperature, the chemical potential μ of a homogeneous system increases with its pressure p (Fig. 11.2) with the slope given by the molar volume, v, the first derivative of chemical potential with respect to pressure at constant temperature,

Fig. 11.2 Schematic dependence of chemical potential on pressure at constant temperature

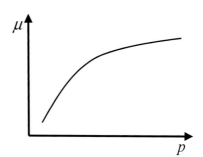

$$\left(\frac{\partial \mu}{\partial p}\right)_T = v \tag{11.4}$$

The curvature of the curve describing the pressure dependence of chemical potential at constant temperature is determined by its second derivative,

$$\left(\frac{\partial^2 \mu}{\partial p^2}\right)_T = -v\beta_T \tag{11.5}$$

where β_T is the isothermal compressibility which is always positive for a stable material. Therefore, the curvature of chemical potential versus pressure is always negative (Fig. 11.2).

It is reminded here that the mixed second derivative of chemical potential with respect to pressure and temperature is related to the volume thermal expansion coefficient α,

$$\left(\frac{\partial^2 \mu}{\partial T \partial p}\right) = \alpha v \tag{11.6}$$

which can be positive or negative although most materials exhibit a positive thermal expansion coefficient.

11.2   Thermodynamic Driving Force for Phase Transition

To determine whether a phase transition is thermodynamically possible, one should first determine the thermodynamic driving force for the transition as discussed in Chap. 8. Let us consider a transition of alpha (α) phase to beta (β) phase without chemical composition change,

$$\alpha \rightarrow \beta \tag{11.7}$$

The chemical potentials of α and β as a function of temperature and pressure are denoted as μ^α and μ^β which describe the thermodynamics of the α and β phases, respectively. The total amount of substance is N, which is assumed to be constant. Let us assume that in a two-phase mixture of α and β phases, the number of moles of α phase is N^α, and the number of moles of β phase is N^β, and thus $N = N^\alpha + N^\beta$. The Gibbs free energy of such a two-phase mixture is

$$G = G^\alpha + G^\beta = N^\alpha \mu^\alpha + N^\beta \mu^\beta \tag{11.8}$$

where G^α and G^β are the Gibbs free energies of α and β phases representing the fundamental equations of thermodynamics for α and β phases, respectively. The differential forms for G^α and G^β can be written as

$$dG^\alpha = -s^\alpha N^\alpha dT + v^\alpha N^\alpha dp + \mu^\alpha dN^\alpha \tag{11.9}$$

$$dG^\beta = -s^\beta N^\beta dT + v^\beta N^\beta dp + \mu^\beta dN^\beta \tag{11.10}$$

where s^α and s^β are molar entropies of α and β phases, and v^α and v^β are molar volumes of α and β phases.

In Eq. (11.8), G can be considered as the fundamental equation for the heterogeneous system containing a mixture of α and β phases ignoring the contribution from the interfacial energy. The differential form of G is

$$dG = -\left(s^\alpha N^\alpha + s^\beta N^\beta\right)dT + \left(v^\alpha N^\alpha + v^\beta N^\beta\right)dp + \mu^\alpha dN^\alpha + \mu^\beta dN^\beta \tag{11.11}$$

At constant temperature and pressure,

$$dG = \mu^\alpha dN^\alpha + \mu^\beta dN^\beta \tag{11.12}$$

We define a parameter called the degree of transition, ξ, from α phase to β phase

$$\xi = \frac{N^\beta}{N} \tag{11.13}$$

and

$$1 - \xi = \frac{N^\alpha}{N} \tag{11.14}$$

i.e., ξ represents the mole fraction (or volume fraction if the two phases have the same molar volume) of β. We can consider ξ as a phase order parameter of the system representing an internal degree of freedom. We can now express the fundamental equation of thermodynamics for the whole heterogeneous system taking into account the possibility of a phase transition by including the phase order parameter ξ, i.e.,

$$G(N, T, p, \xi) = N(1 - \xi)\mu^\alpha(T, p) + N\xi\mu^\beta(T, p) \tag{11.15}$$

Or

$$G(N, T, p, \xi) = N\mu^\alpha(T, p) + N\xi\left[\mu^\beta(T, p) - \mu^\alpha(T, p)\right] \tag{11.16}$$

The corresponding equation for the chemical potential for the heterogeneous system is

$$\mu(T, p, \xi) = G(N, T, p, \xi)/N = \mu^{\alpha}(T, p) + \xi\Delta\mu(T, p) \tag{11.17}$$

where

$$\Delta\mu(T, p) = \mu^{\beta}(T, p) - \mu^{\alpha}(T, p) \tag{11.18}$$

The differential form for the chemical potential at constant temperature and pressure is given by

$$d\mu(T, p) = \Delta\mu(T, p)d\xi \tag{11.19}$$

The driving force for the phase transition is

$$D = -\left(\frac{d\mu}{d\xi}\right)_{T,p} = -\Delta\mu \tag{11.20}$$

The schematic dependences of chemical potentials μ of the α and β phases on temperature T at a constant pressure are shown in Fig. 11.3. T_e is the phase transition temperature. Based on Eq. (11.20) and Fig. 11.3, we can easily deduce that

If $T < T_e$, $\mu^{\beta} > \mu^{\alpha}$, $\Delta\mu = \mu^{\beta} - \mu^{\alpha} > 0$, $D = -\Delta\mu < 0$, the α phase is thermodynamically stable against its transition to the β phase.

If $T = T_e$, $\mu^{\beta} = \mu^{\alpha}$, $\Delta\mu = \mu^{\beta} - \mu^{\alpha} = 0$, $D = -\Delta\mu = 0$, the α phase is at thermodynamic equilibrium with the β phase.

If $T > T_e$, $\mu^{\beta} < \mu^{\alpha}$, $\Delta\mu = \mu^{\beta} - \mu^{\alpha} < 0$, $D = -\Delta\mu > 0$, the α phase is thermodynamically unstable against its transition to the β phase.

When a phase transition takes place with a driving force of D, an amount of chemical energy of D is dissipated due to the transition and converted to thermal

Fig. 11.3 Schematic illustration of temperature dependences of chemical potential and driving force for a phase transition

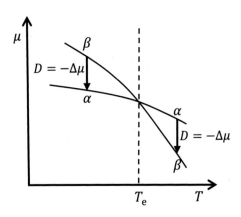

energy. By definition, a phase transition that takes place with a finite driving force is an irreversible process. The amount of entropy produced due to a phase transition is simply related to its driving force, i.e.,

$$\Delta s^{ir} = \frac{D}{T} = \frac{\left(\mu^\alpha - \mu^\beta\right)}{T} = -\frac{\Delta\mu}{T}$$

If the driving force D is zero, Δs^{ir} is zero, i.e., there is no entropy production, and hence a transition that takes place with zero driving force at the equilibrium transition temperature is a reversible process.

It should be emphasized that the knowledge of the thermodynamic driving force for a transition is required for any kinetic theory that describes the rate of the phase transition. For example,

- $\frac{d\xi}{dt} \propto \frac{D}{RT}$ in a linear kinetic theory or
- $\frac{d\xi}{dt} \propto \exp\left(\frac{D}{RT}\right)$ in a nonlinear kinetic theory where R is gas constant.

The rate of entropy production for the transition per unit volume is given by the product of force and the rate of transition, i.e.,

$$\frac{ds^{ir}}{vdt} = \frac{D}{T}\frac{d\xi}{vdt} \tag{11.21}$$

where ds^{ir}/dt is the molar entropy production rate, v is molar volume, and $d\xi/dt$ is the phase transition rate. We can also write Eq. (11.21) in terms of energy dissipation based on the relation between entropy production and energy dissipation in a phase transition,

$$\frac{d\mu}{vdt} = -D\frac{d\xi}{vdt} \tag{11.22}$$

where μ is chemical potential, and $d\mu/vdt$ is the chemical energy density dissipation rate.

11.3 Temperature and Pressure Dependences of Driving Force

To obtain the temperature and pressure dependences of the driving force for a phase transition, we can simply use the information about the temperature and pressure dependences of the chemical potentials of the parent α and new β phases. According to their differential forms, we have

$$d\mu^\alpha = -s^\alpha dT + v^\alpha dp \tag{11.23}$$

and

$$d\mu^\beta = -s^\beta dT + v^\beta dp \tag{11.24}$$

Subtract Eq. (11.23) from Eq. (11.24), we get

$$d(\mu^\beta - \mu^\alpha) = -(s^\beta - s^\alpha)dT + (v^\beta - v^\alpha)dp \tag{11.25}$$

Or

$$d\Delta\mu = -\Delta s dT + \Delta v dp \tag{11.26}$$

where $\Delta\mu = \mu^\beta - \mu^\alpha$, $\Delta s = s^\beta - s^\alpha$, and $\Delta v = v^\beta - v^\alpha$. Therefore, the change in chemical potential for a phase transition as a function of temperature and pressure is given by

$$\Delta\mu(T, p) = - \int_{T_e, 1\,\text{bar}}^{T, 1\,\text{bar}} \Delta s(T, p)dT + \int_{T, 1\,\text{bar}}^{T, p} \Delta v(T, p)dp \tag{11.27}$$

where $\Delta s(T, p)$ and $\Delta v(T, p)$ are the temperature and pressure dependences of changes in molar entropy and molar volume for the transition. The temperatures and pressures that satisfy the following equation form the phase boundary along which the two phases are at equilibrium,

$$\Delta\mu(T, p) = - \int_{T_e, 1\,\text{bar}}^{T, 1\,\text{bar}} \Delta s(T, p)dT + \int_{T, 1\,\text{bar}}^{T, p} \Delta v(T, p)dp = 0 \tag{11.28}$$

Assuming that for a phase transition, the heat capacity change Δc_p is a constant, and the isothermal compressibility $\beta_T(T, p)$ and the thermal expansion coefficient α are zero, we have

$$\Delta s(T, p) = \Delta s^\circ(T_e, 1\,\text{bar}) + \Delta c_p \ln \frac{T}{T_e} \tag{11.29}$$

and

$$\Delta v(T, p) = \Delta v(T, 1\,\text{bar}) = \Delta v^\circ(T_e, 1\,\text{bar}) \tag{11.30}$$

where Δs° and Δv° are the changes in molar entropy and molar volume associated with the phase transition at the equilibrium transition temperature T_e under pressure 1 bar. Therefore,

$$\Delta\mu(T, p) = -\int_{T_e, 1\,bar}^{T, 1\,bar}\left[\Delta s^o + \Delta c_p \ln\frac{T}{T_e}\right]dT + \Delta v^o \Delta p \tag{11.31}$$

Integrating the first term on the right-hand side of the above equation, we have

$$\Delta\mu(T, p) = \left[\Delta c_p - \Delta s^o\right](T - T_e) - \Delta c_p T \ln\frac{T}{T_e} + \Delta v^o(p - 1\,bar) \tag{11.32}$$

which can be rewritten as

$$\Delta\mu(T, p) = \left[\Delta c_p - \Delta s^o\right]\Delta T - \Delta c_p T \ln\left(1 + \frac{\Delta T}{T_e}\right) + \Delta v^o \Delta p \tag{11.33}$$

where $\Delta T = T - T_e$, and $\Delta p = p - 1$ bar.
If $\Delta T \ll T_e$

$$\Delta\mu(T, p) = \left[\Delta c_p - \Delta s^o\right]\Delta T - \Delta c_p T\frac{\Delta T}{T_e} + \Delta v^o \Delta p \tag{11.34}$$

Therefore, the dependences of the chemical potential difference between the two phases as a result of changes in temperature and pressure can be approximated as

$$\Delta\mu(T, p) = -\left[\frac{\Delta h^o + \Delta c_p \Delta T}{T_e}\right]\Delta T + \Delta v^o \Delta p \tag{11.35}$$

11.4 Classification and Order of Phase Transitions

The thermodynamic properties of a material typically undergo dramatic changes near a phase transition or phase transformation. Different types of phase transitions, such as melting of a solid to become a liquid, or transition of a paramagnetic solid to become a ferromagnetic, or transition of a metal to become a superconductor, result in very different types of changes in thermodynamic properties near a transition. Therefore, depending on the behavior of thermodynamic properties near a phase transition, phase transitions have been classified into different types such as first-order phase transitions, second-order phase transitions, high-order phase transitions, discontinuous phase transitions, continuous phase transitions, or infinite-order phase transitions.

Since all the thermodynamic properties of a material can be expressed as the derivatives of a fundamental equation with respect to its natural variables, a phase transition has been classified as a first order, second order or high order according to the behaviors of the derivatives of chemical potentials of the phases involved in a phase transition with respect to temperature and pressure at a phase transition temperature and pressure. For example, according to Ehrenfest's classification[1]:

- A first-order phase transition is one at which the first-order derivatives of chemical potential with respect to temperature and pressure, i.e., molar entropy and molar volume, are discontinuous. The jumps or changes in molar entropy and molar volumes at the transition temperature and pressure are simply called entropy and volume of transition. Since the enthalpy change at the transition is equal to the product of entropy change and transition temperature, and hence the enthalpy change at the transition temperature is also finite. Since the enthalpy change at a transition at constant pressure is equal to the heat absorbed or released, the enthalpy change at the transition is often called latent heat of transition. An order parameter is a parameter for distinguishing the two phases involved in a phase transition, e.g., the difference in mass density between the two phases, spontaneous polarization, atomic order, crystallinity, etc. The order parameters are also discontinuous at the temperature and pressure of a first-order phase transition. The majority of phase transitions in materials such as melting, solidification, phase separation, precipitation, structural transitions, and most of the ferroelectric transitions are first-order phase transitions.
- A second-order phase transition is one at which the first derivatives of chemical potential with respect to temperature and pressure, i.e., the molar entropy and molar volume, are continuous, but the second derivatives of chemical potential with respect to temperature and pressure, i.e., heat capacity, or compressibility, are discontinuous. In a second-order phase transition, the order parameter characterizing the phase transition is continuous. A good example of a second-order phase transition is a ferromagnetic phase transition.
- A high-order phase transition is one at which the first and second derivatives of chemical potential with respect to temperature and pressure are continuous, but the higher-order derivatives of chemical potential with respect to temperature and pressure are discontinuous.

There is a class of phase transitions such as ferromagnetic and superconducting phase transitions at which a certain property or a certain set of properties diverge rather than being discontinuous. For example, the heat capacity diverges at a ferromagnetic transition point. Therefore, there is an increasing trend in the literature to simply use the latent heat of a phase transition to classify phase transitions into two types: (1) first order or discontinuous transitions and (2) second order or continuous phase transitions. A phase transition exhibiting a latent heat is a discontinuous transition, while a phase transition without latent heat is a continuous phase transition. The properties represented by the second derivatives of chemical potential with respect to temperature and pressure or with respect to its order parameter diverge at second-order or *continuous phase transitions*. A glass transition can also be called a

---

[1] Jaeger, Gregg (1 May 1998). "The Ehrenfest Classification of Phase Transitions: Introduction and Evolution". Archive for History of Exact Sciences. 53 (1): 51–81. https://doi.org/10.1007/s00407 0050021.

continuous phase transition, but the transition is more described in terms of dynamics rather than thermodynamics. In physics, there are also transitions that can be classified as *infinite-order phase transitions* since they are continuous with no symmetry breaking, e.g., the Kosterlitz–Thouless transition in the two-dimensional XY model.

11.5 Temperature–Pressure Phase Diagram

At a fixed chemical composition, depending on the temperature and pressure, different phase states (solid, liquid, gas, or their mixtures) may become the stable equilibrium state. According to the thermodynamic equilibrium principle, at constant temperature and pressure, an equilibrium state is obtained by minimizing the Gibbs free energy. However, since when we compare the stability of different thermodynamic states, we have to always employ the same overall chemical composition for the different states. Therefore, to determine the stable states at different temperatures and pressures, we should simply compare the chemical potentials of different phases as a function of temperature and pressure, and the state with the lowest chemical potential is the thermodynamically stable state.

A diagram that collects all the phase boundaries separating the regions of stable (sometimes even metastable) phases in a material is called a phase diagram. For a material with a fixed chemical composition, the graphical representation of stable phases at different pressures and temperatures is also called a $p–T$ diagram. It is also possible to construct other types of phase diagrams, e.g., phase diagrams of equilibrium states represented on diagrams using temperature versus molar volume, pressure versus molar volume, temperature versus molar entropy, and pressure versus molar entropy. Two schematic $p–T$ phase diagrams are shown in Fig. 11.4 for the case that the molar volume of liquid (v^l) is greater than the solid (v^s) and Fig. 11.5 for the case that the molar volume of liquid (v^l) is smaller than the solid (v^s).

- Single-phase regions: The ranges of temperature and pressure within which one of the phases has the lowest chemical potential are called single-phase regions. For example, the regions labeled as solid, liquid, or gas in Figs. 11.4 and 11.5

Fig. 11.4 Schematic diagram of a single-component system with molar volume of liquid larger than solid. T_{tr} and p_{tr} label the temperature and presure of a triple point at which all three phases, solid, liquid, and vapor are at equilibrium with reach other

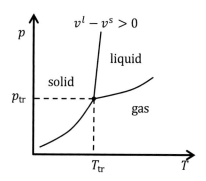

Fig. 11.5 Schematic
diagram of a
single-component system
with molar volume of liquid
smaller than solid

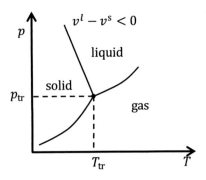

represent the temperature and pressure ranges within which a single-phase solid,
or liquid, or gas is the most stable state, respectively.

- Two-phase coexistence boundaries: If there are two phases that have the same
  chemical potential at certain temperatures and pressures, these two phases are at
  thermodynamic equilibrium with each other at those temperatures and pressures.
  These temperatures and pressures form the lines in the pressure versus temperature
  plot, which are called phase boundaries. For example, the lines separating the
  solid and liquid regions in Figs. 11.4 and 11.5 represent the temperatures and
  pressures at which the solid and liquid are at thermodynamic equilibrium; the lines
  separating the liquid and gas represent the temperatures and pressures at which
  the liquid and gas are at thermodynamic equilibrium; and the lines separating
  the solid and gas represent the temperatures and pressures at which the solid and
  gas phases are at thermodynamic equilibrium. It should be noted that the two-
  phase equilibrium lines on the p–T diagrams will become regions of two-phase
  coexistence fields on the single-component phase diagrams of temperature versus
  molar volume, or pressure versus molar volume, temperature versus or molar
  entropy, or pressure versus molar entropy.

- Triple points (T_{tr}, p_{tr}): If there are three different phases that have the same
  chemical potential at a given temperature and pressure, all three phases are at
  equilibrium with each other at this particular combination of temperature and
  pressure. The points on a pressure versus temperature plot at which three phases
  are at equilibrium and coexist are called triple points (T_{tr}, p_{tr}). For example, the
  points that three-phase boundaries meet in Figs. 11.4 and 11.5 are triple points
  at which the three phases, solid, liquid and gas, are at equilibrium with each
  other. It should also be noted that the triple points on the p–T diagrams will
  become regions of three-phase coexistence fields on the single-component phase
  diagrams of temperature versus molar volume, or pressure versus molar volume,
  or temperature versus molar entropy, or pressure versus molar entropy.

11.6 Gibbs Phase Rule

There are restrictions on the maximum number of phases that can coexist at a given temperature and pressure or on the number of thermodynamic variables that can be independently varied at equilibrium. The expression to count the number of independent variables for a given system is called the Gibbs phase rule.

The derivation of the Gibbs phase rule is simple as it essentially involves counting the number of independent variables. Let us consider a multicomponent multiphase system with n number of components and ψ number of phases. There are $(n - 1)\psi$ number of independent chemical potentials because there is one Gibbs–Duhem relation that relates the n chemical potentials of the n components within each phase. The total number of intensive variables is then $(n - 1)\psi + 2$ where the number 2 corresponds to temperature and pressure. Both temperature and pressure are assumed to be uniform and the same for all phases required by the thermodynamic equilibrium condition. At equilibrium, the chemical potentials of each component must be uniform in the system, and thus there are $n(\psi - 1)$ relations that relate the chemical potentials of each component in all ψ phases. It should be pointed out that not all components have to exist in each phase. For those components that are absent from a phase, the chemical potentials in that phase should be higher than the uniform chemical potentials for those species in a multiphase mixture. Therefore, the total number of independent intensive variables, or the number of degrees of freedom (NDF), for such a multicomponent multiphase mixture is

$$NDF = (n - 1)\psi + 2 - n(\psi - 1) \tag{11.36}$$

Simplifying the above equation, we have

$$NDF = n - \psi + 2 \tag{11.37}$$

NDF represents the number of independent variables that one can vary for a multiphase mixture without changing the number and type of phases in the multiphase mixture. For example, at a triple point of a $p - T$ phase diagram for a material with a fixed chemical composition,

$$NDF = n - \psi + 2 = 1 - 3 + 2 = 0 \tag{11.38}$$

i.e., the number of degrees of freedom is zero, implying that one cannot vary either the temperature or pressure without changing the three-phase mixture to either two-phase mixture or a single phase. The phase rule also sets the maximum number of phases that can coexist in a system at equilibrium,

$$NDF = n - \psi + 2 \geq 0 \Rightarrow \psi \leq n + 2 \tag{11.39}$$

Or

$$\psi_{\max} = n + 2 \tag{11.40}$$

For example, for a single-component system, the maximum number of phases that can coexist at thermodynamic equilibrium is 3.

Along the phase boundaries of a one-component p–T phase diagram,

$$\text{NDF} = n - \psi + 2 = 1 - 2 + 2 = 1 \tag{11.41}$$

Therefore, one can vary either temperature or pressure independently, but not both, without changing the two-phase nature of the system.

Within the single-phase field of a phase diagram,

$$\text{NDF} = n - \psi + 2 = 1 - 1 + 2 = 2 \tag{11.42}$$

which implies that one can vary both temperature T and pressure p simultaneously without leaving a single-phase field.

11.7   Clapeyron Equation

The Clapeyron equation is a differential equation that describes the slopes of phase boundaries in a pressure–temperature diagram.

If we have the chemical potentials for each phase as functions of temperature and pressure for a given material, we can compute its p–T phase diagram by using the equilibrium condition that the phase with the lowest chemical potential is the equilibrium phase. To construct a phase diagram, we first determine the equations describing the phase boundaries in a phase diagram. We then obtain the phase diagram by graphically representing the phase boundary lines on a p–T plot. If we have the chemical potentials of the two phases α and β as a function of temperature T and pressure p, we can obtain the equation describing the α/β phase boundary by simply solving the following equation,

$$\mu^{\alpha}(T, p) = \mu^{\beta}(T, p) \tag{11.43}$$

However, to simply obtain the slope of the α/β phase boundary, we do not need the full knowledge of chemical potentials of both phases as shown by the Clapeyron equation. To derive the Clapeyron equation, we start with the variation of the chemical potential difference between the two phases,

$$d\Delta\mu = -\Delta s dT + \Delta v dp \tag{11.44}$$

Along a phase boundary on the p–T diagram (Fig. 11.6), the α and β phases are at equilibrium with each other, and we call this line the phase boundary separating the

Fig. 11.6 Schematic
illustration of a two-phase
phase boundary and its slope

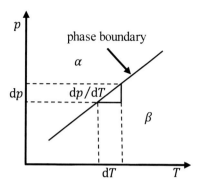

two regions in which either the α phase or the β phase is thermodynamically stable.
Therefore, along the phase boundary between α and β, $\Delta\mu \equiv 0$, or $\mu^\alpha \equiv \mu^\beta$ and
thus

$$-\Delta s \mathrm{d}T + \Delta v \mathrm{d}p = 0$$

Rearranging the above equation, we obtain the local slope of the line, $\mathrm{d}p/\mathrm{d}T$, as
a function of temperature,

$$\frac{\mathrm{d}p}{\mathrm{d}T} = \frac{\Delta s^{\alpha\rightarrow\beta}}{\Delta v^{\alpha\rightarrow\beta}} = \frac{s^\beta - s^\alpha}{v^\beta - v^\alpha} \tag{11.45}$$

where $\Delta s^{\alpha\rightarrow\beta} = s^\beta - s^\alpha$ and $\Delta v^{\alpha\rightarrow\beta} = v^\beta - v^\alpha$ are the changes in the molar entropy
and molar volume for the transition or simply called the entropy and volume of
transition. Therefore, the slope of the α/β phase boundary at a given tempreature and
pressure is simply given by the ratio of entropy of transition to volume of transition.
   If the $\alpha \rightarrow \beta$ phase transition takes place at equilibrium temperature and pressure,
i.e., along a phase boundary, the transition process is a reversible process since the
driving force for this transition, $D = \mu^\alpha - \mu^\beta = 0$, so no entropy is produced.
According to the second law of thermodynamics, for a reversible process,

$$\Delta s^{\alpha\rightarrow\beta} = \frac{Q^{\alpha\rightarrow\beta}}{T_\mathrm{e}^{\alpha\rightarrow\beta}} \tag{11.46}$$

where $Q^{\alpha\rightarrow\beta}$ is the heat released or absorbed for the transition of one mole of α
phase to β phase, and $T_\mathrm{e}^{\alpha\rightarrow\beta}$ represents the equilibrium transition temperature at a
given pressure. According to the first law of thermodynamics applied to a constant
pressure process,

$$Q^{\alpha\rightarrow\beta} = \Delta h^{\alpha\rightarrow\beta} \tag{11.47}$$

Fig. 11.7 Illustration of
relationship between the
slope of a phase boundary
and the molar volume
difference between the two
phases

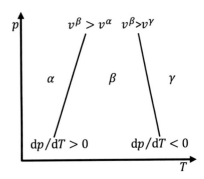

where $\Delta h^{\alpha \to \beta}$ is the molar enthalpy change for the transition of one mole of α phase to β phase or simply called the enthalpy of transition. The slope of the α/β phase boundary can then be rewritten as

$$\frac{dp}{dT} = \frac{\Delta h^{\alpha \to \beta}}{T \Delta v^{\alpha \to \beta}} \qquad (11.48)$$

which is the Clapeyron equation. Therefore, the slope of the α/β phase boundary at a given temperature and pressure can be obtained with the knowledge of heat of transition and volume of transition.

In general, for a transition of a lower temperature phase to a higher temperature phase, the heat of the transition is positive, i.e.,

$$\Delta h^{\alpha \to \beta} > 0$$

Therefore, the sign for the slope of a phase boundary, dp/dT, is determined by the sign of the volume change, Δv. A positive volume change implies a positive slope, or one can determine the sign for the volume change by simply examining the sign for the slope of a phase boundary in a phase diagram (Fig. 11.7). For most materials, the volume change involved in a transition of a lower temperature phase to a higher temperature phase, e.g., melting of a solid to become liquid, is positive, and the slopes of the corresponding phase boundaries are positive. However, for materials whose crystal structures are quite open, melting of a solid might lead to a decrease in volume. The examples happen to be the most familiar materials systems, e.g., ice (H_2O), silicon (Si), and diamond (C). Melting of ice, silicon, and diamond leads to a decrease in volume, i.e., a negative volume of melting. As a result, the slopes for the phase boundaries separating the liquid and solid phases of ice, silicon, and diamond are negative.

11.8 Clausius–Clapeyron Equation and Vapor Pressure of Condensed Phase

If a phase transition involves a vapor phase, e.g., a solid to vapor transition or a liquid to vapor transition, the pressure in the Clapeyron equation is the equilibrium vapor pressure over a solid or liquid if the solid or liquid is at equilibrium with the vapor phase. A single component in a mixture can contribute vapor pressure to the total pressure in the system. The vapor pressure contributed by a particular component is called partial pressure.

As an example, let us consider the transition of a liquid (l) to vapor (v),

$$\text{liquid}(l) \rightarrow \text{vapor}(v)$$

The molar volume of vapor v^v is typically much larger than the molar volume of a condensed phase, liquid v^l or solid v^s, i.e.,

$$\Delta v^{l \rightarrow v} = v^v - v^l \approx v^v$$

The Clapeyron equation can be approximated as

$$\frac{dp}{dT} \approx \frac{\Delta h^{l \rightarrow v}}{T v^v} \tag{11.49}$$

The vapor pressure over a liquid and solid is typically low, so the behavior of a vapor phase can be approximated by the ideal gas equation of state. Therefore, the molar volume of the vapor phase can be expressed in terms of temperature and pressure using the ideal gas law,

$$v^v = \frac{RT}{p} \tag{11.50}$$

Substituting Eqs. (11.50) into (11.49), we have

$$\frac{d \ln p}{dT} = \frac{\Delta h^{l \rightarrow v}}{RT^2} \tag{11.51}$$

which is the Clausius–Clapeyron equation. It should be emphasized that the Clausius–Clapeyron equation is only applied to a transition in which one of the two phases is a vapor phase. Equation (11.51) also expresses the equilibrium vapor pressure above a liquid as a function of temperature.

Normal boiling point: The normal boiling point T_b of a liquid is the temperature at which the liquid is at equilibrium with its vapor phase at 1 bar. The heat of evaporation at the normal boiling point is labeled as Δh_b^o. The dependence of the heat of evaporation Δh_b^o on vapor pressure can be ignored since the vapor pressure is usually very low. The heat of evaporation $\Delta h^{l \to v}$ as a function of temperature can be calculated using the heat capacities of liquid and its vapor phase, i.e.,

$$\Delta h^{l \to v}(T) = \Delta h_b^o + \int_{T_b}^{T} \left(c_p^v(T) - c_p^l(T) \right) dT = \Delta h_b^o + \int_{T_b}^{T} \Delta c_p(T) dT \qquad (11.52)$$

where $c_p^v(T)$ and $c_p^l(T)$ are the heat capacity of vapor phase and liquid phase, respectively. If Δc_p is zero, $\Delta h^{l \to v} = \Delta h_b^o$, i.e., the heat of evaporation is independent of temperature. With this assumption, integrating the Clausius–Clapeyron equation, we have

$$\ln p = -\frac{\Delta h_b^o}{RT} + c \qquad (11.53)$$

where c is an integration constant. The integration constant can be determined using the normal boiling point, i.e.,

$$\ln(1) = -\frac{\Delta h_b^o}{RT_b} + c \to c = \frac{\Delta h_b^o}{RT_b} \qquad (11.54)$$

and thus

$$\ln p = -\frac{\Delta h_b^o}{RT} + \frac{\Delta h_b^o}{RT_b} = \frac{\Delta h_b^o}{R} \left(\frac{1}{T_b} - \frac{1}{T} \right) \qquad (11.55)$$

If Δc_p is assumed to be a constant rather than 0,

$$\Delta h^{l \to v} = \Delta h_b^o + \Delta c_p (T - T_b) \qquad (11.56)$$

The Clausius–Clapeyron equation becomes

$$\frac{d \ln p}{dT} = \frac{\Delta h_b^o + \Delta c_p (T - T_b)}{RT^2} = \frac{\left(\Delta h_b^o - \Delta c_p T_b \right) + \Delta c_p}{RT^2} = \frac{a}{T^2} + \frac{b}{T} \qquad (11.57)$$

where

$$a = \frac{\Delta h_b^o - \Delta c_p T_b}{R} \qquad (11.58)$$

$$b = \frac{\Delta c_p}{R} \qquad (11.59)$$

Integrating the above Eq. (11.57), we have the equilibrium vapor pressure above a liquid,

$$\ln p = -\frac{a}{T} + b \ln T + c \tag{11.60}$$

where a, b, and c are available for some of the liquids in databases in the literature. It should be remembered that this form is derived from the assumption that the heat capacity difference between a vapor and its corresponding liquid or solid is a constant independent of temperature. Equation (11.60) is a typical form for describing equilibrium vapor pressure above a solid or liquid as a function of temperature. This equation also describes the phase boundary between the liquid and vapor phases in a phase diagram.

Triple point: At triple point, the chemical potentials of solid, liquid, and vapor are equal, i.e.,

$$\mu^s(T, p) = \mu^l(T, p) = \mu^v(T, p) \tag{11.61}$$

If we have the full knowledge of the temperature and pressure dependences of chemical potentials of solid, liquid, and vapor phases, we can obtain the triple point by solving the two equations in Eq. (11.61).

If the vapor pressures of both a liquid and the corresponding solid as a function of temperature are known, one can also obtain the triple point temperature and pressure. For example, one can obtain the triple point temperature and pressure by solving the following two coupled equations describing the equilibrium vapor pressure above a solid and liquid,

$$\ln p_{tr} = -\frac{a^{l \to v}}{T_{tr}} + b^{l \to v} \ln T_{tr} + c^{l \to v} \tag{11.62}$$

$$\ln p_{tr} = -\frac{a^{s \to v}}{T_{tr}} + b^{s \to v} \ln T_{tr} + c^{s \to v} \tag{11.63}$$

On the other hand, with known temperature dependence of vapor pressure, one can calculate the heat of evaporation and the heat capacity difference between the vapor and the liquid or solid,

$$\Delta h^{l \to v} = RT^2 \frac{d \ln p^{l \to v}}{dT}, \quad \Delta h^{s \to v} = RT^2 \frac{d \ln p^{s \to v}}{dT} \tag{11.64}$$

$$\Delta c_p^{l \to v} = \frac{d(\Delta h^{l \to v})}{dT}, \quad \Delta c_p^{s \to v} = \frac{d(\Delta h^{s \to v})}{dT} \tag{11.65}$$

With the knowledge of $\Delta h^{l \to v}$, $\Delta h^{s \to v}$, $\Delta c_p^{l \to v}$, and $\Delta c_p^{s \to v}$, we can obtain $\Delta h^{s \to l}$ and $\Delta c^{s \to l}$ using

$$\Delta h^{s \to l} = \Delta h^{s \to v} - \Delta h^{l \to v} \tag{11.66}$$

and

$$\Delta c_p^{s \to l} = \Delta c_p^{s \to v} - \Delta c_p^{l \to v} \tag{11.67}$$

based on the fact that the change of a state function is independent of the process path.

The normal boiling point T_b can be obtained from the equilibrium vapor pressure above a liquid as a function of temperature, for example, by solving the following equation,

$$\ln(1) = -\frac{a}{T_b} + b \ln T_b + c = 0 \tag{11.68}$$

11.9   Size Effect on Phase Transition Temperature

If the size of a material becomes nanoscale, the contribution of surface energy to the phase transition temperature cannot be ignored. Let us consider the melting of a spherical solid particle (Fig. 11.8). Using the result from Sect. 9.6.3 on the size dependence of chemical potential,

$$\mu_r^s = \frac{2\gamma_{sl} v}{r} + \mu_\infty^s \tag{11.69}$$

where μ_r is the chemical potential of a particle with radius r, μ_∞ is the chemical potential of a solid with infinite size, γ_{sl} is solid–liquid interfacial energy, and v is the molar volume of the solid. The treatment of chemical potentials of atoms inside particles of shapes other than spherical such as cuboidal or ellipsoidal is similar.

Fig. 11.8 Illustration of a spherical particle with radius r

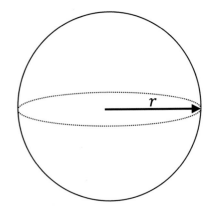

However, chemical potentials of atoms inside particles of arbitrarily complicated shapes can only be obtained numerically.

At the solid–liquid equilibrium, the chemical potentials of solid and liquid are equal. We label the melting temperature T_m for infinite particle size as T_∞ at which

$$\mu_\infty^s = \mu^l$$

where μ^l is the chemical potential of atoms in the liquid.

The entropy of melting is given by

$$\Delta s_m = \frac{\Delta h_m}{T_\infty} \tag{11.70}$$

where Δh_m is the enthalpy or heat of melting.

Therefore, for a solid particle of size r, the chemical potential of atoms inside the particle must be equal to the chemical potentials of atoms in the liquid, i.e.,

$$\mu_r^s = \mu_\infty^s + \frac{2\gamma_{sl} v}{r} = \mu^l \tag{11.71}$$

Assuming that the heat Δh_m and entropy Δs_m of melting are independent of temperature, we have at melting temperature T_r of particle size r,

$$\mu^l - \mu_\infty^s = \Delta h_m - T\frac{\Delta h_m}{T_\infty} = \Delta h_m \frac{T_\infty - T_r}{T_\infty} \tag{11.72}$$

Using Eq. (11.69),

$$\mu^l - \mu_\infty^s = \frac{2\gamma_{sl} v}{r} = \frac{(T_\infty - T_r)\Delta h_m}{T_\infty} \tag{11.73}$$

If we solve Eq. (11.73) for the melting temperature T_r, we get

$$T_r = \left(1 - \frac{2\gamma_{sl} v}{r\Delta h_m}\right) T_\infty \tag{11.74}$$

The transition temperature as a function of size is schematically shown in Fig. 11.9 as an illustration.

11.10 Landau Theory of Phase Transitions

In principle, any phase transition can be characterized by physically well-defined order parameters that distinguish the parent and new transformed phases. An order

Fig. 11.9 Illustration of melting temperature of a particle as a function of particle radius

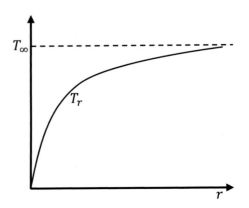

parameter is typically defined to be zero in the high-temperature phase and has a finite value in the low-temperature phase. For example, the relative density of a liquid relative to its vapor can be used as an order parameter to describe a vapor to liquid transition. Other examples include relative compositional difference between two phases for compositional phase separation, long-range order parameter for order–disorder transitions, spontaneous polarization for ferroelectric transitions, spontaneous strain for ferroelastic transitions, and spontaneous magnetization for ferromagnetic transitions.

The thermodynamics of a phase transition can be described by a free energy density function which is expressed as a polynomial of order parameters based on the Landau theory of phase transitions. All the terms in the free energy density function are required to be invariant with respect to the symmetry operations of the high-temperature phase. For example, for a phase transition described by a single order parameter η, we can express the free energy density as a function of the order parameter η as

$$
\begin{aligned}
f(\eta) = f(0) + \left(\frac{\partial f}{\partial \eta}\right)_{\eta=0} \eta + \frac{1}{2}\left(\frac{\partial^2 f}{\partial \eta^2}\right)_{\eta=0} \eta^2 \\
+ \frac{1}{3!}\left(\frac{\partial^3 f}{\partial \eta^3}\right)_{\eta=0} \eta^3 + \frac{1}{4!}\left(\frac{\partial^4 f}{\partial \eta^4}\right)_{\eta=0} \eta^4 + \cdots
\end{aligned}
\tag{11.75}
$$

Let us also consider the case that the symmetry of the system requires that all odd terms in the free energy function vanish,

$$
f(\eta) = f(0) + \frac{1}{2}\left(\frac{\partial^2 f}{\partial \eta^2}\right)_{\eta=0} \eta^2 + \frac{1}{4!}\left(\frac{\partial^4 f}{\partial \eta^4}\right)_{\eta=0} \eta^4 + \cdots
\tag{11.76}
$$

Keeping terms up to fourth order, we have

$$f(\eta) - f(0) = \frac{1}{2}\left(\frac{\partial^2 f}{\partial \eta^2}\right)_{\eta=0} \eta^2 + \frac{1}{4!}\left(\frac{\partial^4 f}{\partial \eta^4}\right)_{\eta=0} \eta^4 \tag{11.77}$$

If we assume only the coefficient in the first term is temperature-dependent, and that in the second term is independent of temperature, we have

$$f(\eta) - f(0) = \frac{A(T - T_c)}{2}\eta^2 + \frac{B}{4}\eta^4 \tag{11.78}$$

where A and B are positive coefficients, and T_c is the critical temperature for the phase transition. The behavior of the first term in the right-hand side of Eq. (11.78) is shown in Fig. 11.10. At $T < T_c$, the parent phase with $\eta = 0$ is unstable since a finite value of order parameter η has a lower energy density. However, with only the first term, the order parameter in the ordered state goes to infinite since the free energy density keeps decreasing as the order parameter increases. Therefore, if we ignore all terms higher than the fourth order, the fourth-order coefficient B must be positive in order to have finite values for the order parameter in the ordered state at different temperatures below the critical temperature, T_c. With a positive B, the behavior of the free energy density function as a function of order parameter at different temperatures is schematically shown in Fig. 11.11. At temperatures higher than T_c, the state with $\eta = 0$ has the lowest free energy density and thus the stable state. Below T_c the states with finite values of the order parameter, η_0^- and η_0^+, are the stable states. The magnitudes of η_0^- and η_0^+ can be found by minimizing the free energy density with respect to the order parameter, i.e.,

$$\left.\frac{\partial f(\eta)}{\partial \eta}\right|_{\eta_0} = A(T - T_c)\eta_0 + B\eta_0^3 = 0 \tag{11.79}$$

$$\eta_0 = \pm\sqrt{-\frac{A(T - T_c)}{B}} \tag{11.80}$$

Fig. 11.10 Illustration of thermodynamic stability of a material at different temperature

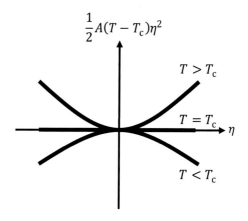

Fig. 11.11 Illustration of the free energy density as a function of order parameter at three temperatures and the equilibrium order parameter values

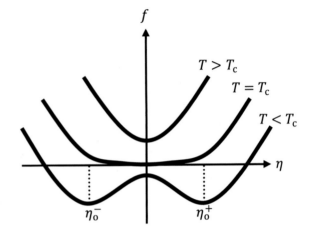

or

$$\eta_o^+ = \sqrt{-\frac{A(T - T_c)}{B}}$$

$$\eta_o^- = -\sqrt{-\frac{A(T - T_c)}{B}}$$

At temperature T_c, the value for the parameter becomes zero, i.e., $\eta_o^- = \eta_o^+ = 0$, and

$$\left.\frac{\partial^2 f(\eta)}{\partial \eta^2}\right|_{\eta=0, T=T_c} = 0$$

Figure 11.12 displays the schematic dependence of η_o^+ as a function of temperature. It is shown that the magnitude of order parameter gradually goes to zero as the

Fig. 11.12 Illustration of equilibrium order parameter values as a function of temperature

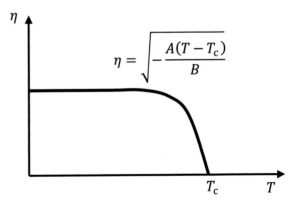

temperature approaches to the critical temperature, i.e., there is no jump in the order parameter value at the transition temperature. A phase transition at which the order parameter value is continuous is called a second-order phase transition or simply a continuous transition. At the transition, both the entropy and enthalpy of transition are zero.

To describe a first-order phase transition, it is necessary to add either a cubic or a sixth-order term to above free energy density function. Let us first look at the $\eta^2 - \eta^3 - \eta^4$ free energy density model,

$$f(\eta) - f(0) = \frac{A(T - T_c)}{2}\eta^2 - \frac{B}{3}\eta^3 + \frac{C}{4}\eta^4 \qquad (11.81)$$

where the coefficients A and C are positive. The sign of B determines the sign of the order parameter value. To find the equilibrium order parameter value at a given temperature, we set

$$\frac{\partial f(\eta)}{\partial \eta}\bigg|_{\eta_o} = A(T - T_c)\eta_o - B\eta_o^2 + C\eta_o^3 = 0 \qquad (11.82)$$

The solution to the above equation gives

$$\eta_{o1} = 0$$

$$\eta_{o2} = \frac{B + \sqrt{B^2 - 4AC(T - T_c)}}{2C}$$

$$\eta_{o3} = \frac{B - \sqrt{B^2 - 4AC(T - T_c)}}{2C}$$

At the critical temperature T_c

$$\eta_{o1} = 0$$

$$\eta_{o2} = \frac{B}{C}$$

$$\eta_{o3} = 0$$

At the transition temperature T_o,

$$B^2 - 4AC(T_o - T_c) = 0$$

$$T_o = T_c + \frac{B^2}{4AC}$$

Therefore, in this case the critical temperature and the transition temperature are different, and there is a jump in the order parameter value at the transition temperature (see Fig. 11.13). A phase transition at which there is a finite jump of the order

Fig. 11.13 Illustration of order parameter as a function of temperature for a first-order phase transition

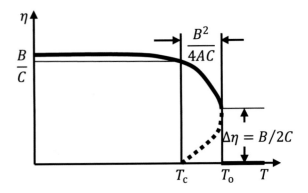

parameter is a first-order phase transition or simply discontinuous transition. There are also finite changes in entropy and enthalpy at the first-order phase transition.

A first-order phase transition can also be described by a free energy density function with a six-order polynomial of order parameters, i.e.,

$$f(\eta) - f(0) = \frac{A(T - T_c)}{2}\eta^2 - \frac{B}{4}\eta^4 + \frac{C}{6}\eta^6 \qquad (11.83)$$

where the coefficients A, B, and C are positive. To find the equilibrium order parameter value at a given temperature, we set

$$\left.\frac{\partial f(\eta)}{\partial \eta}\right|_{\eta_o} = A(T - T_c)\eta_o - B\eta_o^3 + C\eta_o^5 = 0 \qquad (11.84)$$

The solution to the above equation gives five different order parameter values (see Fig. 11.14)

$$\eta_{o1} = 0$$

$$\eta_{o2}^{\pm} = \pm\sqrt{\frac{B + \sqrt{B^2 - 4AC(T - T_c)}}{2C}}$$

$$\eta_{o3}^{\pm} = \pm\sqrt{\frac{B - \sqrt{B^2 - 4AC(T - T_c)}}{2C}}$$

At T_c

$$\eta_{o1} = 0$$

$$\eta_{o2}^{\pm} = \pm\sqrt{\frac{B}{C}}$$

$$\eta_{o3}^{\pm} = 0$$

Fig. 11.14 Illustration of order parameter as a function of temperature obtained from a free energy represented by a six-order polynomial as a function of order parameter

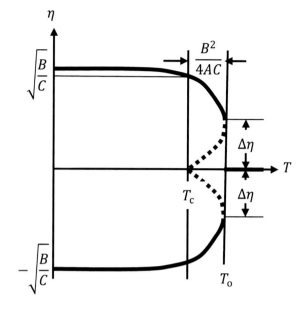

At the transition temperature T_o,

$$B^2 - 4AC(T_o - T_c) = 0$$

$$T_o = T_c + \frac{B^2}{4AC}$$

$$\eta_{o1} = 0$$

$$\eta_{o2}^{\pm} = \pm\sqrt{\frac{B}{2C}}$$

$$\eta_{o3}^{\pm} = \pm\sqrt{\frac{B}{2C}}$$

11.11 Examples

Example 1 The atomic weight of Si is 28.085 g/mol, and the density of liquid (l) Si at the melting temperature 1687 K is 2.57 g/cm³. The lattice parameter a_o at 298 K is 0.5431 nm, and the linear thermal expansion coefficient of Si at room temperature is 2.6 μm/(m K). The chemical potentials of Si in diamond (d) structure and liquid as a function of temperature at 1 bar are given as follows using the enthalpy of Si in diamond structure as the reference (from SGTE element database):

$$\mu_{Si}^d = -8162.6 + 137.24T - 22.832T \ln T - 1.9129 \times 10^{-3}T^2$$
$$- 0.003552 \times 10^{-6}T^3 + 176667T^{-1} \quad \text{(G)} \quad (298\,\text{K} < T < 1687\,\text{K})$$
$$\mu_{Si}^d = -9457.6 + 167.28T - 27.196T \ln T - 420.369$$
$$\times 10^{28}T^{-9} \quad \text{(G)} \quad (1687\,\text{K} < T < 3600\,\text{K})$$
$$\mu_{Si}^l = 42533.8 + 107.14T - 22.831T \ln T - 1.9129 \times 10^{-3}T^2$$
$$- 0.003552 \times 10^{-6}T^3 + 176667T^{-1} + 209.31$$
$$\times 10^{-23}T^7 \quad \text{(G)} \quad (298\,\text{K} < T < 1687\,\text{K})$$
$$\mu_{Si}^l = 40370.5 + 137.72T - 27.196T \ln T \quad \text{(G)} \quad (1687\,\text{K} < T < 3600\,\text{K})$$

(a)　What is the entropy of melting for Si in diamond structure at the melting temperature?

(b)　What is the heat of melting for Si in diamond structure at the melting temperature?

(c)　What is the heat capacity difference between liquid Si and solid Si in diamond structure at the melting temperature?

(d)　Estimate the molar volume change for the melting of Si at the melting temperature by assuming that the thermal expansion coefficient is a constant independent of temperature.

(e)　Estimate the slope of dp/dT for the solid–liquid phase boundary on the p–T phase diagram of Si at 1 bar by assuming both the solid and liquid Si are incompressible.

Solution

(a)　To obtain the entropy of melting for Si in diamond structure at the melting temperature, we first obtain the entropy of solid Si in diamond structure and entropy of liquid at the melting temperature. The entropy of solid Si is given by

$$s_{Si}^d = -\left(\frac{\partial \mu_{Si}^d}{\partial T}\right)_{1\,\text{bar}}$$
$$s_{Si}^d = -137.24 + 22.832 \ln T + 22.832 + 2 \times 1.9129$$
$$\times 10^{-3}T + 3 \times 0.003552 \times 10^{-6}T^2 + 176667T^{-2}$$
$$= -114.41 + 22.832 \ln T + 3.8258 \times 10^{-3}T$$
$$+ 0.010656 \times 10^{-6}T^2 + 176667T^{-2} \quad \text{(J/mol K)}$$
$$(298\,\text{K} < T < 1687\,\text{K})$$

The entropy of liquid Si is given by

$$s_{Si}^l = -\left(\frac{\partial \mu_{Si}^l}{\partial T}\right)_{1\,\text{bar}}$$

$$s_{Si}^l = -137.72 + 27.196 \ln T + 27.196$$
$$= -110.524 + 27.196 \ln T \ \text{(J/mol K)}$$
$$(1687 \text{ K} < T < 3600 \text{ K})$$

Therefore, the entropy of melting of Si is given by the entropy difference between the entropy of liquid Si and solid Si at the melting temperature,

$$\Delta s_m^o = s_{Si}^l - s_{Si}^d$$
$$= 3.88 + 4.364 \ln T - 3.8258 \times 10^{-3}T - 0.010656$$
$$\times 10^{-6}T^2 - 176667T^{-2}$$
$$= 29.763 \text{ J/(mol K)}$$

(b) The heat of melting for Si in diamond structure at the melting temperature is

$$\Delta h_m^o = T_m \Delta s_m^o = 1687 \times 29.763 = 50210 \text{ J/mol}$$

(c) The heat capacity difference between liquid Si and solid Si in diamond structure at the melting temperature is

$$\Delta c_p = T\left(\frac{\partial \Delta s_m}{\partial T}\right) = 4.3642 - 3.8258 \times 10^{-3}T - 0.021312 \times 10^{-6}T^2 + 353334T^{-2}$$

At the melting temperature,

$$\Delta c_p = -2.0264 \text{ J/(mol K)}$$

(d) To estimate the molar volume change for the melting of Si at the melting temperature by assuming that the thermal expansion coefficient is a constant independent of temperature, we first determine the molar volumes of solid and liquid Si at the melting temperature.

The lattice parameter of solid Si as a function of temperature is given by

$$a_{Si}^d(T) \approx a_o\left[1 + \alpha_{Si}^d(T - 298)\right] \text{nm}$$

Hence the lattice parameter of solid Si at the melting temperature is

$$a_{Si}^d(1687 \text{ K}) = 0.5431 \times \left[1 + 2.6 \times 10^{-6}(1687 - 298)\right] \text{nm} = 0.5451 \text{ nm}$$

The molar volume of liquid Si at the melting temperature is

$$v_{Si}^l(1687 \text{ K}) = m_{Si}/\rho_{Si}^l = 28.085 \text{ g/mol}/(2.57 \text{ g/cm}^3)$$
$$= 10.928 \text{ cm}^3/\text{mol}$$

Therefore, the molar volume of melting is given by

$$\Delta v_m^o = v_{Si}^l(1687\,\text{K}) - v_{Si}^d(1687\,\text{K}) = 10.928 - 12.192 = -1.264\,\text{cm}^3/\text{mol}$$

(e) Estimate the slope of dp/dT for the solid–liquid phase boundary on the p–T phase diagram of Si at 1 bar by assuming both the solid and liquid Si are incompressible.

$$\left(\frac{dp}{dT}\right)_{T_m,\,1\,\text{bar}} = \frac{\Delta h_m^o}{T_m \Delta v_m^o} \approx \frac{50210\,\text{J/mol}}{1687\,\text{K} \times \left(-1.264 \times 10^{-6}\,\text{m}^3/\text{mol}\right)}$$
$$\approx -2.355 \times 10^7\,\text{Pa/K}$$

Example 2 When an initially homogeneous phase becomes unstable with respect to a first-order phase transition to a new phase with lower chemical potential as the thermodynamic conditions, e.g., temperature and pressure, change, the new phase often appears through a nucleation and growth mechanism. The nucleation of a new phase particle involves the creation of an interface between the new phase particle and the initial parent matrix. Therefore, although the transition of the initially homogeneous parent phase to the new phase particle results in a decrease in chemical potential and thus the decrease in the Gibbs free energy of the particle, the creation of the interface increases the energy of the system. Since the Gibbs free energy reduction is proportional to the volume of the particle while the interfacial energy increase is proportional to the interfacial area, the interfacial energy increase dominates over the Gibbs free energy reduction when the particle size is very small, and the Gibbs free energy reduction dominates over the interfacial energy increase at large sizes. As a result, there is a critical size for the nucleating particle, above which the total free energy, the Gibbs free energy and interfacial energy, decreases with size while below which increases with size. Given interfacial energy, γ, and Gibbs free energy density difference between the new phase and the initial phase, Δg_v, derive the expressions for the critical size and the critical free energy formation of a nucleus in terms of γ and Δg_v by assuming the nuclei are spherical particles.

Solution
The Gibbs free energy change ΔG due to the transition of a spherical region of radius r from the initial phase to the new phase is given by

$$\Delta G = \frac{4\pi}{3}r^3 \Delta g_v + 4\pi r^2 \gamma$$

To find the critical radius r^*, we set

$$\left(\frac{d\Delta G}{dr}\right)_{r=r^*} = 4\pi \left(r^*\right)^2 \Delta g_v + 8\pi r^* \gamma = 0$$

Solving the above equation for r^*, we get

$$r^* = -\frac{2\gamma}{\Delta g_v}$$

The corresponding critical free energy of formation for the critical particle ΔG^* is then

$$\Delta G^* = \frac{4\pi}{3}(r^*)^3 \Delta g_v + 4\pi (r^*)^2 \gamma = \frac{16\pi \gamma^3}{3(\Delta g_v)^2}$$

Example 3 Compute the melting temperature of ice particles as a function of particle size. For simplicity, assume they are spherical particles. For melting of ice to become water, we have the following data: ice water interfacial energy: $\gamma_{sl} = 0.033$ J/m^2, equilibrium melting temperature of ice ignoring the size effect: $T_\infty = 273$ K, heat or enthalpy of melting of ice at the equilibrium temperature: $\Delta h_m^o = 6007$ J/mol, and molar volume of ice: $v = 19.65 \times 10^{-6}$ m^3/mol.

Solution
Using the data in the problem, we can compute

$$\frac{2\gamma_{sl}v}{\Delta h_m^o} = \frac{2 \times 0.033 \text{ J/m}^2 \times 19.65 \times 10^{-6} \text{ m}^3/\text{mol}}{6007 \text{ J/mol}}$$
$$\approx 2.0 \times 10^{-10} \text{ m} = 0.2 \text{ nm}$$

Substituting the data into Eq. (11.73), we have

$$T_r = \left(1 - \frac{0.2 \text{ nm}}{r}\right)T_\infty$$

For example, for $r = 50$ nm, we have

$$T_{50 \text{ nm}} \approx 272 \text{ K}$$

or

$$T_{50 \text{ nm}} - T_\infty \approx -1 \text{ K}$$

11.12 Exercises

1. Is the chemical potential of one mole of oxygen gas at 1 bar and 1000 K: (i) <, or (ii) =, or (iii) > the chemical potential of one mole of oxygen gas at 1 bar and 300 K?

2. The chemical potential of oxygen gas at 1 bar and 298 K is higher than that at 100 bar and 298 K, true or false? Explain.

3. The chemical potential of a solid always increases as the pressure increases at constant temperature: true or false?

4. The chemical potential of a pure substance always decreases with temperature at constant pressure, true or false?

5. The curvature of a chemical potential of a substance vs. temperature plot at constant pressure is always negative, true or false?

6. At 0 °C and 1 bar, is the chemical potential of ice (i) higher than, (ii) equal to or (iii) lower than that of water?

7. When a pure solid is at equilibrium with its vapor phase at a given temperature, do the atoms in the solid have (i) lower, (ii) equal, or (iii) higher chemical potential than those in the vapor?

8. The maximum number of phases that can coexist at any temperature and pressure for a single-component system at equilibrium is 2, true or false?

9. Rank the chemical potentials of solid, liquid, and vapor Cu at room temperature from the lowest to the highest at 1 bar.

10. When you take an ice cube out from a freezer and let it sit at room temperature and melt, is the Gibbs free energy change for the melting process positive or negative?

11. If a melting process results in a volume decrease, would you expect the melting temperature to increase or decrease when the system pressure increases?

12. In a pressure–temperature phase diagram, if the slope of a solid–liquid phase boundary is positive, it implies that the liquid has a larger molar volume than the corresponding solid at the melting temperature, true or false?

13. The higher the heat of vaporization of a liquid, the higher the equilibrium vapor pressure above the liquid at a given temperature under 1 bar total pressure: true or false?

14. If the chemical potential difference between the solid state and the liquid state of a substance is a linear function of temperature, the specific heat capacity difference between solid and liquid is zero: true or false?

15. One mole of undercooled liquid (l) water at 268 K solidifies with a surrounding, which is also at 268 K, to become solid (s) ice at 268 K. If the chemical potential difference between ice and liquid water at 268 K is $\mu^s - \mu^l = \Delta\mu$, then the entropy produced during this solidification process is $(-\Delta\mu/268\ \text{K})$: true or false?

16. For the melting of solid (s) ice to become liquid (l) water at the equilibrium melting temperature of 273 K under 1 bar, the heat of melting is 6007 J/mol. The entropy of water at 298 K and 1 bar is 70 J/(mol K), and the enthalpy of water at 298 K and 1 bar is -286 kJ/mol. The heat capacities of ice and water at 1 bar are 38 J/(mol K) and 75 J/(mol K), respectively. The molar volumes of ice and water at 273 K are 19.65×10^{-6} m^3/mol and 18.02×10^{-6} m^3/mol, respectively

    (a) Which one has higher enthalpy, one mole of ice at 273 K and 1 bar or one mole of water at 273 K and 1 bar?
    (b) What is the entropy difference ($s^l - s^s$) between one mole of ice and one mole of water both at 273 K and 1 bar?
    (c) Which one has higher entropy, one mole of ice or one mole of water both at 273 K and 1 bar?
    (d) One mole of ice and one mole of water have different levels of enthalpy at 273 K and 1 bar, why are ice and water at equilibrium at 273 K and 1 bar?
    (e) What is the chemical potential difference between one mole of ice and one mole of water both at 273 K and 1 bar?
    (f) Is the chemical potential difference $\left(\mu^l - \mu^s\right)$ between one mole of water and one mole of ice at 273 K and 1001 bar positive or negative?
    (g) Estimate the chemical potential difference $\left(\mu^s_{1001\ \text{bar}} - \mu^s_{1\ \text{bar}}\right)$ between one mole of ice under 1 bar and 1001 bar both at 273 K (assuming ice is incompressible).
    (h) Estimate the chemical potential difference $\left(\mu^l - \mu^s\right)$ between one mole of water and one mole of ice both at 273 K and 1001 bar (assuming water and ice are incompressible).

17. Given data of graphite (g) and diamond (d):
    Molar enthalpy of diamond: $h^{\text{o,d}}_{298\ \text{K, 1 bar}} = 1900$ J/mol.
    Molar entropy of graphite and diamond: $s^{\text{o,g}}_{298\ \text{K, 1 bar}} = 5.6$ J/(K mol),
    $s^{\text{o,d}}_{298\ \text{K, 1 bar}} = 2.4$ J/(K mol),
    Molar volume of graphite and diamond: $v^{\text{o,g}}_{298\ \text{K, 1 bar}} = 6.0$ cm^3/mol,
    $v^{\text{o,d}}_{298\ \text{K, 1 bar}} = 3.4$ cm^3/mol.
    Molar heat capacity of graphite and diamond: $c^{\text{g}}_p = 8.5$ J/(K mol), $c^{\text{d}}_p = 6.0$ J/(K mol).
    Isothermal compressibility of graphite and diamond: $\beta^{\text{g}}_T = 3 \times 10^{-6}$/bar, $\beta^{\text{d}}_T = 2 \times 10^{-7}$/bar,

Volume thermal expansion coefficient of graphite and diamond: $\alpha^g = 3 \times 10^{-5}/K$, $\alpha^d = 10^{-5}/K$.

Consider the hypothetical phase transition from graphite to diamond:

$$C(g) = C(d)$$

Please do the following (assume all the properties α^g, α^d, β_T^g, β_T^d, c_p^g, and c_p^d are constant independent of temperature and pressure):

(a) Calculate the heat of transition at 298 K and 1 bar.

(b) Calculate the entropy change for the transition at 298 K and 1 bar.

(c) Calculate the chemical potential change for the transition at 298 K and 1 bar.

(d) Based on your calculation in (c), determine which is more stable, graphite or diamond at 298 K and 1 bar?

(e) Calculate the entropy change for the surrounding if one mole of graphite were transformed to one mole of diamond in the system at 298 K and 1 bar.

(f) Calculate the total entropy change of the system and surrounding at 298 K and 1 bar if one mole of graphite were transformed to diamond at 298 K and 1 bar.

(g) Based on the above calculations, determine whether the transition from graphite to diamond at 298 K and 1 bar is (i) spontaneous or (ii) not possible.

(h) Calculate the heat of transition at 1000 K and 1 bar.

(i) Calculate the entropy change for the transition at 1000 K and 1 bar.

(j) Calculate the chemical potential change for the transition at 1000 K and 1 bar.

(k) Calculate the enthalpy change for the transition at 298 K and 10^6 bar.

(l) Calculate the entropy change for the transition at 298 K and 10^6 bar.

(m) Calculate the chemical potential change for the transition at 298 K and 10^6 bar.

(n) Based on the above calculations, determine whether transition from graphite to diamond at 298 K and 10^6 bar is (i) spontaneous or (ii) impossible.

(o) Calculate the pressure at which the diamond and graphite are at thermodynamic equilibrium at room temperature.

18. Calcium carbonate ($CaCO_3$) has two polymorphs, calcite and aragonite. The volume change for the calcite to aragonite transition is $-2.784 \, cm^3/mol$ and is assumed to be independent of temperature and pressure. The chemical potential change for the calcite to aragonite transition as a function of temperature at 1 bar is given by

$$\Delta\mu = -210 + 4.2T \ (G = J/mol)$$

(a) What is the molar heat of transition?
(b) What is the molar entropy change for the transition?
(c) Determine the stable phase at room temperature 298 K under 1 bar.
(d) Calculate the temperature at which calcite and aragonite are at equilibrium under 1 bar.
(e) Determine whether an increase in pressure at a fixed temperature will increase or decrease the thermodynamic stability of aragonite with respect to calcite.

19. The equilibrium melting temperature of copper (Cu) is 1358 K. The heat of fusion (or melting) is 13.26 kJ/mol. The densities of solid and liquid Cu at the melting temperature are $8.82 \, g/cm^3$ and $8.02 \, g/cm^3$, respectively. The atomic weight of Cu is 63.55 g/mol. Assuming that the constant pressure heat capacities of solid Cu and liquid Cu are the same, 24.0 J/(mol K). Please answer the following questions:

(a) What is the difference in chemical potentials between solid Cu and liquid Cu at 1358 K.
(b) Estimate what is the slope (dp/dT) of the phase boundary between solid and liquid Cu on the p–T phase diagram at melting temperature 1358 K.

20. The equilibrium vapor pressure of solid Cu as a function of temperature is given by

$$\ln\left(\frac{p}{1\,\text{bar}}\right) = -\frac{45,650}{T} - 0.306 \ln T + 10.81$$

(a) Determine the equilibrium vapor pressure above solid Cu at its equilibrium melting temperature of 1358 K.
(b) Calculate the heat of sublimation of solid Cu at 1358 K.
(c) What is the heat capacity difference between solid and vapor of Cu?

21. The solid $\beta - Ti_3O_5$ to solid $\lambda - Ti_3O_5$ phase transition takes place around 375 K at 0.1 bar with heat of transition of 230 kJ/L. The unit cell volumes of $\beta - Ti_3O_5$ and $\lambda - Ti_3O_5$ are $\sim 350 \, \text{Å}^3$ and $\sim 371 \, \text{Å}^3$, respectively. Assuming that both the heat capacity and the compressibility of the two structures are the same

(a) Determine the molar volumes of $\beta - Ti_3O_5$ and $\lambda - Ti_3O_5$ in unit of m^3/mol.
(b) Determine the molar heat of transition in unit of J/mol.
(c) Determine the molar entropy of transition in unit of J/(mol K).
(d) Determine the chemical potential of transition as a function of temperature.
(e) Determine the slope of the $\beta - Ti_3O_5/\lambda - Ti_3O_5$ two-phase boundary (dp/dT) on the pressure (p) and temperature (T) phase diagram.

(f) Determine the equation describing the transition pressure as a function of transition temperature.

(g) What pressure is required in unit of bar to obtain a transition temperature of 400 K for the transition from solid $\beta - Ti_3O_5$ to solid $\lambda - Ti_3O_5$?

Chapter 12
Chemical Potentials of Solutions

A pure material with one component is called a pure, single-component, or unary system. A material consisting of two, three, or more components is called a binary, ternary or multicomponent system. Solutions, by definition, have more than one chemical components uniformly mixed together. The previous chapters are mainly focused on simple, homogeneous materials and establishing the relationships among the thermodynamic properties such as temperature, entropy, pressure, volume, chemical potential, amount of substance, and energy. This chapter focuses on the chemical compositional dependence of thermodynamic properties, in particular, the compositional dependence of chemical potential of a solution and its components at constant temperature and pressure. Therefore, we first review the representation of the chemical composition of a material.

12.1 Representation of Chemical Composition

The chemical composition of a solution can be expressed in several forms. The most straightforward representation of composition is the amount $(N_1, N_2, \ldots, N_i, \ldots, N_n)$ for each component $1, 2, \ldots, n$. Since we usually use one mole as the basis for performing thermodynamic calculations, the chemical composition is most often expressed by the mole fraction x_i of component i in a binary or multicomponent material, i.e.,

$$x_i = \frac{N_i}{\sum_{i=1}^{n} N_i} = \frac{N_i}{N} \tag{12.1}$$

where N is the total number of moles of an n-component material.

From the definition of mole fraction, it is easy to see that

© Springer Nature Singapore Pte Ltd. 2022
L.-Q. Chen, *Thermodynamic Equilibrium and Stability of Materials*,
https://doi.org/10.1007/978-981-13-8691-6_12

$$\sum_{i=1}^{n} x_i = 1 \tag{12.2}$$

$$\sum_{i=1}^{n} dx_i = 0 \tag{12.3}$$

where dx_i is the differential of x_i. Therefore, there are only $n - 1$ independent mole fractions in an n-component system. We typically call a component with a small x_i as solute and one with a large x_i as solvent.

We distinguish composition x_i, which is the mole fraction, from concentration c_i, which is referred to as the amount of component i per unit volume, e.g., moles per cubic meter (mol/m^3),

$$c_i = \frac{N_i}{V} \tag{12.4}$$

$$c = \frac{N}{V} \tag{12.5}$$

$$x_i = \frac{c_i}{c} \tag{12.6}$$

where V is the total volume of the solution, and c is the total concentration of n components. The concentration c_i is often called molar concentration or molarity of component i.

There are other representations of chemical compositions such as weight fraction (w_i)

$$w_i = \frac{m_i}{\sum_{i=1}^{n} m_i} \tag{12.7}$$

or molality (M_i),

$$M_i = \frac{N_i(\text{mol})}{\sum_{j=1, j \neq i}^{n} m_j(\text{kg})} \tag{12.8}$$

where m_j is the mass of species j.

In polymer solutions, different types of molecule components often have very different molecular sizes, and hence, the composition of a polymer is often represented by the volume fraction of each type of polymer, i.e.,

$$\varphi_i = \frac{N_i v_i}{\sum_{i=1}^{n} N_i v_i} = \frac{x_i v_i}{\sum_{i=1}^{n} x_i v_i} = \frac{V_i}{V} \tag{12.9}$$

where N_i and v_i are the number of moles and the molar volume of molecular type i, and V_i and V are the volume occupied by i type of molecules in a solution and the total volume of the solution.

12.2 Fundamental Equation of Thermodynamics of a Multicomponent Solution

For a simple system with n number of components, the differential form for the fundamental equation of thermodynamics in terms of internal energy U is given by

$$dU = TdS - pdV + \mu dN = TdS - pdV + \mu_1 dN_1 + \mu_2 dN_2 + \cdots \mu_n dN_n$$
(12.10)

where T is temperature, S entropy, p pressure, V volume, $N = (N_1 + N_2 + \cdots + N_n)$ the total number of moles, and $N_i, N_2, \ldots, N_n$ the number of moles of component $1, 2, \ldots, n$. The μ in the above equation is the chemical potential or the molar Gibbs free energy of the whole multicomponent system,

$$\mu = \left(\frac{\partial U}{\partial N}\right)_{S,V}$$
(12.11)

and $\mu_1, \mu_2, \ldots,$ and μ_n are the chemical potentials of each individual component $1, 2, \ldots, n$,

$$\mu_i = \left(\frac{\partial U}{\partial N_i}\right)_{S,V,N_{j \neq i}}$$
(12.12)

Based on the above equation, the chemical potential of a component i in a homogeneous phase can be interpreted as the increase in the internal energy of the phase per mole of i added, with the addition being made at constant S and V and fixed numbers of moles of all the other components.

The integrated form for Eq. (12.10) is

$$U = TS - pV + \mu N = TS - pV + \mu_1 N_1 + \mu_2 N_2 + \cdots \mu_n N_n \qquad (12.13)$$

The above integrated fundamental equation of thermodynamics can be rewritten in terms of Gibbs free energy G as

$$G = U - TS + pV = H - TS = F + pV = \sum_{i=1}^{n} \mu_i N_i$$
(12.14)

where H is enthalpy, and F is Helmholtz free energy. The differential form for the Gibbs free energy is given by

$$dG = -SdT + Vdp + \sum_{i=1}^{n} \mu_i dN_i \qquad (12.15)$$

where the chemical potentials μ_i is defined as

$$\mu_i = \left(\frac{\partial G}{\partial N_i} \right)_{T,p,N_{j \neq i}} \qquad (12.16)$$

Therefore, the chemical potential of a component i in a homogeneous phase can also be interpreted as the increase in the Gibbs free energy of the system per mole of i added, with the addition being made at constant T and p and fixed numbers of moles of all the other components.

12.3   Chemical Potential as Fundamental Equation for a Multicomponent System

Gibbs pointed out that for the purpose of defining chemical potentials, any chemical element or any combination of elements in given proportions may be considered as a substance, regardless of whether it may or may not exist itself as a homogeneous body. To see this, let us rewrite the integrated form (Eq. 12.14) for the Gibbs free energy again,

$$G = U - TS + pV = \mu N = \mu_1 N_1 + \mu_2 N_2 + \cdots \mu_n N_n \qquad (12.17)$$

Therefore, we can write down the chemical potential μ of a homogeneous multi-component system in terms of Gibbs free energy per mole of substance (G/N) or the molar Gibbs free energy (g), or a combination of molar internal energy u, molar thermal energy Ts, and molar mechanical energy pv, or the chemical potentials (sometimes referred to as partial molar free energy) of n individual components or species, $\mu_1, \mu_2, \ldots,$ and μ_n, i.e.,

$$\mu = \frac{G}{N} = g = u - Ts + pv = \mu_1 x_1 + \mu_2 x_2 + \cdots \mu_n x_n \qquad (12.18)$$

where s and v are molar entropy and molar volume of the solution, and $x_i (= N_i/N)$ is the mole fraction of component i in the solution. Therefore, the chemical potential of a homogeneous system is simply the Gibbs free energy or chemical energy per mole of the homogeneous system. The chemical potential of a component is the Gibbs free energy per mole of that component in the homogeneous solution.

It should be emphasized that $\mu(T, p, x_1, x_2, \ldots, x_n)$ in Eq. (12.18) is a fundamental equation of thermodynamics for the multicomponent system and thus contains all the thermodynamic information about the multicomponent system except the size of the system. The chemical potential of a specific component as a function of temperature, pressure, and mole fractions of all components, $\mu_i(T, p, x_1, x_2, \ldots, x_n)$, is an equation of state and thus contains less thermodynamic information than $\mu(T, p, x_1, x_2, \ldots, x_n)$. Therefore, one can derive all the μ_i for each component from μ, but one cannot derive μ from one or a partial list of μ_i. However, the knowledge about all the chemical potentials, $\mu_1, \mu_2, \ldots$, and μ_n, together contains the same information as μ as it can be easily seen from Eq. (12.18).

The differential form for the chemical potential of a multicomponent system is

$$d\mu = -sdT + vdp + \mu_1 dx_1 + \mu_2 dx_2 + \cdots \mu_n dx_n \qquad (12.19)$$

and the differential form for the chemical potentials of each individual component in a multicomponent system with composition $x_1, x_2, \ldots, x_n$ are

$$d\mu_i = -s_i dT + v_i dp \qquad (12.20)$$

where s_i and v_i are the molar entropy and molar volume of component i in a multicomponent solution, also called the partial molar entropy and partial molar volume of component i, respectively, in the literature.

It is worth pointing out that although the Gibbs free energy has been the most often mentioned energy function in most existing thermodynamics literature, almost all materials processes are more conveniently described by the changes in chemical potentials. As a matter of fact, the molar Gibbs free energies of elemental substances, compounds, and solution phsaes in all existing thermodynamic databases are actually the chemical potentials of the elements, compounds, and solutions. Therefore, we employ chemical potential as a function of temperature, pressure, and composition as the fundamental equation of thermodynamics for the rest of the discussion of equilibrium states in a multicomponent system.

12.4 Chemical Potential of a Mixture of Pure Components

The chemical potential of a mixture of pure components is simply given by

$$\mu^\circ(T, p, x_1, x_2, \ldots, x_n) = x_1\mu_1^\circ(T, p) + x_2\mu_2^\circ(T, p) + \cdots + x_n\mu_n^\circ(T, p) \qquad (12.21)$$

where $x_1, x_2, \ldots, x_n$ are the mole fractions of pure component 1, 2, ..., n, and $\mu_1^\circ, \mu_2^\circ,$ ..., μ_n° are the chemical potentials of pure component 1, 2, ..., n at temperature T and p. For the examples without specifying the pressure, we assume p is at 1 bar, and

Fig. 12.1 Illustration of chemical potential of a mixture of pure components as a function of composition x_B in a A − B binary system

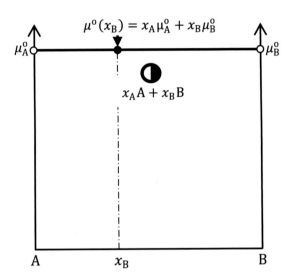

thus, μ_1^o, μ_2^o, ..., μ_n^o are the chemical potentials of the standard states for components 1, 2, ..., n at temperature T and 1 bar.

For a binary mixture of pure A and pure B, the chemical potential as a function of overall composition x_B is given by

$$\mu^o(T, x_B) = x_A \mu_A^o(T) + x_B \mu_B^o(T)$$

If we plot $\mu^o(x_B, T)$ as a function of x_B, it is represented by a straight line, e.g., the horizontal straight line on Fig. 12.1. Based on the schematic plot, one can easily read that the chemical potential of a mixture of x_A fraction of pure A and x_B fraction of pure B is given by $\mu^o(x_B)$. It should be noted that the vertical axis at A ($x_B = 0$) represents the chemical potential of A, and the vertical axis B ($x_B = 1$) represents the chemical potential of B. These two vertical axes are shifted so that the chemical potentials of A and B at standard states, $\mu_A^o(T)$ and $\mu_B^o(T)$, are on the same horizontal line although they have different values.

12.5 Chemical Potential of a Component in a Multicomponent Solution

The chemical potential μ_i of a given component i in a solution is often expressed in terms of the activity a_i of each component in the solution, i.e.,

$$\mu_i(T, x_1, x_2, \ldots, x_n) = \mu_i^o(T) + RT \ln a_i(T, x_1, x_2, \ldots, x_n) \tag{12.22}$$

where μ_i^o is the chemical potential of i at its standard state (stable state of pure i at temperature T and pressure 1 bar), which is only a function of temperature.

In general, the activity a_i is a function of temperature and composition $x_1, x_2, \ldots, x_n$. It is a common practice to write the activity a_i as the product of an activity coefficient γ_i and mole fraction x_i,

$$a_i = \gamma_i x_i \tag{12.23}$$

Therefore, Eq. (12.22) can be rewritten as

$$\mu_i(T, x_1, x_2, \ldots, x_n) = \mu_i^o(T, 1\,\text{bar}) + RT \ln \gamma_i(T, x_1, x_2, \ldots, x_n) + RT \ln x_i \tag{12.24}$$

where $RT \ln x_i$ represents the chemical potential change due to the random mixing of component i into the solution with composition x_i, i.e., due to the configurational entropy change, and $RT \ln \gamma_i$ represents the chemical potential change due to the chemical bonding energy changes of component i in a solution as compared to i in its pure state as well as the contribution from the thermal energy changes associated with changes in the atom movements or lattice vibrations in a solid due to the mixing of one mole of i into a solid solution.

12.6   Chemical Potential of Homogeneous Solution from Component Chemical Potentials

One can generalize the definition of chemical potential for a given component to any combination of different components with a fixed composition. As a matter of fact, one can define an infinite number of chemical potentials for a given solution since we can have an infinite number of different compositions for a solution. For example, the chemical potential $\mu(x_B)$ of a homogeneous binary solution as a function of composition x_B can be expressed in terms of chemical potentials $\mu_A(x_B)$ and $\mu_B(x_B)$ of the two components A and B as a function of x_B,

$$\mu(x_B) = x_A \mu_A(x_B) + x_B \mu_B(x_B)$$

where $\mu(x_B)$ is shown by the curve on the left schematic drawing of Fig. 12.2. In Fig. 12.2, μ_A^o and μ_B^o are the chemical potentials of A and B at their standard states, i.e., pure A and B in their stable states at the particular temperature and 1 bar. The chemical potentials of A and B in the solutions are related to the chemical potentials of A and B at the standard state by the following equations:

$$\mu_A(T, x_B) = \mu_A^o(T) + RT \ln a_A$$
$$\mu_B(T, x_B) = \mu_B^o(T) + RT \ln a_B$$

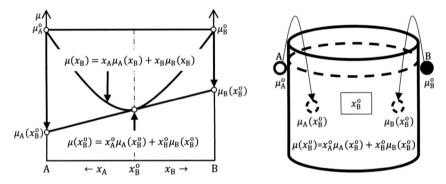

Fig. 12.2 Left: Illustration of chemical potential $\mu(T, x_B)$ of a solution as a function of composition x_B (represented by the schematic curve), the chemical potentials μ_A^o and μ_B^o of pure A and B, chemical potential of the solution $\mu(x_B^o)$ at composition x_B^o, and chemical potentials $\mu_A(x_B^o)$ and $\mu_B(x_B^o)$ of A and B in the solution with composition x_B^o; Right: Illustration of formation of a solution of composition x_B^o

If we mix x_A^o moles of pure A and x_B^o moles of pure B to form a homogeneous solution with composition x_A^o and x_B^o (right illustration of Fig. 12.2), the chemical potential of the solution $\mu(x_B^o)$ at composition x_A^o and x_B^o is illustrated on the left schematic of Fig. 12.2. The relationship among $\mu(x_B^o)$, $\mu_A(x_B^o)$, and $\mu_B(x_B^o)$ is shown in the left illustration of Fig. 12.2 by the tangent line and its intercepts with the $x_B = 0$ and $x_B = 1$ vertical axes. At equilibrium, both the chemical potential of each component and the chemical potential of the solution itself are uniform.

It is straightforward to extend the chemical potential of a binary solution to an n-component system with each component in an n-component solution associated with a chemical potential $\mu_i(x_1, x_2, \ldots, x_n)$. The chemical potential of a solution as a function of composition, $x_1, x_2, \ldots, x_n$, is given by chemical potentials of individual components in the solution weighted by their mole fractions,

$$\mu(T, x_1, x_2, \ldots, x_n) = \sum_{i=1}^{n} x_i \mu_i(T, x_1, x_2, \ldots, x_n) \qquad (12.25)$$

Using Eq. (12.22), we can rewrite Eq. (12.25) as

$$\mu(T, x_1, x_2, \ldots, x_n) = \sum_{i=1}^{n} x_i \mu_i^o(T) + RT \sum_{i=1}^{n} x_i \ln a_i(T, x_1, x_2, \ldots, x_n) \quad (12.26)$$

The chemical potential of a solution as a function of temperature and composition is the fundamental equation of thermodynamics from which one can derive all the thermodynamic properties of the solution with respect to temperature and chemical composition.

12.7 Obtaining Chemical Potentials of Components from Chemical Potential of Solution

The chemical potential of a solution as a function of composition contains all the information about the molar thermodynamic properties of a solution as a function of temperature and composition. For example, one can determine all the chemical potentials of each individual component, μ_i, as a function of composition from the knowledge of the chemical potential μ of the solution as a function of composition. The integrated and differential forms for the fundamental equation of thermodynamics for a multicomponent system at constant temperature and pressure are given by

$$\mu = \sum_{i=1}^{n} \mu_i x_i \tag{12.27}$$

and

$$d\mu = \sum_{i=1}^{n} \mu_i dx_i \tag{12.28}$$

Since $\sum_{i=1}^{n} x_i = 1$, we have $x_i = 1 - \sum_{j=1, j\neq i}^{n} x_j$, so we can rewrite Eq. (12.27) as

$$\mu = \mu_i + \sum_{j=1, j\neq i}^{n} x_j (\mu_j - \mu_i) \tag{12.29}$$

From $\sum_{i=1}^{n} dx_i = 0$, we can write $dx_i = -\sum_{j=1, j\neq i}^{i=n} dx_j$. Using this relation in Eq. (12.28), we can obtain the following relation,

$$\left(\frac{\partial \mu}{\partial x_j} \right)_{j\neq i} = (\mu_j - \mu_i) \tag{12.30}$$

By substituting Eq. (12.30) into Eq. (12.29), we have

$$\mu = \mu_i + \sum_{j=1, j\neq i}^{n} x_j (\mu_j - \mu_i) = \mu_i + \sum_{j=1, j\neq i}^{n} x_j \frac{\partial \mu}{\partial x_j} \tag{12.31}$$

Expressing μ_i in terms of μ, we get

$$\mu_i = \mu - \sum_{j=1, j\neq i}^{n} x_j \frac{\partial \mu}{\partial x_j} \tag{12.32}$$

Therefore, the knowledge of all the chemical potentials of each individual component of a solution is equivalent to the knowledge of the chemical potential of the solution as a function of composition. We can obtain the chemical potential of the solution using Eq. (12.27) if we have the knowledge of the chemical potentials of all the components, and we can calculate the chemical potentials of all the components using Eq. (12.32) if we have the chemical potential of the solution as a function of composition.

For example, in a binary system, the chemical potential of the solution and chemical potentials of each component as a function of composition are simply related through

$$\mu = x_A \mu_A + x_B \mu_B \tag{12.33}$$

Therefore, if we have the knowledge of μ_A and μ_B as a function of composition, we can easily obtain the chemical potential of the solution as a function of composition.

On the other hand, we can obtain the chemical potentials of the two components from the chemical potential of the solution as a function of composition. To see this, let us rewrite Eq. (12.33) and its differential form as

$$\mu = x_A \mu_A + x_B \mu_B = (1 - x_B)\mu_A + x_B \mu_B = \mu_A + x_B(\mu_B - \mu_A) \tag{12.34}$$

$$d\mu = \mu_A dx_A + \mu_B dx_B = (\mu_B - \mu_A)dx_B \tag{12.35}$$

Expressing μ_A and μ_B in terms of μ, we have

$$\mu_A = \mu - x_B(\mu_B - \mu_A) = \mu - x_B \frac{\partial \mu}{\partial x_B} \tag{12.36}$$

$$\mu_B = \mu - x_A(\mu_A - \mu_B) = \mu - x_A \frac{\partial \mu}{\partial x_A} \tag{12.37}$$

The above two equations describe the same straight-line tangent to the chemical potential μ of the solution at x_B. The intercept of this tangent line at $x_B = 0$ is μ_A, and the intercept of the same tangent line at $x_B = 1$ is μ_B. Therefore, suppose that we have an A − B solution at composition x_B^o, the chemical potential of the solution is $\mu(x_B^o)$, and the chemical potentials of A and B in the solution at x_B^o are $\mu_A(x_B^o)$ and $\mu_B(x_B^o)$, which can be obtained by the intercepts of the tangent line to the $\mu(x_B)$ at x_B^o with the vertical axis at $x_B = 0$ and $x_B = 1$, respectively (see Fig. 12.2 for the illustration).

12.8 Chemical Potential of a Uniform Solution at a Fixed Composition

Suppose that we have an $A - B$ solution at a fixed composition x_B^o, and the chemical potential of the solution is $\mu(x_B^o)$. Here we can define the chemical potential of any combination of A and B, represented by composition x_B, with the chemical potentials of each component defined by the uniform solution with composition x_B^o, i.e.,

$$\mu'(x_B) = x_A \mu_A(x_B^o) + x_B \mu_B(x_B^o) \tag{12.38}$$

i.e., $\mu'(x_B)$ is the chemical potential of a substance with x_A fraction of A atoms and x_B fraction of B atoms with the component chemical potentials defined by the uniform solution with composition x_B^o (Fig. 12.3)).

For example, adding or taking a small amount of solution with a composition x_B to or from a large quantity of solution with composition x_B^o, the chemical potential of this small amount of solution with composition x_B is given by the tangent line $\mu'(x_B)$. Therefore, not only the chemical potentials of A and B are uniform at equilibrium, the chemical potential $\mu'(x_B)$ of a substance with any combination of A atoms and B atoms is also uniform. There is an infinite number of such chemical potentials corresponding to each composition with their chemical potential values given by the tangent line $\mu'(x_B)$. According to Gibbs, "*In the same homogeneous mass, therefore, we may distinguish the potentials for an indefinite number of substances, each of which has a perfectly determined value.*" However, the total number of independent chemical potentials in a multicomponent system is finite because of the Gibbs–Duhem relation relating all the potentials.

Fig. 12.3 Tangent line $\mu'(x_B)$ represents the chemical potential of the solution as a function of composition x_B with the chemical potentials of each component defined by the solution at composition x_B^o

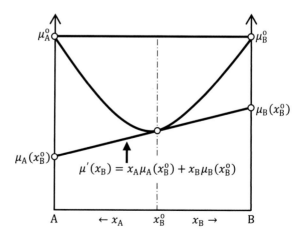

12.9   Gibbs–Duhem Equation for Multicomponent Systems

In an n-component system, one can define an infinite number of chemical potentials since the composition of a homogeneous system can vary continuously, and each homogeneous solution with a given composition can be associated with a chemical potential. However, there are only $n - 1$ independent chemical potentials since the chemical potential of a n-component solution is related to the n chemical potentials for each component (Eq. 12.18), and n chemical potentials for each component in the solution are related by the Gibbs–Duhem relation.

To derive the Gibbs–Duhem relation, we start from the differential form of Gibbs free energy,

$$dG = -SdT + Vdp + \mu_1 dN_1 + \mu_2 dN_2 + \cdots \mu_n dN_n \qquad (12.39)$$

Taking a differential of Eq. (12.17), we have

$$dG = \mu_1 dN_1 + \mu_2 dN_2 + \cdots \mu_n dN_n + N_1 d\mu_1 + N_2 d\mu_2 + \cdots N_n d\mu_n \qquad (12.40)$$

Comparing Eqs. (12.39) and (12.40), we get

$$SdT - Vdp + N_1 d\mu_1 + N_2 d\mu_2 + \cdots N_n d\mu_n = 0 \qquad (12.41)$$

which is the Gibbs–Duhem relation for a multicomponent system.

For a single component system, the Gibbs–Duhem relation is simply the differential form of the chemical potential,

$$SdT - Vdp + Nd\mu = 0 \text{ or } d\mu = -sdT + vdp \qquad (12.42)$$

For a binary system at constant temperature and pressure, the Gibbs–Duhem relation is

$$N_1 d\mu_1 + N_2 d\mu_2 = 0 \qquad (12.43)$$

The general Gibbs–Duhem relation for a multicomponent system can also be directly obtained from the integrated and differential forms of chemical potential,

$$\mu = \sum_{i=1}^{n} x_i \mu_i \qquad (12.44)$$

$$d\mu = -sdT + vdp + \sum_{i=1}^{n} \mu_i dx_i \qquad (12.45)$$

By differentiating both sides of Eq. (12.44), we have

$$d\mu = \sum_{i=1}^{n} (\mu_i dx_i + x_i d\mu_i) \tag{12.46}$$

Comparing Eqs. (12.45) and (12.46), we obtain the Gibbs–Duhem relation for a multicomponent system,

$$-sdT + vdp - \sum_{i=1}^{n} x_i d\mu_i = 0 \tag{12.47}$$

This Gibbs–Duhem relation relates all the intensive parameters $(T, p, \mu_1, \mu_2, \ldots, \mu_n)$ of a system, and thus for an n-component system, the number of independent intensive parameters is $n + 1$. At constant temperature and pressure,

$$\sum_{i=1}^{n} x_i d\mu_i = 0 \tag{12.48}$$

and hence, the number of independent chemical potentials for an n-component system at constant temperature and pressure is $n - 1$.

For a binary system at constant temperature and pressure, the Gibbs–Duhem relation is reduced to

$$x_1 d\mu_1 + x_2 d\mu_2 = 0 \tag{12.49}$$

Expressing the chemical potentials in terms of activities a_i or activity coefficients γ_i and compositions leads to other forms of Gibbs–Duhem relations at constant temperature and pressure for a binary system,

$$x_1 d\ln a_1 + x_2 d\ln a_2 = 0 \tag{12.50}$$

or

$$x_1 d\ln \gamma_1 + x_2 d\ln \gamma_2 = 0 \tag{12.51}$$

where we have used the fact that

$$x_1 d\ln x_1 + x_2 d\ln x_2 \equiv 0 \tag{12.52}$$

One of the applications of the Gibbs–Duhem relation is to find the chemical potential, activity, or activity coefficient of a component in a binary system if the chemical potential, activity, or activity coefficient of the other component is known. If the chemical potentials, activities, and activity coefficients of both components are known, the accuracies and consistencies of their values can be checked with the Gibbs–Duhem relation.

12.10   Thermodynamic Stability of a Solution and Thermodynamic Factor

In Chap. 7, we focus on the stability of an equilibrium state with respect to perturbations in thermal and mechanical variables (S, T, V, and p). Here we discuss the stability of an equilibrium state with respect to chemical variables (μ_i and N_i, $i = 1, 2, \ldots, n$) at a given temperature and pressure. We formulate chemical stability criteria of an equilibrium state similar to the thermal and mechanical stability. The entropy produced from the perturbations in chemical variables (μ_i and N_i, $i = 1, 2, \ldots, n$) of the stable equilibrium state at constant temperature and pressure can be written as,

$$\Delta S^{\text{ir}} = \delta^2 S = -\frac{\sum_{i=1}^{n} \mathrm{d}\mu_i \mathrm{d}N_i}{T} < 0 \tag{12.53}$$

At constant temperature and pressure, the entropy produced is related to the Gibbs free energy ΔG,

$$\Delta G = -T\Delta S^{\text{ir}} = \sum_{i=1}^{n} \mathrm{d}\mu_i \mathrm{d}N_i > 0 \tag{12.54}$$

We now express perturbations $\mathrm{d}\mu_i$ in terms of perturbation $\mathrm{d}N_j$,

$$\mathrm{d}\mu_i = \sum_{j=1}^{n} \left(\frac{\partial \mu_i}{\partial N_j}\right) \mathrm{d}N_j \tag{12.55}$$

Substituting Eq. (12.55) into Eq. (12.54), we have

$$\Delta G = \sum_{i=1}^{n} \sum_{j=1}^{n} \left(\frac{\partial \mu_i}{\partial N_j}\right) \partial N_i \partial N_j = \sum_{i=1}^{n} \sum_{j=1}^{n} \left(\frac{\partial^2 G}{\partial N_i \partial N_j}\right) \partial N_i \partial N_j \tag{12.56}$$

For a thermodynamically stable solution,

$$\Delta G = \sum_{i=1}^{n} \sum_{j=1}^{n} \left(\frac{\partial^2 G}{\partial N_i \partial N_j}\right) \partial N_i \partial N_j > 0 \tag{12.57}$$

Therefore, the stability of a multicomponent solution requires that the following determinant is positive-definite, i.e.,

$$\begin{vmatrix} \dfrac{\partial^2 G}{\partial N_1^2} & \cdots & \dfrac{\partial^2 G}{\partial N_1 \partial N_n} \\ \vdots & \ddots & \vdots \\ \dfrac{\partial^2 G}{\partial N_n \partial N_1} & \cdots & \dfrac{\partial^2 G}{\partial N_n^2} \end{vmatrix} > 0 \tag{12.58}$$

To illustrate the thermodynamic stability conditions of a homogeneous solution, for simplicity, let us consider a binary system A-B. For a stable or metastable region of a binary system,

$$\left(\frac{\partial^2 G}{\partial N_A^2}\right)_{T,p,N_B} = \left(\frac{\partial \mu_A}{\partial N_A}\right)_{T,p,N_B} > 0, \left(\frac{\partial^2 G}{\partial N_B^2}\right)_{T,p,N_A} = \left(\frac{\partial \mu_B}{\partial N_B}\right)_{T,p,N_A} > 0 \tag{12.59}$$

and

$$\begin{vmatrix} \dfrac{\partial^2 G}{\partial N_A^2} & \dfrac{\partial^2 G}{\partial N_A \partial N_B} \\ \dfrac{\partial^2 G}{\partial N_B \partial N_A} & \dfrac{\partial^2 G}{\partial N_B^2} \end{vmatrix} = \begin{vmatrix} \dfrac{\partial \mu_A}{\partial N_A} & \dfrac{\partial \mu_B}{\partial N_A} \\ \dfrac{\partial \mu_A}{\partial N_B} & \dfrac{\partial \mu_B}{\partial N_B} \end{vmatrix} > 0 \tag{12.60}$$

which implies that the chemical potential of a component increases with the amount of the component in a thermodynamically stable solution.

If the total number of moles N is fixed,

$$\Delta G = -T \Delta S^{ir} = d\mu_A dN_A + d\mu_B dN_B = d(\mu_B - \mu_A)dN_B > 0 \tag{12.61}$$

For $N = 1$ mol

$$\Delta \mu = d\mu_A dx_A + d\mu_B dx_B = d(\mu_B - \mu_A)dx_B > 0 \tag{12.62}$$

Since

$$\left(\frac{\partial \mu}{\partial x_B}\right)_{T,p} = \mu_B - \mu_A \tag{12.63}$$

we can rewrite Eq. (12.62) as

$$\Delta \mu = d\mu_A dx_A + d\mu_B dx_B = d\left(\frac{\partial \mu}{\partial x_B}\right)_{T,p} dx_B > 0 \tag{12.64}$$

or

$$\Delta \mu = d\mu_A dx_A + d\mu_B dx_B = \left(\frac{\partial^2 \mu}{\partial x_B^2}\right)_{T,p} (dx_B)^2 > 0 \tag{12.65}$$

We can derive a similar expression for the change in chemical potential $\Delta\mu$ in terms of fluctuation dx_A in x_A. Therefore, for a stable binary solution,

$$\left(\frac{\partial^2\mu}{\partial x_B^2}\right)_{T,p} > 0, \left(\frac{\partial^2\mu}{\partial x_A^2}\right)_{T,p} > 0 \tag{12.66}$$

The second derivative of chemical potential of the solution is

$$\left(\frac{\partial^2\mu}{\partial x_B^2}\right)_{T,p} = \left(\frac{\partial(\mu_B - \mu_A)}{\partial x_B}\right)_{T,p} = \left(\frac{\partial\mu_B}{\partial x_B}\right)_{T,p} + \left(\frac{\partial\mu_A}{\partial x_A}\right)_{T,p} \tag{12.67}$$

We express the chemical potentials of a component in terms of activity coefficient and composition,

$$\mu_i = \mu_i^\circ + RT \ln \gamma_i + RT \ln x_i \tag{12.68}$$

It follows that

$$\left(\frac{\partial\mu_A}{\partial \ln x_A}\right)_{T,p} = RT\left(1 + \frac{\partial \ln \gamma_A}{\partial \ln x_A}\right) = RT\psi$$

$$\left(\frac{\partial\mu_B}{\partial \ln x_B}\right)_{T,p} = RT\left(1 + \frac{\partial \ln \gamma_B}{\partial \ln x_B}\right) = RT\psi$$

Therefore, we have

$$\left(\frac{\partial\mu_A}{\partial x_A}\right)_{T,p} = \frac{RT}{x_A}\psi \tag{12.69}$$

$$\left(\frac{\partial\mu_B}{\partial x_B}\right)_{T,p} = \frac{RT}{x_B}\psi \tag{12.70}$$

The second derivative of chemical potential with respect to composition is then given by

$$\left(\frac{\partial^2\mu}{\partial x_B^2}\right)_{T,p} = \left(\frac{\partial\mu_A}{\partial x_A}\right)_{T,p} + \left(\frac{\partial\mu_B}{\partial x_B}\right)_{T,p} = \frac{RT}{x_A x_B}\psi \tag{12.71}$$

where

$$\psi = \left(1 + \frac{\partial \ln \gamma_A}{\partial \ln x_A}\right) = \left(1 + \frac{\partial \ln \gamma_B}{\partial \ln x_B}\right) = \frac{x_A x_B}{RT}\left(\frac{\partial^2\mu}{\partial x_B^2}\right)_{T,p} \tag{12.72}$$

is called the thermodynamic factor in chemical diffusion kinetics, which is related to the second derivative of chemical potential of a solution with respect to composition.

The chemical diffusion coefficient in a solution is proportional to this thermodynamic factor, and hence, the chemical diffusion coefficient is negative within the unstable region for the solution.

For a stable solution,

$$\psi > 0$$

which implies that

$$\left(\frac{\partial \mu_A}{\partial x_A}\right)_{T,p} > 0, \text{ and } \left(\frac{\partial \mu_B}{\partial x_B}\right)_{T,p} > 0, \text{ and } \left(\frac{\partial^2 \mu}{\partial x_B^2}\right)_{T,p} > 0 \qquad (12.73)$$

The thermodynamic stability limit is reached if

$$\psi = 0, \left(\frac{\partial \mu_A}{\partial x_A}\right)_{T,p} = 0, \left(\frac{\partial \mu_B}{\partial x_B}\right)_{T,p} = 0, \left(\frac{\partial^2 \mu}{\partial x_B^2}\right)_{T,p} = 0 \qquad (12.74)$$

The compositions that satisfy the conditions given in Eq. (12.74) are called spinodal points which separate the compositions for which the solutions are metastable and those for which the solutions are unstable.

In a thermodynamically unstable region,

$$\psi < 0, \left(\frac{\partial \mu_A}{\partial x_A}\right)_{T,p} < 0, \left(\frac{\partial \mu_B}{\partial x_B}\right)_{T,p} < 0, \left(\frac{\partial^2 \mu}{\partial x_B^2}\right)_{T,p} < 0 \qquad (12.75)$$

The above conditions are schematically shown in Fig. 12.4 for the range of compositions in which a solution is stable (S), metastable (MS), and unstable (US).

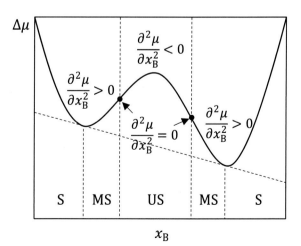

Fig. 12.4 Illustration of stable (S), metastable (MS), and unstable (US) regions based on the sign of the second derivative of chemical potential with respect to composition

Finally, at a critical point, the critical temperature and composition at which two spinodal points merge into one, the chemical potential of component i satisfies the following conditions,

$$\left(\frac{\partial^2 \mu}{\partial x_B^2}\right)_{T,p} = 0, \quad \left(\frac{\partial^3 \mu}{\partial x_B^3}\right)_{T,p} = 0 \tag{12.76}$$

12.11   Lever Rule

Lever rule derives from the mass conservation condition. It can be used to determine the fractions of phases if we have the information about the overall average composition and the equilibrium compositions of the individual phases.

Suppose that we have N moles of a binary solution with overall composition $\left(x_A^o, x_B^o\right)$. We now suppose that this homogeneous solution separates into two solutions: a solution with N^α moles and composition $\left(x_A^\alpha, x_B^\alpha\right)$, and the other with N^β moles and composition $\left(x_A^\beta, x_B^\beta\right)$. The amount of A, N_A, and amount of B, N_B, should be the same before and after the phase separation, so

$$N_A = x_A^o N = x_A^\alpha N^\alpha + x_A^\beta N^\beta \tag{12.77}$$

$$N_B = x_B^o N = x_B^\alpha N^\alpha + x_B^\beta N^\beta \tag{12.78}$$

Let us define the fractions of α and β phases in the two-phase mixture as

$$\varphi^\alpha = \frac{N^\alpha}{N}, \quad \varphi^\beta = \frac{N^\beta}{N} \tag{12.79}$$

The two conservation conditions, Eqs. (12.77) and (12.78), become

$$x_A^o = x_A^\alpha \varphi^\alpha + x_A^\beta \varphi^\beta, \quad x_B^o = x_B^\alpha \varphi^\alpha + x_B^\beta \varphi^\beta \tag{12.80}$$

Solving the above two equations for φ^α and φ^β, we have

$$\varphi^\alpha = \frac{x_B^\beta - x_B^o}{x_B^\beta - x_B^\alpha}, \quad \varphi^\beta = \frac{x_B^o - x_B^\alpha}{x_B^\beta - x_B^\alpha} \tag{12.81}$$

Equation (12.81) allows one to determine the fractions φ^α and φ^β of the two phases α and β based on the overall composition x_B^o of the two-phase mixture and the equilibrium compositions of two phases x_B^α and x_B^β. Equation (12.81) is the mathematical expression of the Lever Rule. Illustration of Fig. 12.5 explains why this mass conservation equation is called the Lever Rule.

Fig. 12.5 Illustration of
lever rule

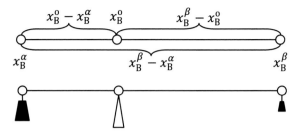

12.12   Chemical Potential Change of Solution Formation from Pure Components

The process of mixing n pure components, $A_1, A_2, \ldots$, and A_n, to form a homogeneous solution with composition $x_{A_1}, x_{A_2}, \ldots$, and x_{A_n} can be described as

$$x_{A_1} A_1 + x_{A_2} A_2 + \cdots + x_{A_n} A_n \rightarrow \left(x_{A_1} A_1, x_{A_2} A_2, \ldots, x_{A_n} A_n \right) \tag{12.82}$$

The chemical potential change for the formation of a homogeneous solution from pure components at the same temperature is given by

$$\Delta\mu\left(x_{A_1}, x_{A_2}, \ldots, x_{A_n}\right) = \mu\left(x_{A_1}, x_{A_2}, \ldots, x_{A_n}\right) - \sum_{i=1}^{n} x_{A_i} \mu_{A_i}^{o} \tag{12.83}$$

The chemical potential change due to the formation of a solution from pure components is often called the molar Gibbs free energy of mixing or forming a solution in existing literature. Equation (12.83) can be rewritten in terms of activities a_i,

$$\Delta\mu = RT \sum_{i=1}^{n} x_i \ln a_i \tag{12.84}$$

and

$$\Delta\mu_i = RT \ln a_i \tag{12.85}$$

where $\Delta\mu_i$ is the chemical potential change for component i and is often called partial molar free energy change of component i in existing literature. It represents the chemical potential change when one mole of pure species i is dissolved into the solution with composition $x_1, x_2, \ldots, x_n$.

The relation between $\Delta\mu_i$ and $\Delta\mu$ is given by an equation similar to Eq. (12.32). It is graphically shown in Fig. 12.6 using a binary solution as an example. The formation of a binary solution $x_A A x_B B$ from x_A moles of pure A and x_B moles of pure B can be expressed as

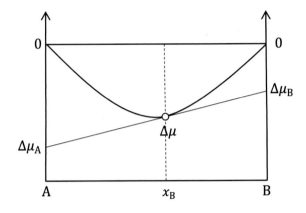

Fig. 12.6 Relation between chemical potential change of solution formation $\Delta\mu$ and the chemical potential change of each component $\Delta\mu_A$ and $\Delta\mu_B$

$$x_A A + x_B B \rightarrow x_A A x_B B \tag{12.86}$$

The chemical potential change from x_A moles of pure A and x_B moles of pure B to form one mole of $x_A A x_B B$ solution with composition x_A and x_B is

$$\Delta\mu = x_A \Delta\mu_A + x_B \Delta\mu_B \tag{12.87}$$

i.e., one can use the chemical potential change for each individual species to obtain the chemical potential change for the solution. On the other hand, one could obtain the chemical potential change for each individual component using the chemical potential change for the solution. We can rewrite Eq. (12.87) as

$$\Delta\mu = x_A \Delta\mu_A + x_B \Delta\mu_B = \Delta\mu_A + x_B(\Delta\mu_B - \Delta\mu_A) = \Delta\mu_A + x_B \frac{\partial \Delta\mu}{\partial x_B} \tag{12.88}$$

or

$$\Delta\mu = x_A \Delta\mu_A + x_B \Delta\mu_B = \Delta\mu_B + x_A(\Delta\mu_A - \Delta\mu_B) = \Delta\mu_B + x_A \frac{\partial \Delta\mu}{\partial x_A} \tag{12.89}$$

Therefore, we can easily see that

$$\Delta\mu_A = \Delta\mu - x_B \frac{\partial \Delta\mu}{\partial x_B} \tag{12.90}$$

$$\Delta\mu_B = \Delta\mu - x_A \frac{\partial \Delta\mu}{\partial x_A} \tag{12.91}$$

which are similar to Eqs. (12.36) and (12.37). $\Delta\mu_A$ and $\Delta\mu_B$ can be graphically obtained from the intercepts at $x_B = 0$ and $x_B = 1$ of the tangent line to $\Delta\mu$ at x_B (Fig. 12.6).

12.13 Chemical Potential Change for Adding Pure Components into a Solution

Let us consider a process of adding a small quantity of pure A into a large amount of solution with composition x_B^o. The process can be described as

$$A(\text{pure}) \rightarrow A\left(\text{in A} - \text{B solution with composition } x_B^o\right)$$

The chemical potential change for the above process is

$$\Delta\mu_A = \mu_A\left(x_B^o\right) - \mu_A^o = RT \ln a_A \tag{12.92}$$

Similarly, the process of adding a small quantity of pure B into the solution can be written as

$$B(\text{pure}) \rightarrow B\left(\text{in A} - \text{B solution with composition } x_B^o\right)$$

and the chemical potential change for this process is

$$\Delta\mu_B = \mu_B\left(x_B^o\right) - \mu_B^o = RT \ln a_B \tag{12.93}$$

The process of adding a small amount, e.g., 1 mol, of a mixture with x_A fraction of pure A and x_B fraction of pure B into the solution with composition x_B^o can be written as

$$x_A A(\text{pure}) + x_B B(\text{pure}) \rightarrow (x_A A x_B B)\left(\text{in A} - \text{B solution with composition } x_B^o\right)$$

The chemical potential change is schematically illustrated in the right illustration of Fig. 12.7,

$$\Delta\mu = \mu'(x_B) - \mu^o(x_B) = \left[x_A\mu_A\left(x_B^o\right) + x_B\mu_B\left(x_B^o\right)\right] - \left(x_A\mu_A^o + x_B\mu_B^o\right) \tag{12.94}$$

which can be rewritten as

$$\Delta\mu = x_A\left[\mu_A\left(x_B^o\right) - \mu_A^o\right] + x_B\left[\mu_B\left(x_B^o\right) - \mu_B^o\right] \tag{12.95}$$

or

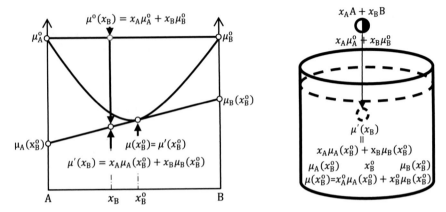

Fig. 12.7 Left: Illustration of chemical potential change from $x_A \mu_A\left(x_B^o\right) + x_B \mu_B\left(x_B^o\right)$ to $x_A \mu_A\left(x_B^o\right) + x_B \mu_B\left(x_B^o\right)$ due to the addition of x_A moles of pure A and x_B moles of pure B into a large amount of solution with composition solution x_B^o; Right: Illustration of the addition process

$$\Delta\mu = RT\left[x_A \ln a_A\left(x_B^o\right) + x_B \ln a_B\left(x_B^o\right)\right] \qquad (12.96)$$

12.14 Chemical Potential Change of Adding a Solution into Another Solution

The process of adding a small amount of A − B solution with composition x_B into a solution with composition x_B^o (see Fig. 12.8 for the illustration of this process) can be written as

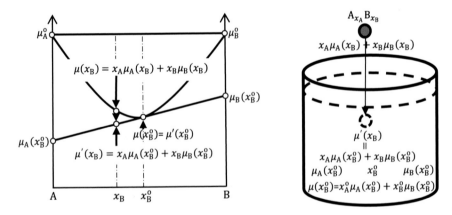

Fig. 12.8 Left: Illustration of chemical potential change from $x_A \mu_A(x_B) + x_B \mu_B(x_B)$ to $x_A \mu_A\left(x_B^o\right) + x_B \mu_B\left(x_B^o\right)$ due to the addition of a solution with composition x_A and x_B into a large amount of solution with composition solution x_B^o; Right: Illustration of the addition process

$$x_A A x_B B(\text{solution}) \rightarrow (x_A A x_B B)\left(\text{in A} - \text{B solution with composition } x_B^o\right)$$

Therefore, the chemical potential change for this process is

$$\Delta \mu = \left[x_A \mu_A\left(x_B^o\right) + x_B \mu_B\left(x_B^o\right)\right] - \left[x_A \mu_A(x_B) + x_B \mu_B(x_B)\right] \qquad (12.97)$$

or

$$\Delta \mu = x_A\left[\mu_A\left(x_B^o\right) - \mu_A(x_B)\right] + x_B\left[\mu_B\left(x_B^o\right) - \mu_B(x_B)\right] \qquad (12.98)$$

Expressing chemical potentials in terms of activities, we have

$$\Delta \mu = RT\left[x_A \ln \frac{a_A\left(x_B^o\right)}{a_A(x_B)} + x_B \ln \frac{a_B\left(x_B^o\right)}{a_B(x_B)}\right] \qquad (12.99)$$

The chemical potential change for this process is shown in Fig. 12.8.

12.15   Driving Force for Precipitation in a Solution

As discussed above, the chemical potential of a small amount of solution with composition $x_1^P, x_2^P, \ldots, x_n^P$ in a large amount of solution with composition $x_1^m, x_2^m, \ldots, x_n^m$ is given by

$$\mu'\left(x_1^P, x_2^P, \ldots, x_n^P\right) = \sum_{i=1}^{n} x_i^P \mu_i\left(x_1^m, x_2^m, \ldots, x_n^m\right) \qquad (12.100)$$

which is a linear function of composition, i.e., the multidimensional plane tangent to $\mu(x_i)$ at x_i^m. $\mu'(x_i)$ can be considered as the chemical potential of x_1 moles of component 1, x_2 moles of component 2, etc., taken from a solution with composition x_i^m. It should be emphasized that to determine the most stable state, one has to compare the chemical potentials among all the states of a solution with the same overall composition.

For example, given a large quantity of a binary A-B solution with composition $\left(x_A^m, x_B^m\right)$, if we take x_A^P moles of A and x_B^P moles of B from the solution, its chemical potential is given by

$$\mu\left(x_A^P, x_B^P\right) = x_A^P \mu_A\left(x_A^m, x_B^m\right) + x_B^P \mu_B\left(x_A^m, x_B^m\right) \qquad (12.101)$$

i.e., the chemical potential for the solution is fixed by the solution with composition $\left(x_A^m, x_B^m\right)$. This is shown in Fig. 12.9m. The chemical potential of the solution with composition $\left(x_A^m, x_B^m\right)$ is given by the open circle while the chemical potential of the solution with composition $\left(x_A^P, x_B^P\right)$ taken from the solution with composition

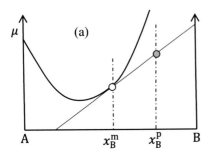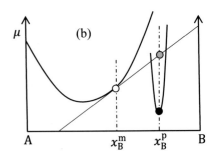

Fig. 12.9 a Chemical potential of a small amount of solution with composition $\left(x_A^P, x_B^P\right)$ taken from a large amount of matrix solution with composition $\left(x_A^m, x_B^m\right)$ is represented by the gray circle, and **b** the driving force for precipitation is given by the chemical potential difference between the chemical potential of the precipitate (dark circle) and that of the solution with composition $\left(x_A^P, x_B^P\right)$ from matrix

$\left(x_A^m, x_B^m\right)$ is represented by the light-shaded circle. This is particularly relevant to the precipitation process of a phase with composition $\left(x_A^P, x_B^P\right)$ from a matrix with composition $\left(x_A^m, x_B^m\right)$. As shown in Fig. 12.9b, the driving force D for precipitation is given by the chemical potential difference between the chemical potentials represented by the light-shaded and black circles. The chemical potential of the light-shaded circle represents the chemical potential of the solution with composition $\left(x_A^P, x_B^P\right)$ before it is transformed into the precipitate phase with the same composition, and the black-shaded circle represents the chemical potential of the stable precipitate phase.

12.16 Chemical Potential of a Two-Phase Mixture

It is important to remember that the relative stabilities of different phases or thermodynamic states in a multicomponent system should be determined under the same set of thermodynamic conditions, $T, p, x_1, x_2, \ldots, x_n$. A good example is the determination of driving forces for nucleation and growth of a new phase in a diffusional phase transition in which the composition of a nucleus is typically different from that of the matrix. One has to use the same composition for the initial and final states to determine the driving force for a process.

A very useful criterion for determining the stability of a homogeneous solution or compound with regard to phase separation or decomposition into two phases is the chemical potential of a two-phase mixture with each phase having its own chemical potential and composition. However, in order to compare the stability of a homogeneous state and a two-phase mixture, we have to make sure that the overall average composition of the two-phase mixture is the same as the homogeneous state.

Let us assume that the chemical potential of the homogeneous solution with composition $\left(x_A^o, x_B^o\right)$ is μ and the chemical potentials of the two phases, α with

Fig. 12.10 Illustration of chemical potential μ^o of a mixture of pure A and pure B, chemical potential of μ^{tp} of a two-phase mixture $\alpha + \beta$, and chemical potential of homogeneous solution μ

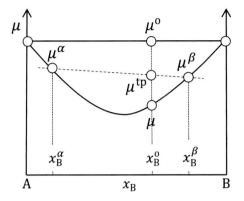

composition $\left(x_A^\alpha, x_B^\alpha\right)$ and β with composition $\left(x_A^\beta, x_B^\beta\right)$, are μ^α and μ^β, respectively (see Fig. 12.10). The chemical potential of a two-phase state μ^{tp} can be obtained using a weighted average with the phase fractions from the Lever Rule,

$$\mu^{tp} = \varphi^\alpha \mu^\alpha + \varphi^\beta \mu^\beta \qquad (12.102)$$

Using Eqs. (12.80) and (12.81) for the fractions φ^α and φ^β, we have

$$\mu^{tp} = \frac{x_B^\beta - x_B^0}{x_B^\beta - x_B^\alpha} \mu^\alpha + \frac{x_B^0 - x_B^\alpha}{x_B^\beta - x_B^\alpha} \mu^\beta \qquad (12.103)$$

which represents the straight line (the dotted line in Figs. 12.11, 12.12, and 12.13) connecting the two points in the two-dimensional space of μ and x_B,

$$\left(x_B^\alpha, \mu^\alpha\right) \text{ and } \left(x_B^\beta, \mu^\beta\right) \qquad (12.104)$$

Fig. 12.11 Illustration of an unstable solution with respect to phase separation

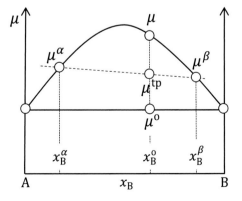

Fig. 12.12 An unstable
solution with respect to
phase separation within a
compositional range

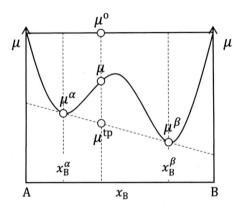

The compositions $x_B^\alpha = 0$ and $x_B^\beta = 1$ represent the pure A and pure B, and hence

$$\mu^o = x_A^o \mu_A^o + x_B^o \mu_B^o \qquad (12.105)$$

Figure 12.10 shows the relative chemical potentials of a homogeneous solution, a two-phase mixture with compositions x_B^α and x_B^β, and a mixture of pure A and B. The compositional dependence of μ has a positive curvature; i.e., it is a convex function of composition. It is easy to see that

$$\mu^o > \mu^{tp} > \mu$$

Therefore, a homogeneous solution with composition $\left(x_A^o, x_B^o\right)$ is more stable than a mixture of two phases with composition x_B^α and x_B^β, which itself is more stable than the mixture of pure A and pure B.

However, if the chemical potential of a homogeneous composition has a negative curvature or the shape is concave as shown in Fig. 12.11,

$$\mu^o < \mu^{tp} < \mu$$

In this case, a homogeneous solution is unstable, and the most stable state is a mixture of pure A and pure B.

If the dependence of chemical potential of the homogeneous solution on composition shows a double-well shape as shown in Fig. 12.12,

$$\mu^{tp} < \mu < \mu^o$$

within the composition range $\left(x_B^\alpha, x_B^\beta\right)$. In this case, a mixture of two phases with compositions x_B^α and x_B^β has the lowest chemical potential and is thus the most stable state. The chemical potentials of each component are uniform in the two-phase mixture,

$$\mu_A^\alpha = \mu_A^\beta \text{ and } \mu_B^\alpha = \mu_B^\beta \qquad (12.106)$$

12.17 Stability at 0 K (Convex Hull)

For single-component systems or for multicomponent compounds, or a multicomponent system with a fixed chemical composition, the stability of different states at 0 K can be obtained by simply comparing their energies or enthalpies. Figure 12.13 is a schematic illustration of how one could determine the equilibrium states as a function of composition in a binary A − B system at 0 K. The dotted circles represent the energies of different spatial arrangements of A and B atoms at different fixed chemical compositions. For a single-phase state with a fixed composition, the structure with the lowest energy is the most stable single-phase state at each composition.

To determine the stability of a single-phase state at a given composition with respect to a two-phase mixture with different compositions, we draw a line connecting any two points in the chemical potential versus composition plot and compare the chemical potential of a two-phase mixture described by this line and the chemical potentials of single-phase states between the two points to determine whether a single-phase or a mixture of two phases is more stable at the same overall composition. For example, at composition x_B^{II}, the lowest chemical potential among all possible structures is represented by circle 2, and thus, it is the most stable single-phase structure at this particular composition. However, the value of the chemical potential represented by the circle at point 2 is lying above the line connecting the two circles represented by point 1 and 3 at composition x_B^I and x_B^{III}, respectively. Therefore, the structure represented by circle 2 is unstable with respect to decomposition into the two structures represented by circle 1 and 3. On the other hand, the value of chemical potential represented by circle 3 is lying below the line connecting the two circles 1 and 4, and thus, the single-phase structure represented by circle 3 is stable with respect to the decomposition to two phases represented by circle 1

Fig. 12.13 Illustration of the stability of compounds at different compositions of a binary system A − B. The dotted circles represent the energies of compounds with different crystal structures at different compositions. The solid lines and the solid circles represent the stable structures at different compositions, forming so-called the convex hull

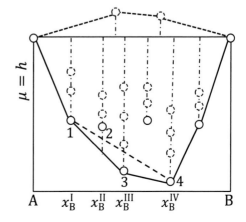

and 4 at composition x_B^I and x_B^{IV}, respectively. Using the same analysis, one can conclude that the compounds represented by the solid line circles on Fig. 12.13 are stable at 0 K, and two-phase mixtures are stable for compositions which deviate from the stoichiometric compositions. The stable states represented by the solid lines connecting the solid circles make up the so-called convex hull.

The same principle works for multicomponent systems to determine the relative stabilities of single-phases and phase mixtures.

12.18 Chemical Potentials of Ideal Gas Mixtures

For ideal gases, a pure gas at 1 bar is usually chosen as the standard state, and the chemical potential of component i with partial pressure p_i can be expressed as:

$$\mu_i(T, p_i) = \mu_i^\circ(T, 1 \text{ bar}) + RT \ln p_i \qquad (12.107)$$

or

$$\mu_i(T, p_i) = \mu_i^\circ(T, 1 \text{ bar}) + RT \ln p^{\text{tot}} + RT \ln x_i \qquad (12.108)$$

$\mu_i^\circ(T, 1 \text{ bar})$ is the chemical potential at the standard state of component i. The second term in Eq. (12.108) represents the chemical potential change when the pressure of the unmixed gas species i is changed from 1 bar to p^{tot}, and the last term in Eq. (12.108) is the chemical potential change due to the mixing of x_i moles of component i with the other $1 - x_i$ moles of gases in the mixture.

The chemical potential of a mixture of ideal gases of composition x_i is

$$\mu = \sum_{i=1}^{n} x_i \mu_i = \sum_{i=1}^{n} \left[x_i \mu_i^\circ(T, 1 \text{ bar}) + RT x_i \ln p_i \right] \qquad (12.109)$$

The chemical potential change from pure gas components to a homogeneous mixture of gases is

$$\Delta \mu = RT \sum_{i=1}^{n} [x_i \ln p_i] \qquad (12.110)$$

12.19   Relation Between Activity of a Component in Solution and Its Vapor Pressure

It was shown in Chap. 11 that the equilibrium vapor pressure above a pure solid or liquid A is a function of temperature and can be approximated as

$$\ln p_A^o = -\frac{a}{T} + b \ln T + c \qquad (12.111)$$

where a, b, and c are constants related to heat of vaporization of a solid or liquid and the heat capacity difference between the vapor phase and its corresponding solid or liquid. On the other hand, the chemical potential of a component can be expressed in terms of the chemical potential of the component at the standard state and the activity of the component in a solid or liquid solution, i.e.,

$$\mu_A = \mu_A^o + RT \ln a_A \qquad (12.112)$$

Let us use a liquid (l) as an example. For a liquid, we choose the pure liquid at a given temperature T and 1 bar as the standard state, hence for a pure liquid A at 1 bar,

$$\mu_A^l = \mu_A^{l,o}(T, 1 \text{ bar}) \text{ and } a_A = 1 \qquad (12.113)$$

If we treat the vapor phase as an ideal gas, the standard state for a gas is the pure gas at a given temperature T and 1 bar; thus, the chemical potential of A in the vapor phase with the equilibrium vapor pressure can be written as,

$$\mu_A^v(T, p_A^o) = \mu_A^{v,o}(T, 1 \text{ bar}) + RT \ln p_A^o \qquad (12.114)$$

When the vapor phase over the pure liquid A is at equilibrium with the liquid, the chemical potential of A atoms in the vapor phase is equal to the chemical potential of A atoms in the pure liquid phase, i.e.,

$$\mu_A^{v,o}(T, 1 \text{ bar}) + RT \ln p_A^o = \mu_A^{l,o}(T, p_A^o) \qquad (12.115)$$

Assuming incompressibility, the chemical potential of the liquid phase under p_A^o can be approximated as,

$$\mu_A^{l,o}(T, p_A^o) \approx \mu_A^{l,o}(T, 1 \text{ bar}) + v_A(p_A^o - 1) \qquad (12.116)$$

Since p_A^o is generally low, the second term on the right can be neglected. Therefore,

$$\mu_A^{v,o}(T, 1 \text{ bar}) + RT \ln p_A^o = \mu_A^{l,o}(T, 1 \text{ bar}) \qquad (12.117)$$

Consequently, the equilibrium vapor pressure is given by

$$p_A^o(T) = e^{-\frac{\left[\mu_A^{v,o}(T,1\,bar)-\mu_A^{l,o}(T,1\,bar)\right]}{RT}}$$ (12.118)

At the normal boiling temperature T_b,

$$\mu_A^{v,o}(T_b, 1\,bar) = \mu_A^{l,o}(T_b, 1\,bar), \; p_A^o(T) = 1\,bar$$ (12.119)

Sometimes, the deviation from equilibrium is measured relative to the equilibrium vapor pressure. For example, the chemical potential for water vapor is given by

$$\mu_{water}(T, p_{water}) = \mu_{water}^o\left(T, p_{water}^o\right) + RT \ln\left(\frac{p_{water}}{p_{water}^o}\right)$$ (12.120)

The humidity is then the logarithm of water vapor pressure relative to the supersaturation temperature which can be computed thermodynamically in terms of chemical potential or partial pressure of water vapor,

$$\text{Humidity} = \ln\frac{p_{water}}{p_{water}^o} = \frac{\mu_{water} - \mu_{water}^o}{RT}$$ (12.121)

Now we consider a liquid solution of A-B (Fig. 12.14). The chemical potential of a given component A in the solution is,

$$\mu_A^l = \mu_A^{l,o}(T) + RT \ln a_A$$ (12.122)

where $\mu_A^{l,o}$ is the chemical potential of liquid A in the standard state, and a_A is the activity of A in the A-B binary liquid solution. Similarly, the chemical potential of A atoms in the vapor phase is given by

Fig. 12.14 Illustration of liquid–vapor equilibrium for a pure liquid and a liquid solution

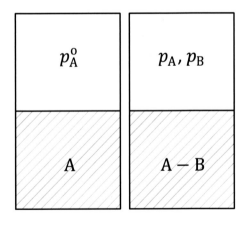

$$\mu_A^v = \mu_A^{v,o}(T) + RT \ln p_A \tag{12.123}$$

where $\mu_A^{v,o}$ is the chemical potential of vapor A in the standard state. We assume that the chemical potentials in the liquid solution under the total equilibrium vapor pressure can be approximated by those under 1 bar. Therefore, if A atoms in the liquid solution A-B are at equilibrium with the A atoms in the vapor phase, the chemical potential of A atoms in the vapor phase is equal to the chemical potential of A atoms in the liquid phase, i.e.,

$$\mu_A^{v,o}(T) + RT \ln p_A = \mu_A^{l,o}(T) + RT \ln a_A \tag{12.124}$$

Using Eq. (12.117),

$$\mu_A^{l,o}(T) = \mu_A^{v,o}(T) + RT \ln p_A^o,$$

in Eq. (12.124), we have

$$RT \ln\left(\frac{p_A}{p_A^o}\right) = RT \ln a_A \text{ or } a_A = \left(\frac{p_A}{p_A^o}\right)$$

Therefore, the activity of A in the liquid solution is the ratio of the partial pressure of A in the corresponding vapor phase above the liquid solution to the equilibrium vapor pressure over pure liquid A at the same temperature.

The relation between the activity of B in the liquid solution and the partial pressure of B above the liquid solution and the equilibrium partial pressure of B above pure liquid is the same as that for A, i.e.,

$$a_B = \left(\frac{p_B}{p_B^o}\right)$$

12.20   Raoult's Law and Henry's Law

A component is said to obey the Raoult's law if its activity a_i and the corresponding activity coefficient γ_i follow the relation,

$$a_i = x_i \text{ or } \gamma_i = 1 \tag{12.125}$$

where x_i is the mole fraction of i in the solution. A solution is called an ideal solution if all components in the solution obey Raoult's law.

Raoult's law is a good approximation when the composition of a component is approaching 1. When the composition of a component is small, Henry's law is a good approximation, i.e.,

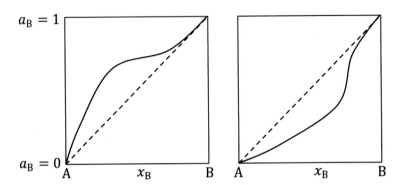

Fig. 12.15 Illustration of activity as a function of composition

$$a_A = k_h x_A \tag{12.126}$$

where k_h is Henry's coefficient, which is independent of composition. The schematic dependencies of activity versus composition are shown in Fig. 12.15 for negative (left) or positive (right) deviation from Raoult's law.

12.21 Chemical Potentials of Ideal Solutions

For an ideal solution, by definition,

$$a_i = x_i \tag{12.127}$$

i.e., every component obeys Raoult's Law.

Therefore, the chemical potential of component i in an ideal solution is

$$\mu_i = \mu_i^o + RT \ln x_i \tag{12.128}$$

or

$$x_i = \exp\left(\frac{\mu_i - \mu_i^o}{RT}\right) \tag{12.129}$$

The chemical potential change for one mole of species i from the pure state to solution is

$$\Delta\mu_i = RT \ln x_i \tag{12.130}$$

The compositionally averaged chemical potential of an ideal solution with composition x_i is

$$\mu = \sum_{i=1}^{n} x_i \mu_i = \sum_{i=1}^{n} \left(x_i \mu_i^o + RT x_i \ln x_i \right) \tag{12.131}$$

The chemical potential change of forming a solution from its corresponding pure components is

$$\Delta\mu = RT \sum_{i=1}^{n} (x_i \ln x_i) \tag{12.132}$$

The heat of mixing for an ideal solution is zero,

$$\Delta h = 0 \tag{12.133}$$

which is expected since there is assumed to be no change in the effective interatomic interactions in an ideal solution compared to its unmixed state.

The entropy change for one mole of component i from the pure state to solution is

$$\Delta s_i = - \left(\frac{\partial \Delta\mu_i}{\partial T} \right)_{p,x_i} = -R \ln x_i \tag{12.134}$$

The entropy of mixing for an ideal solution is

$$\Delta s = - \left(\frac{\partial \Delta\mu}{\partial T} \right)_{p,x_i} = -R \sum_{i=1}^{n} (x_i \ln x_i) \tag{12.135}$$

The volume of mixing for an ideal solution is zero,

$$\Delta v = \left(\frac{\partial \Delta\mu}{\partial p} \right)_{T,x_i} = 0 \tag{12.136}$$

In summary, we have the following relations for a binary ideal solution:

$$\mu = x_A \mu_A^o + x_B \mu_B^o + RT (x_A \ln x_A + x_B \ln x_B) \tag{12.137}$$

$$\Delta\mu = \mu - \left(x_A \mu_A^o + x_B \mu_B^o \right) = RT (x_A \ln x_A + x_B \ln x_B) \tag{12.138}$$

$$\Delta\mu_A = \mu_A - \mu_A^o = RT \ln x_A \tag{12.139}$$

$$\Delta\mu_B = \mu_B - \mu_B^o = RT \ln x_B \tag{12.140}$$

Fig. 12.16 Schematic
illustration of property
changes from pure
components to an ideal
solution

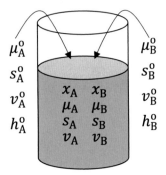

$$\Delta s = -\left(\frac{\partial \Delta \mu}{\partial T}\right)_{p,x_B} = -R(x_A \ln x_A + x_B \ln x_B) \qquad (12.141)$$

$$\Delta s_A = -\left(\frac{\partial \Delta \mu_A}{\partial T}\right)_{p,x_B} = -R \ln x_A \qquad (12.142)$$

$$\Delta s_B = -\left(\frac{\partial \Delta \mu_B}{\partial T}\right)_{p,x_B} = -R \ln x_B \qquad (12.143)$$

$$\Delta v_A = \left(\frac{\partial \Delta \mu_A}{\partial p}\right)_{T,x_B} = 0 \qquad (12.144)$$

$$\Delta v_B = \left(\frac{\partial \Delta \mu_B}{\partial p}\right)_{T,x_B} = 0 \qquad (12.145)$$

$$\Delta h = 0 \qquad (12.146)$$

where Δs, Δv, and Δh are molar entropy, volume, and enthalpy of formation of an ideal binary solution, respectively.

The ideal solution model is the simplest possible thermodynamic description to model the behavior of a solution. Figure 12.16 shows the thermodynamic properties before and after mixing, and thus their changes during the formation process of a solution.

12.22 Chemical Potentials of Regular Solutions

The regular solution model is the simplest possible model for describing a nonideal solution. It has one adjustable parameter, called the regular solution parameter, which can be used to model a real solid solution. As one can imagine, its application is limited since it has only one adjustable parameter. However, it serves as a simple model to discuss the behavior of real solutions. A regular solution is a nonideal solution with an ideal entropy of mixing, i.e.,

$$a_i \neq x_i, \ \Delta h \neq 0, \ \Delta s = -R \sum_{i=1}^{n} x_i \ln x_i \tag{12.147}$$

For a binary system modeled by a regular solution model, the activity coefficients as a function of composition are described by

$$RT \ln \gamma_A = \alpha' x_B^2, \ RT \ln \gamma_B = \alpha' x_A^2 \tag{12.148}$$

where α' is a constant, called the regular solution parameter, independent of temperature. Sometimes, they are expressed as

$$\ln \gamma_A = \alpha x_B^2, \ \ln \gamma_B = \alpha x_A^2 \tag{12.149}$$

where

$$\alpha = \alpha'/RT \tag{12.150}$$

From the definition of a regular binary solution, one can easily derive its various thermodynamic properties. For example, the chemical potential of component A is

$$\mu_A = \mu_A^o + RT \ln \gamma_A + RT \ln x_A = \mu_A^o + \alpha' x_B^2 + RT \ln x_A \tag{12.151}$$

The chemical potential of component B is

$$\mu_B = \mu_B^o + RT \ln \gamma_B + RT \ln x_B = \mu_B^o + \alpha' x_A^2 + RT \ln x_B \tag{12.152}$$

The chemical potential of a regular solution with composition x_A and x_B is

$$\mu = x_A \mu_A^o + x_B \mu_B^o + \alpha' x_A x_B + RT(x_A \ln x_A + x_B \ln x_B) \tag{12.153}$$

The chemical potential change of forming a binary solution from pure components is given by

$$\Delta \mu = RT(x_A \ln a_A + x_B \ln a_B) \tag{12.154}$$

or

$$\Delta \mu = RT(x_A \ln \gamma_A + x_B \ln \gamma_B) + RT(x_A \ln x_A + x_B \ln x_B) \tag{12.155}$$

Using the definition of regular solutions for activity coefficients, we have

$$\Delta \mu = \alpha' \left(x_A x_B^2 + x_B x_A^2 \right) + RT(x_A \ln x_A + x_B \ln x_B) \tag{12.156}$$

or

$$\Delta\mu = \alpha' x_A x_B + RT(x_A \ln x_A + x_B \ln x_B) \qquad (12.157)$$

From the above equation, it is easy to see that the heat of mixing is

$$\Delta h = \alpha' x_A x_B \qquad (12.158)$$

The entropy of mixing is

$$\Delta s = -R(x_A \ln x_A + x_B \ln x_B) \qquad (12.159)$$

The volume of mixing is zero since α' is a constant.

12.23   Flory–Huggins Model of Polymer Solutions[1,2]

For polymer solutions, the molecule sizes of solvent and solutes are very different. The sizes of the macromolecular polymer chains are usually much larger than the sizes of solvent molecules. As a result, although the mole fraction of polymer chains may be small, they could occupy a large fraction of the volume of a solution. Therefore, the entropy of mixing will need to take into account the large differences in molecular sizes between a polymer and a solvent. According to the Flory–Huggins solution theory, the Gibbs free energy change for forming a polymer solution from a pure polymer and a pure solvent can be approximated as

$$\Delta G = RT[\chi N_1 \varphi_2 + N_1 \ln \varphi_1 + N_2 \ln \varphi_2] \qquad (12.160)$$

where x_1 and φ_1 are the mole and volume fractions of solvent, x_2 and φ_2 are the mole and volume fractions of polymer solutes, and χ is the interaction parameter similar to the regular solution parameter α' in Eq. (12.157). To relate the volume fractions to the number of moles of solvent N_1 and the number of moles of polymer solute N_2 as well as the size of the polymer chains, the size of solvent molecules is assumed to be the same as that of one monomer of the polymer chain. If a polymer chain is assumed to contain x monomers, the volume fractions of the solvent φ_1 and polymer solutes φ_2 can be expressed as

$$\varphi_1 = \frac{N_1}{N_1 + x N_2}$$
$$\varphi_2 = \frac{x N_2}{N_1 + x N_2}$$

---

[1] P. J. Flory, "Thermodynamics of high polymer solutions," J. Chem. Phys. 10, 51–61 (1942).
[2] M. L. Huggins, "Solutions of Long Chain Compounds," Journal of Chemical Physics, 9, 440 (1941).

Equation (12.160) can be rewritten as

$$\Delta G = RT[\chi(N_1 + xN_2)\varphi_1\varphi_2 + N_1 \ln \varphi_1 + N_2 \ln \varphi_2] \qquad (12.161)$$

or

$$\Delta G = (N_1 + xN_2)RT\left[\chi \varphi_1\varphi_2 + \varphi_1 \ln \varphi_1 + \frac{\varphi_2}{x} \ln \varphi_2\right] \qquad (12.162)$$

Therefore, we could also write down the chemical potential change for mixing,

$$\Delta \mu = \frac{\Delta G}{(N_1 + xN_2)} = RT\left[\chi \varphi_1\varphi_2 + \varphi_1 \ln \varphi_1 + \frac{\varphi_2}{x} \ln \varphi_2\right] \qquad (12.163)$$

If $x = 1$, the above equation is reduced to the regular solution described by Eq. (12.157). However, if x is greater than 1, the function is not symmetric with respect to $\varphi_1 = \varphi_2 = 0.5$.

12.24 Redlich–Kister Expansion for Chemical Potential of a Solution[3]

For a binary A-B solution, a more general description for the compositional dependence of the activity coefficients on composition can be described by using the Redlich–Kister Expansion,

$$x_A \ln \gamma_A + x_B \ln \gamma_B = x_A x_B\left[a + b(x_B - x_A) + c(x_B - x_A)^2 + \cdots\right] \qquad (12.164)$$

- For an ideal solution: $a = b = c = \cdots = 0$,

$$x_A \ln \gamma_A + x_B \ln \gamma_B = 0$$

- For a regular solution: $a \neq 0, b = c = \cdots = 0$,

$$x_A \ln \gamma_A + x_B \ln \gamma_B = ax_A x_B$$

- A solution described by $a \neq 0, b \neq 0, c = \cdots = 0$ is called a Margules solution for which

$$x_A \ln \gamma_A + x_B \ln \gamma_B = x_A x_B[a + b(x_B - x_A)] \qquad (12.165)$$

---

[3] Otto Redlich and A. T. Kister. Algebraic representation of thermodynamic properties and the classification of solutions. Industrial & Engineering Chemistry, 40(2):345–348, 1948.

There can be more complicated descriptions, e.g., $a \neq 0, b = 0, c \neq 0, d = \cdots = 0, a \neq 0, b \neq 0, c \neq 0, d = \cdots = 0$, etc.

Similarly, one can express the compositional dependence of activity coefficients for an A-B-C ternary solution,

$$
\begin{aligned}
Y_{AB} &= x_A x_B \left[a_{AB} + b_{AB}(x_B - x_A) + c_{AB}(x_B - x_A)^2 + \cdots \right] \\
Y_{BC} &= x_B x_C \left[a_{BC} + b_{BC}(x_C - x_B) + c_{BC}(x_C - x_B)^2 + \cdots \right] \\
Y_{CA} &= x_C x_A \left[a_{CA} + b_{CA}(x_A - x_C) + c_{CA}(x_A - x_C)^2 + \cdots \right] \\
Y &= x_A \ln \gamma_A + x_B \ln \gamma_B + x_C \ln \gamma_C \\
Y &= Y_{AB} + Y_{BC} + Y_{CA} + x_A x_B x_C [c + d_1(x_A - x_B) + d_2(x_B - x_C) + \cdots]
\end{aligned}
$$
$$(12.166)$$

Therefore, ignoring the high-order terms, one can model a ternary solution using the expression,

$$
\begin{aligned}
Y &= x_A x_B [a_{AB} + b_{AB}(x_B - x_A)] + x_B x_C [a_{BC} + b_{BC}(x_C - x_B)] \\
&\quad + x_C x_A [a_{CA} + b_{CA}(x_A - x_C)] + c x_A x_B x_C
\end{aligned}
$$

12.25 Osmotic Pressure

The Nobel Prize in Chemistry was awarded to J. H. Van't Hoff in 1901 for his discovery and explanation of Osmotic pressure. Osmotic pressure is the pressure difference that is developed between the two sides of a semipermeable membrane separating a pure solvent A (e.g., water) on one side and a solution of solvent A with solute B (e.g., sugar) atoms. The membrane is only permeable to the solvent atoms A but not to the solute atoms B. This concept is very important in biology since many cell membranes are only permeable to certain types of ions (Fig. 12.17).

It is easy to understand that the chemical potential of A atoms in the pure solvent at a given temperature and pressure is higher than the chemical potential of A atoms in the solution at the same temperature and pressure. Therefore, A atoms will move across the membrane to the solution. In order to balance or stop the diffusion of solvent A atoms, the pressure on the solution side must be increased.

Let us consider the isothermal process

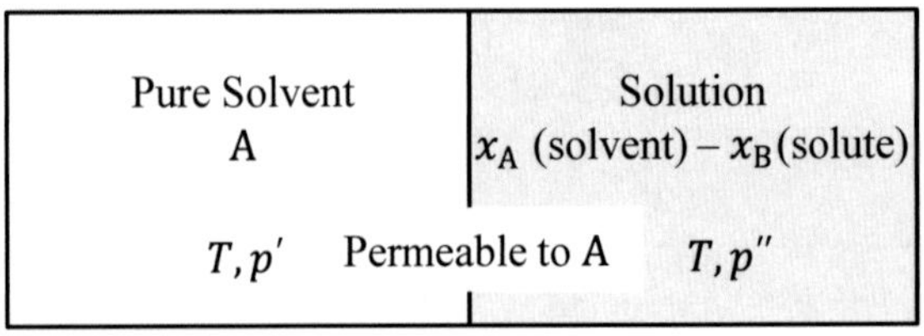

Fig. 12.17 Illustration of equilibrium between a pure solve and a solution and the Osmotic pressure

Pure Solvent A	Solution x_A (solvent) $-$ x_B (solute)
T, p' Permeable to A	T, p''

$$A(\text{pure}, p') \rightleftharpoons A(\text{solution with composition } x_A, p'') \tag{12.167}$$

At equilibrium

$$\mu_{A(\text{pure}, p')} = \mu_{A(\text{solution}, p'')} \tag{12.168}$$

Assuming that the liquid solution is incompressible, the chemical potential of pure solvent A at temperature T and pressure p' is

$$\mu_{A(\text{pure})} = \mu_A^o + v_A(p' - 1) \tag{12.169}$$

The chemical potential of A in the solution at temperature T and pressure p'' is

$$\mu_{A(\text{solution})} = \mu_A^o + v_A(p'' - 1) + RT \ln x_A \tag{12.170}$$

At equilibrium

$$\mu_A^o + v_A(p' - 1) \rightleftharpoons \mu_A^o + v_A(p'' - 1) + RT \ln x_A \tag{12.171}$$

Simplifying the above equation, we have

$$v_A p' \rightleftharpoons v_A p'' + RT \ln x_A \tag{12.172}$$

or

$$v_A(p'' - p') = v_A \Delta p = -RT \ln x_A = -RT \ln(1 - x_B) \tag{12.173}$$

Since x_B is very small,

$$\ln(1 - x_B) \cong -x_B \tag{12.174}$$

Therefore,

$$v_A(p'' - p') \cong RT x_B \tag{12.175}$$

or

$$\Delta p = p'' - p' = \frac{RT x_B}{v_A} = c_B RT \tag{12.176}$$

where Δp is called the Osmotic pressure, and c_B is the molar concentration of solute atoms. Notably, the above equation has a similar form to the ideal gas law.

12.26 Chemomechanical Potentials of Components in a Solution

For solid solutions, if the lattice parameters of a solution depend on composition, the elastic energy contribution to the chemical potential of the solution has to be considered. The chemomechanical potential is defined as the sum of chemical and elastic potentials.

Let us consider a cubic binary solution in which the concentration dependence of lattice parameter $a(c)$ of the solution is assumed to follow Vegard's law, i.e., the lattice parameter $a(c)$ of a cubic solution is linearly dependent on solute concentration c,

$$a(c) = a(c_0) + \frac{da}{dc}(c - c_0) \tag{12.177}$$

where $a(c_0)$ is the lattice parameter of the solution with the reference solute concentration c_0, and da/dc is the rate of change in lattice parameter with respect to concentration. Let us define the concentration-dependent lattice expansion coefficient as,

$$\varepsilon_0 = \frac{1}{a(c_0)}\frac{da}{dc} = \frac{[a(c) - a(c_0)]}{a(c_0)(c - c_0)} = \frac{v_m}{a(c_0)}\frac{da}{dx} \tag{12.178}$$

where x is the mole fraction of solute atoms, and $v_m = x/c$ is molar volume of the solution.

The concentration-dependent eigenstrain tensor ε_{ij}^o can be expressed as

$$\varepsilon_{ij}^o = \varepsilon_0(c - c_0)\delta_{ij} \tag{12.179}$$

where δ_{ij} is the Kronecker delta function defined as

$$\delta_{ij} = \begin{bmatrix} 1 \ \text{if} \ i = j \\ 0 \ \text{if} \ i \neq j \end{bmatrix} \tag{12.180}$$

The elastic strain ε_{ij}^{el} is given by

$$\varepsilon_{ij}^{el} = \varepsilon_{ij} - \varepsilon_{ij}^o = \varepsilon_{ij} - \frac{1}{a_0}\frac{da}{dc}(c - c_0)\delta_{ij} = \varepsilon_{ij} - \varepsilon_0(c - c_0)\delta_{ij} \tag{12.181}$$

where ε_{ij} is the total strain containing both the elastic strain and compositional strain. The local elastic strain, ε_{ij}^{el}, of an inhomogeneous solid solution is obtained by solving the mechanical equilibrium equation. We will assume that the local elastic displacements, elastic strains, and thus the elastic stresses are all available. The corresponding elastic stress σ_{ij}^{el} is given by

$$\sigma_{ij}^{el} = C_{ijkl}\varepsilon_{kl}^{el} \tag{12.182}$$

where C_{ijkl} is the elastic modulus tensor.

Therefore, the chemomechanical potential μ^{cm} of solute atoms is given by

$$\mu^{cm} = \sigma_{ij}^{el}\frac{d\varepsilon_{ij}^{el}}{dc} + \mu = -\sigma_{ij}^{el}\varepsilon_o\delta_{ij} + \mu \tag{12.183}$$

where σ_{ij}^{el} is the local elastic stress, and μ is chemical potential of solute atoms. The term, $-\sigma_{ij}^{el}\varepsilon_o\delta_{ij}$, is therefore the mechanical contribution to the total chemomechanical potential, μ^{cm}.

For example, let us consider a relatively simple example on the segregation of solute B atoms around an edge dislocation for which the elastic solution is available. Let us assume that the edge dislocation is located at $(x = 0, y = 0)$ (see Fig. 12.18) with the x-axis along the horizontal direction, y-axis along the vertical direction, and z-direction is out of the paper. The stress distribution around such an edge dislocation with a Burgers vector magnitude of B is given by

$$\sigma_{xx}^{el} = -\frac{Gb}{2\pi(1-v)}\frac{y(3x^2 + y^2)}{(x^2 + y^2)^2} \tag{12.184}$$

$$\sigma_{yy}^{el} = \frac{Gb}{2\pi(1-v)}\frac{y(x^2 - y^2)}{(x^2 + y^2)^2} \tag{12.185}$$

$$\sigma_{zz}^{el} = \frac{Gvb}{\pi(1-v)}\frac{y}{(x^2 + y^2)} \tag{12.186}$$

where G is the shear modulus, and v is the Poisson ratio.

Fig. 12.18 Illustration of a edge dislocation

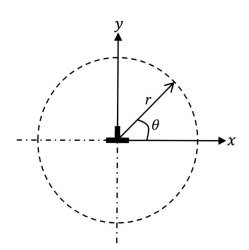

Therefore, the mechanical contribution to the chemomechanical potential of solute atoms around an edge dislocation is

$$-\varepsilon_0 \sigma_{ij}^{el}\delta_{ij} = -\varepsilon_0\left(\sigma_{xx}^{el} + \sigma_{yy}^{el} + \sigma_{zz}^{el}\right) = \frac{\varepsilon_0 Gb}{\pi}\frac{y}{(x^2+y^2)} = \frac{\varepsilon_0 Gb}{\pi}\frac{\sin\theta}{r} \quad (12.187)$$

where r is the distance from the origin, and θ is the angle from the positive x-axis (see Fig. 12.18).

Let us assume that the mole fraction of solute B atoms in the solution in the absence of dislocations or very far away from the dislocation is x_B^o and that the behavior of B atoms can be approximated using Raoult's law for simplicity. The bulk chemical potential of B is

$$\mu_B^b = \mu_B^o + RT\ln x_B^o \quad (12.188)$$

where μ_B^b is the bulk chemical potential without the mechanical contribution, and μ_B^o is the chemical potential of B at standard state, i.e., the chemical potential of pure solid B. In the presence of an edge dislocation, the local chemomechanical $\mu_B^{\sigma,d}$ is given by

$$\mu_B^{\sigma,d} = \mu_B^o + RT\ln x_B + \frac{\varepsilon_0 Gb}{\pi}\frac{\sin\theta}{r} \quad (12.189)$$

At equilibrium,

$$\mu_B^{\sigma,d} = \mu_B^b \quad (12.190)$$

or

$$\mu_B^o + RT\ln x_B + \frac{\varepsilon_0 Gb}{\pi}\frac{\sin\theta}{r} = \mu_B^o + RT\ln x_B^o \quad (12.191)$$

Solving the above equation for x_B, we have

$$RT\ln\frac{x_B}{x_B^o} = -\frac{\varepsilon_0 Gb}{\pi}\frac{\sin\theta}{r} \quad (12.192)$$

or

$$\frac{x_B}{x_B^o} = \exp\left(-\frac{\varepsilon_0 Gb}{RT}\frac{\sin\theta}{\pi r}\right) \quad (12.193)$$

where x_B is the composition distribution of solute atoms around an edge dislocation. If $\varepsilon_0 > 0$, it implies that the size of B atoms is larger than that of A atoms. For a given r, minimum x_B takes place at $\theta = \pi/2$; i.e., solute B atoms are depleted above the dislocation, and maximum x_B takes place at $\theta = -\pi/2$; i.e., solute B atoms are accumulated below the dislocation. If $\varepsilon_o < 0$, it implies that the size of B atoms is smaller than that of A atoms, and the x_B extrema are reversed, with solute accumulation above the dislocation and depletion below the dislocation.

12.27 Electrochemical Potentials of Components in a Solution

Similar to the compositional distribution in the presence of an elastic potential obtained from the equilibrium condition based on the chemomechanical potential, the spatial distribution of charge species in the presence of an electric potential can be obtained from the spatial distribution of electrochemical potential. The electrical potential itself in an inhomogeneous solution has to be obtained by solving the Poisson equation under appropriate boundary conditions just like the knowledge of the elastic potential requiring the solution to the mechanical equilibrium equation. We assume the electric potential is available.

Let us designate x_B^o as the composition distribution of charged species, e.g., ions or electrons, in the absence of an electric potential and $x_B(r)$ as the compositional distribution in the presence of an electric potential $\phi(r)$. We also assume the configurational entropy contribution to the chemical potential of charged species B can be described by "$-R \ln x_B$". The bulk chemical potential μ_B^b of charged species B in the absence of an electric potential is then given by

$$\mu_B^b = \mu_B^o + RT \ln x_B^o \tag{12.194}$$

where μ_B^o is the chemical potential of B in the standard state.

The electrochemical potential of B, $\mu_B^\phi(r)$, in the presence of an electric potential $\phi(r)$ is given by

$$\mu_B^\phi(r) = \mu_B^o + RT \ln x_B(r) + z_B \mathcal{F} \phi(r) \tag{12.195}$$

where $x_B(r)$ is the composition of B in the presence of the electric potential, z_B is the valence of charged species B, and $\mathcal{F}$ is Faraday's constant.

At equilibrium

$$\mu_B^\phi(r) = \mu_B^b \tag{12.196}$$

Therefore,

$$RT \ln \frac{x_B(r)}{x_B^o} = -z_B \mathcal{F} \phi(r) \tag{12.197}$$

or we can express the composition of B as a function of position

$$\frac{x_B(r)}{x_B^o} = \exp\left(-\frac{z_B \mathcal{F}\phi(r)}{RT}\right) \qquad (12.198)$$

12.28   Particle Size Dependence of Solubility

Let us consider a simple case of solubility of a pure solid B in A to form a solid solution (α) with A as solvent and B as solute. If we ignore the pure solid B particle size or assume that the pure solid B particle size is infinite, i.e., the particle/solution interface is flat (see the left illustration of Fig. 12.19), the equilibrium condition for atom B is given by

$$\mu_B^\alpha = \mu_B^o \qquad (12.199)$$

where μ_B^o is the chemical potential of pure solid B, and μ_B^α is the chemical potential of solute atom B in the solution α with composition $\left(x_A^{\alpha,\infty}, x_B^{\alpha,\infty}\right)$. We can rewrite the above equation as

$$\mu_B^o + RT \ln \gamma_B^{\alpha,\infty} + RT \ln x_B^{\alpha,\infty} = \mu_B^o \qquad (12.200)$$

where $x_B^{\alpha,\infty}$ is the solubility of atom B in A at temperature T assuming that the pure solid B particle size is infinite or the size effect of the particle can be ignored, and $\gamma_B^{\alpha,\infty}$ is the activity coefficient of B atoms in the solution at temperature T and composition $x_B^{\alpha,\infty}$.

If we assume pure solid B is in the form of spherical particles with particle radius r (see the right illustration of Fig. 12.19), the chemical potential of B atoms in a particle of size r, $\mu_B^r(r)$, is then given by

$$\mu_B^r(r) = \mu_B^o + \frac{2\gamma_{B\alpha} v_m}{r} \qquad (12.201)$$

Fig. 12.19 Equilibrium between solid B and a solution with a flat interface or circular interface

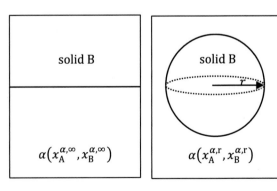

where $\gamma_{B\alpha}$ is the interfacial energy between the pure solid B and solid solution α, v_m is the molar volume of the solid (assumed to be independent of composition for simplicity). The equilibrium condition between the B atoms in a particle with radius r and those in the solid solution is then given by

$$\mu_B^o + RT \ln \gamma_B^{\alpha,r} + RT \ln x_B^{\alpha,r} = \mu_B^o + \frac{2\gamma_{B\alpha} v_m}{r} \qquad (12.202)$$

where $x_B^{\alpha,r}$ is the solubility of B atoms in A at temperature T when the particle size of B is r, and $\gamma_B^{\alpha,r}$ is the activity coefficient at temperature T and composition $x_B^{\alpha,r}$. It is reasonable to assume that $\gamma_B^{\alpha,r} = \gamma_B^{\alpha,\infty}$ since the difference $x_B^{\alpha,r}$ and $x_B^{\alpha,\infty}$ is expected to be small, and both $x_B^{\alpha,r}$ and $x_B^{\alpha,\infty}$ are small.

Comparing Eqs. (12.200) and (12.202), we have,

$$RT \ln \frac{x_B^{\alpha,r}}{x_B^{\alpha,\infty}} = \frac{2\gamma_{B\alpha} v_m}{r} \qquad (12.203)$$

or

$$\frac{x_B^{\alpha,r}}{x_B^{\alpha,\infty}} = \exp\left(\frac{2\gamma_{B\alpha} v_m}{r RT}\right) \qquad (12.204)$$

which expresses the composition of B in the solution α outside a pure solid B particle with radius r relative to the composition of B in the solution next to a flat particle solution interface. A schematic dependence of solubility shift by considering the particle size is shown in Fig. 12.20.

The spatial distribution of chemical potentials of B around differently sized particles is the driving force for the coarsening of particles, a phenomenon known as Ostwald ripening. For example, the difference in the chemical potential of B atoms around a particle of size r' and that of around a particle of size r'' (Fig. 12.21) is

$$\mu_{r'} - \mu_{r''} = 2\gamma_{B\alpha} v_m \left(\frac{1}{r'} - \frac{1}{r''}\right) \qquad (12.205)$$

Fig. 12.20 Phase boundary shift with particle size

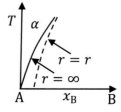

Fig. 12.21 Driving force for coarsening

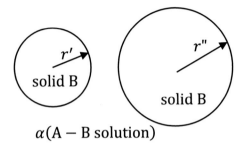

$$\alpha(\text{A} - \text{B solution})$$

12.29 Interfacial Segregation

The interfacial atoms or defects such as vacancies, e.g., atoms or defects at grain boundaries and surfaces inside a solid, typically have different atomic environments and thus have different formation energies or enthalpies at or near the interfaces versus those away from interfaces. However, at equilibrium the chemical potential of atoms of a given species or a given type of defects is uniform throughout the solid, i.e., the chemical potential of each species at or near the interfaces is equal to that inside the corresponding bulk. As a result, the spatial distributions of compositions of different atomic species are expected to be nonuniform.

Let us first consider a simplest possible case in which we have a chemical species, B, with a different formation energy at a grain boundary from that in the bulk of a solid A, for example, hydrogen atoms in a metal such as iron. Then, at equilibrium, we have,

$$\mu_B^{gb} = \mu_B^b \tag{12.206}$$

i.e., the chemical potential μ_B^{gb} of B atoms at the grain boundary is equal to that of B atoms inside the grains μ_B^b. We assume that the formation energy per atom B at a grain boundary of A is $E_B^{gb,o}$ and the corresponding formation energy per atom B in the bulk crystal of A is $E_B^{b,o}$. We also assume that the number density of B atoms is small both in the bulk and at the grain boundary, so the mutual interactions between B atoms can be neglected. If we use x_B^{gb} to represent the fraction of available lattice sites at the grain boundary that are occupied by the B atoms and x_B^o the fraction of available bulk lattice sites that are occupied by the B atoms inside the bulk, according to Eq. (12.206), we have

$$E_B^{gb,o} + k_B T \ln x_B^{gb} = E_B^{b,o} + k_B T \ln x_B^o \tag{12.207}$$

Solving Eq. (12.207) for x_B^{gb}, we have

$$x_B^{gb} = x_B^o \exp\left(-\frac{\left(E_B^{gb,o} - E_B^{b,o}\right)}{k_B T}\right) = x_B^o \exp\left(-\frac{\Delta E_B^{gb}}{k_B T}\right) \qquad (12.208)$$

where ΔE_B^{gb} is the segregation energy of atom B. If we include the vibrational contributions and contributions from the formation volume difference of B atoms at the grain boundary and in the bulk, then ΔE_B^{gb} in Eq. (12.208) can be replaced by $\Delta \mu_B^{gb,o}$, the standard chemical potential difference between B atoms at the grain boundary and in the bulk, i.e.,

$$x_B^{gb} = x_B^o \exp\left(-\frac{\left(\mu_B^{gb,o} - \mu_B^{b,o}\right)}{k_B T}\right) = x_B^o \exp\left(-\frac{\Delta \mu_B^{gb,o}}{k_B T}\right) \qquad (12.209)$$

Let us consider another simple case involving vacancy segregation to a grain boundary in a single-component solid A. When a vacancy V_A^b moves from inside the bulk of a grain to become a vacancy V_A^{gb} at a grain boundary, in exchange, an atom A originally at the grain boundary A_A^{gb} moves to inside the bulk of the grain to become A_A^b. This process can be expressed as a reaction:

$$A_A^{gb} + V_A^b \rightleftharpoons A_A^b + V_A^{gb} \qquad (12.210)$$

At equilibrium, we have

$$\mu_{A_A^{gb}} + \mu_{V_A^b} = \mu_{A_A^b} + \mu_{V_A^{gb}} \qquad (12.211)$$

Let us assume that the formation energy of a vacancy at a grain boundary is $E_{V_A^{gb}}$, the formation energy of a vacancy in the bulk is $E_{V_A^b}$, the formation energy of an A atom at a grain boundary is $E_{A_A^{gb}}$, and the formation energy of an A atom in the bulk is $E_{A_A^b}$. If we also assume that vacancies and A atoms behave like an ideal binary solution, we have

$$E_{A_A^{gb}} + k_B T \ln x_A^{gb} + E_{V_A^b} + k_B T \ln x_V^b = E_{A_A^b} + k_B T \ln x_A^b + E_{V_A^{gb}} + k_B T \ln x_V^{gb}$$

where x_A^{gb} is the fraction of grain boundary sites occupied by A atoms.

We can reorganize the above equation as

$$k_B T \ln \frac{x_V^{gb} x_A^b}{x_V^b x_A^{gb}} = -\left[\left(E_{A_A^b} + E_{V_A^{gb}}\right) - \left(E_{A_A^{gb}} + E_{V_A^b}\right)\right] = -\Delta E_{gb}$$

or

$$\frac{x_V^{gb} x_A^b}{x_V^b x_A^{gb}} = \exp\left(-\frac{\Delta E_{gb}}{k_B T}\right) \tag{12.212}$$

or

$$\frac{x_V^{gb}}{\left(1 - x_V^{gb}\right)} = \frac{x_V^b}{\left(1 - x_V^b\right)} \exp\left(-\frac{\Delta E_{gb}}{k_B T}\right) \tag{12.213}$$

where ΔE_{gb} is the segregation energy. If the vacancy fractions both at the grain boundary and inside the bulk are small, we can simply rewrite the Eq. (12.213) as

$$x_V^{gb} = x_V^b \exp\left(-\frac{\Delta E_{gb}}{k_B T}\right) \tag{12.214}$$

If we include the vibrational contribution and consider the effect of formation volumes, we can replace ΔE_{gb} with ΔG_{gb}^o, the standard segregation free energy.

Finally, let us consider the case of segregation in a binary solid solution with A as solvent and B as solute, which is very similar to the case of vacancy segregation in a pure material by treating atom-vacancy pairs as a binary system. When a solute atom B^b moves from inside the bulk of a grain to become a solute atom B^{gb} at a grain boundary, in exchange, an atom A originally at the grain boundary A^{gb} moves into inside the bulk of the grain to become A^b. This process can be expressed as an exchange reaction:

$$A^{gb} + B^b \rightleftharpoons A^b + B^{gb} \tag{12.215}$$

At equilibrium, we have

$$\mu_A^{gb} + \mu_B^b = \mu_A^b + \mu_B^{gb} \tag{12.216}$$

which can be rewritten as

$$\mu_B^b - \mu_A^b = \mu_B^{gb} - \mu_A^{gb} \tag{12.217}$$

Equation (12.217) indicates that the slope for the tangent line to the chemical potential (or molar Gibbs free energy) of the bulk solid solution at composition x_B^b is equal to the slope for the tangent line to the chemical (or molar Gibbs free energy) of the grain boundary solid solution at x_B^{gb}. To see this, let us write the chemical potential of the bulk solid solution phase as

$$\mu^b = x_A^b \mu_A^b + x_B^b \mu_B^b \tag{12.218}$$

The corresponding differential form is given by

$$d\mu^b = \mu_A^b dx_A^b + \mu_B^b dx_B^b = \left(\mu_B^b - \mu_A^b\right)dx_B^b \tag{12.219}$$

Therefore, the slope of the tangent line to the chemical potential of the bulk solid solution at composition at x_B^b is given by

$$\left(\frac{\partial\mu^b}{\partial x_B^b}\right)_{x_B^b} = \mu_B^b\left(x_B^b\right) - \mu_A^b\left(x_B^b\right) \tag{12.220}$$

Similarly, for the grain boundary solid solution phase, we have

$$\mu^{gb} = x_A^{gb}\mu_A^{gb} + x_B^{gb}\mu_B^{gb} \tag{12.221}$$

The corresponding differential form for the grain boundary solid solution is given by

$$d\mu^{gb} = \mu_A^{gb} dx_A^{gb} + \mu_B^{gb} dx_B^{gb} = \left(\mu_B^{gb} - \mu_A^{gb}\right)dx_B^{gb} \tag{12.222}$$

Therefore, the slope of the tangent line to the chemical potential of the grain boundary solid solution at composition at x_B^{gb} is given by

$$\left(\frac{\partial\mu^{gb}}{\partial x_B^{gb}}\right)_{x_B^{gb}} = \mu_B^{gb}\left(x_B^{gb}\right) - \mu_A^{gb}\left(x_B^{gb}\right) \tag{12.223}$$

At equilibrium,

$$\left(\frac{\partial\mu^b}{\partial x_B^b}\right)_{x_B^b} = \mu_B^b\left(x_B^b\right) - \mu_A^b\left(x_B^b\right) = \left(\frac{\partial\mu^{gb}}{\partial x_B^{gb}}\right)_{x_B^{gb}} = \mu_B^{gb}\left(x_B^{gb}\right) - \mu_A^{gb}\left(x_B^{gb}\right) \tag{12.224}$$

i.e., the two tangent lines are parallel at equilibrium. Therefore, if we know the chemical potentials (molar Gibbs free energies) of the grain boundary solution phase and bulk solution phase as a function of composition, we can use the parallel tangent construction to obtain the composition in the grain boundary that is at equilibrium with the composition in the bulk.

Let us write the chemical potentials of A and B in terms of their activities, a_A^{gb}, a_B^b, a_A^b, and a_B^{gb} in the bulk and at the grain boundaries,

$$\mu_A^{gb} = \mu_A^{gb,o} + k_B T \ln a_A^{gb}$$
$$\mu_B^b = \mu_B^{b,o} + k_B T \ln a_B^b$$
$$\mu_A^b = \mu_A^{b,o} + k_B T \ln a_A^b$$
$$\mu_B^{gb} = \mu_B^{gb,o} + k_B T \ln a_B^{gb} \tag{12.225}$$

where $\mu_A^{gb,o}$, $\mu_B^{b,o}$, $\mu_A^{b,o}$, and $\mu_B^{gb,o}$ are the standard chemical potentials of A and B in the bulk and at grain boundaries. Bulk standard chemical potentials $\mu_A^{b,o}$ and $\mu_B^{b,o}$ are essentially the chemical potentials of pure A and B assuming the crystal structure of the binary solution, which include the contribution from the interaction potential energy, thermal vibration free energy, and electron systems. The standard chemical potentials of A and B at grain boundaries, $\mu_A^{gb,o}$ and $\mu_B^{gb,o}$, represent the formation energies, or formation free energies excluding the configuration entropy contributions, of A and B atoms at the grain boundary.

Substituting Eq. (12.225) to Eq. (12.216), we have

$$\mu_A^{gb,o} + k_B T \ln a_A^{gb} + \mu_B^{b,o} + k_B T \ln a_B^{b} = \mu_A^{b,o} + k_B T \ln a_A^{b} + \mu_B^{gb,o} + k_B T \ln a_B^{gb}$$

We can rewrite the above equation as

$$k_B T \ln \frac{a_B^{gb} a_A^{B}}{a_A^{gb} a_B^{b}} = -\left[\left(\mu_A^{b,o} + \mu_B^{gb,o} \right) - \left(\mu_A^{gb,o} + \mu_B^{b,o} \right) \right] = -\Delta G_{gb}^{o}$$

or

$$\frac{a_B^{gb}}{a_A^{gb}} = \frac{a_B^{b}}{a_A^{b}} \exp\left(-\frac{\Delta G_{gb}^{o}}{k_B T} \right) \qquad (12.226)$$

In terms of compositions and activity coefficients γ_A^{gb}, γ_B^{b}, γ_A^{b}, and γ_B^{gb}, the above equation can be rewritten as

$$\frac{x_B^{gb} \gamma_B^{gb}}{x_A^{gb} \gamma_A^{gb}} = \frac{x_B^{b} \gamma_B^{b}}{x_A^{b} \gamma_A^{b}} \exp\left(-\frac{\Delta G_{gb}^{o}}{k_B T} \right) \qquad (12.227)$$

or

$$\frac{x_B^{gb} \gamma_B^{gb}}{\left(1 - x_B^{gb}\right) \gamma_A^{gb}} = \frac{x_B^{b}}{\left(1 - x_B^{b}\right)} \frac{\gamma_B^{b}}{\gamma_A^{b}} \exp\left(-\frac{\Delta G_{gb}^{o}}{k_B T} \right) \qquad (12.228)$$

where ΔG_{gb}^{o} is the segregation free energy excluding the configurational entropy contribution. If we assume that solutions both in the bulk and at the grain boundaries can be approximated by the regular solution model, then we have

$$k_B T \ln \gamma_A^{b} = \alpha^{b} \left(x_B^{b}\right)^2, \quad k_B T \ln \gamma_B^{b} = \alpha^{b} \left(x_A^{b}\right)^2$$

$$k_B T \ln \gamma_A^{gb} = \alpha^{gb} \left(x_B^{gb}\right)^2, \quad k_B T \ln \gamma_B^{gb} = \alpha^{gb} \left(x_A^{gb}\right)^2$$

where α^{b} and α^{gb} are the regular solution parameters for the bulk and the grain boundary solid solution phases, respectively. Therefore, we can write Eq. (12.228) in terms of regular solution parameters,

$$\frac{x_B^{gb}}{\left(1 - x_B^{gb}\right)} \exp\left[\frac{\alpha^{gb}\left(1 - 2x_B^{gb}\right)}{k_B T}\right] = \frac{x_B^b}{\left(1 - x_B^b\right)} \exp\left[\frac{\alpha^b\left(1 - 2x_B^b\right)}{k_B T} - \frac{\Delta G_{gb}^\circ}{k_B T}\right]$$

(12.229)

Consequently, if we have the knowledge about the regular solution parameters both in the bulk and at the grain boundaries, α^b and α^{gb}, as well as the standard segregation free energy, ΔG_{gb}°, we can numerically solve Eq. (12.229) for the grain boundary composition x_B^{gb} as a function of bulk composition x_B^b and temperature T.

12.30 Examples

Example 1 Consider an A − B binary ideal solution at 300 K,

(a) Calculate the amount of heat released or absorbed when one mole of pure A at 300 K is added to a large quantity of A − B solution with $x_A = 0.25$ at 300 K.
(b) Calculate the entropy change when one mole of pure A at 300 K is added to a large quantity of A − B solution with $x_A = 0.25$ at 300 K.
(c) What is the difference between the chemical potential of A in the solution with composition $x_A = 0.25$ and that of pure A both at 300 K.

Solution

(a) The amount of heat released or absorbed when one mole of pure A at 300 K is added to a large quantity of A − B solution with $x_A = 0.25$ at 300 K is zero according to the definition of an ideal solution.
(b) The entropy change when one mole of pure A at 300 K is added to a large quantity of A − B solution with $x_A = 0.25$ at 300 K is

$$\Delta s_A = -R \ln x_A = -8.314 \times \ln 0.25 = 11.52 \ (J/(K \, mol))$$

(c) The difference between the chemical potential of A in the solution with composition $x_A = 0.25$ and that of pure A both at 300 K is

$$\Delta \mu_A = RT \ln x_A = 8.314 \times 300 \times \ln 0.25 = -3457.70 \ (G)$$

Example 2 Let us consider the electron occupation at the conduction band edge using the thermodynamics of solutions. Assuming there are n electrons occupying at the conduction band edge with energy E_c and density of states, N_c, find the electron density at the conduction band edge at temperature T under electron chemical potential or Fermi level E_f.

Solution
The number of ways that n number of electrons distribute among the N_C number of states at the edge of the conduction band is given by

$$\Omega = \frac{N_c!}{n!(N_c - n)!}$$

The entropy of the electron system at the conduction band is then given by

$$S = k_B \ln \Omega$$

Using Stirling's approximation, $\ln N! \approx N \ln N - N$, for large N, we have

$$S = k_B \ln \Omega \approx k_B \{N_c \ln N_c - N_c - [(N_c - n) \ln(N_c - n) - (N_c - n)] - (n \ln n - n)\}$$

The above equation can be simplified as

$$S = k_B [N_c \ln N_c - (N_c - n) \ln(N_c - n) - n \ln n]$$

or

$$S = -N_c k_B \left[\frac{(N_c - n)}{N_c} \ln \frac{(N_c - n)}{N_c} - \frac{n}{N_c} \ln \frac{n}{N_c} \right]$$

which is essentially the entropy of a binary ideal solution of electrons and unoccupied states.

The entropy per electron in the conduction band, which is called the partial molar entropy in many textbooks, is given by

$$s = \left(\frac{\partial S}{\partial n} \right)_{T,V} = -k_B \ln \frac{n}{(N_c - n)}$$

The energy per electron at the conduction band is given by

$$E = E_c$$

Therefore, the chemical potential of electrons at the conduction band at constant temperature and volume is given by

$$\mu = E - Ts = E_c + k_B T \ln \frac{n}{(N_c - n)}$$

At equilibrium, all the electrons in a crystal have the same chemical potential called the Fermi level E_f, i.e.,

$$\mu = E_c + k_B T \ln \frac{n}{(N_c - n)} = E_f$$

Solving the above equation for n, we get the equilibrium electron concentration at the conduction band edge at temperature T under Fermi level E_f,

$$n = \frac{N_c}{1 + \exp\left(\frac{E_c - E_f}{k_B T}\right)}$$

If n is much smaller than N_c, we can simplify the above equation to

$$n = N_c \exp\left(-\frac{E_c - E_f}{k_B T}\right)$$

which is a familiar equation in semiconductor physics.

Example 3 Show that if the activity coefficient of component A in a binary solution as a function of composition is given by

$$RT \ln \gamma_A = \Omega x_B^2,$$

the activity coefficient of component B in the same binary solution as a function of composition is given by

$$RT \ln \gamma_B = \Omega x_A^2$$

Solution
The Gibbs–Duhem relation for relating the activity coefficients of a binary solution is

$$x_A d \ln \gamma_A + x_B d \ln \gamma_B = 0$$

Solving the above equation for the activity coefficient γ_B, we have

$$d \ln \gamma_B = -\frac{x_A}{x_B} d \ln \gamma_A$$

Integrating both sides of the above equation, we get

$$\int_{x_B=1}^{x_B} d \ln \gamma_B = -\int_{x_A=0}^{x_A} \frac{x_A}{x_B} d \ln \gamma_A$$

Substituting the expression for the activity coefficient of A as a function of composition into the right-hand side of the above equation, we obtain

$$\int_{x_B=1}^{x_B} d\ln\gamma_B = -\int_{x_A=0}^{x_A} \frac{x_A}{x_B} d\left(\frac{\Omega x_B^2}{RT}\right) = -\frac{2\Omega}{RT}\int_{x_A=0}^{x_A} x_A dx_B$$

Using the relation $dx_B = -dx_A$ and the fact that $\gamma_B = 1$ at $x_B = 1$, we arrive at the final expression for the activity coefficient of B as a function of composition,

$$\ln\gamma_B = \frac{2\Omega}{RT}\int_{x_A=0}^{x_A} x_A dx_A = \Omega x_A^2$$

12.31   Exercises

1. Is the chemical potential of O_2 in the air (i) higher, (ii) equal to, or (iii) lower than the chemical potential of pure O_2 at 1 bar?

2. The number of independent chemical potentials in a binary system at constant temperature and pressure is (i) 1, (ii) 2, or (iii) infinite.

3. Does the activity of Li in a stable Al–Li solution (i) increase or (ii) decrease with its composition at constant temperature and ambient pressure?

4. The activity coefficient of a component in a solution can be larger than 1.0, true or false?

5. The activity coefficient of a component in a solution can be less than 0, true or false?

6. Rank the chemical potential of alcohols in beer, wine, and scotch whisky.

7. Is the chemical potential of salt dissolved in water (i) higher, (ii) equal to, or (iii) lower than the chemical potential of pure solid salt?

8. Comparing the entropy of one mole of pure A s_A, and the entropy of one mole of A and B mixture with 3/4 mol of A and 1/4 mol B, s_{AB}, at the same temperature and pressure, which of the following is your best guess?
    (A) $s_A > s_{AB}$, (B) $s_A = s_{AB}$, (C) $s_A < s_{AB}$

9. The chemical potential of a dopant atom in the host lattice below the solubility limit is (i) <, (ii) =, or (iii) > the chemical potential of a dopant atom in its pure solid.

10. The chemical potential of a vacancy in an elemental solid at equilibrium is (i) < 0, (ii) = 0, or (iii) > 0.

11. For a gas species with mole fraction 0.25 in an ideal gas mixture with total pressure of 1 bar, what is the activity of the gas species?

12. True or false questions:

(a) A real binary system cannot exist as a homogeneous solution at an equilibrium state at 0 K, true or false?

(b) For a binary solution at a given temperature and pressure, the two chemical potentials for the two components depend on each other, true or false?

(c) If the second derivative of the chemical potential of a solution with respect to composition is negative at a certain composition, the solution at that composition is always unstable with respect to phase separation to two phases with different compositions, true or false?

(d) For a binary two-phase mixture at equilibrium at a given temperature and pressure, if both components are present in the two phases, the chemical potential of each component must be uniform throughout the two phases, true or false?

(e) For a two-phase mixture at equilibrium at a given temperature and pressure, if both components are present in both phases, the activity of each component must be uniform throughout the two phases, true or false?

(f) For a two-phase mixture at equilibrium at a given temperature and pressure, if both components are present in both phases, the activity coefficient of each component must be uniform throughout the two phases, true or false?

13. Consider the mixing of A and B atoms to form an A-B solution, A (pure) + B (pure) = A-B (solution).

(a) If both A atoms and B atoms obey Raoult's law, determine whether the enthalpy change for the mixing is (i) 0; (ii) positive; or (iii) negative.

(b) If both A atoms and B atoms obey Raoult's law, determine whether the heat of mixing is (i) 0; (ii) positive; or (iii) negative.

(c) If both A atoms and B atoms obey Raoult's law, the entropy change for the mixing is (i) 0; (ii) positive; or (iii) negative.

(d) If A atoms and B atoms do not like each other, i.e., the atomic bonding between A and B atoms is weaker than those among the A atoms and among the B atoms, determine whether the enthalpy change for the mixing is (i) 0; (ii) positive; or (iii) negative.

(e) If A atoms and B atoms like each other, i.e., the atomic bonding between A and B atoms is stronger than those among the A atoms and among the B atoms, the heat of mixing is (i) 0; (ii) positive; or (iii) negative.

14. Consider an ideal binary liquid solution with components A and B. Please do the following:

(a) Write down the relationship between activities and mole fractions of A and B, respectively.

(b) What are the activity coefficients of A and B?

(c) What are the chemical potentials of A and B in a homogeneous solution of composition $x_A = 0.5$ at 1000 K relative to pure A and B, respectively?

(d)   What are the chemical potential, entropy, and enthalpy changes when 0.25 mol of pure liquid A is mixed with 0.75 mol of pure liquid B to form 1.0 mol of homogeneous liquid solution at 1000 K?

(e)   What are the chemical potential, entropy, and enthalpy changes when 0.75 mol of pure liquid A is mixed with 0.25 mol of pure liquid B to form 1.0 mol of homogeneous liquid solution at 1000 K?

(f)   What are the chemical potential, entropy, and enthalpy changes when 0.5 mol of a solution with composition $x_A = 0.25$ is mixed with 0.5 mol of a solution with composition $x_A = 0.75$ to form 1.0 mol of homogeneous solution at 1000 K?

(g)   What are the chemical potential, entropy, and enthalpy changes when 1.0 mol of pure A liquid is added to and mixed with a large quantity of homogeneous liquid solution with composition of $x_A = 0.5$ at 1000 K?

(h)   If the molar heat of melting of solid A is 10,000 J at the equilibrium melting temperature of 1100 K, and the molar heat capacity of both pure solid and liquid A is 30 J/K, what are the Gibbs free energy, entropy, and enthalpy changes when 1.0 mol of pure A solid at 1000 K is added to and mixed with a large quantity of homogeneous liquid solution with a composition of $x_A = 0.5$ at 1000 K?

15.   For a binary solution in which the solvent obeys Raoult's law and the solute obeys Henry's law, show that the thermodynamic factor ψ given below is equal to 1.0.

$$\psi = \frac{x_A x_B}{RT}\left(\frac{\partial^2 \mu}{\partial x_B^2}\right)_{T,p}$$

where μ is the chemical potential of the binary solution.

16.   Consider the formation of A-B solutions, $A(\text{pure}) + B(\text{pure}) = AB(\text{solution})$. The thermodynamics of the binary solution can be approximated using a regular solution model:

$$RT \ln \gamma_A = 4000 x_B^2 \ (\text{J/mol})$$

(a)   Calculate the activity coefficient of A at 300 K at $x_A = 0.5$.

(b)   Calculate the activity of A at 300 K at $x_A = 0.5$.

(c)   Calculate the amount of heat released or absorbed when one mole of pure A is added to a large quantity of A − B solution with $x_A = 0.5$.

(d)   Calculate the entropy change when one mole of pure A is added to a large quantity of A − B solution with $x_A = 0.5$.

(e)   What is the chemical potential difference of one mole of A in the solution with composition $x_A = 0.5$ and one mole of pure A both at 300 K.

(f)   Write down an expression for activity coefficient of B as a function of composition of A.

(g) Determine whether the enthalpy change for the mixing is (i) 0; (ii) positive; or (iii) negative.

(h) Determine whether heat during mixing is (i) 0; (ii) released; or (iii) absorbed.

(i) Determine whether the entropy change for the mixing is (i) 0; (ii) positive; or (iii) negative.

(j) The equilibrium state for this binary system at 0 K is a two-phase mixture of pure A and pure B rather than a homogeneous random solution of A and B, true or false?

(k) Calculate the chemical potential change when 0.25 mol of pure A is combined with 0.75 mol of pure B to form 1 mol of A − B solution with composition $x_A = 0.25$ at 300 K.

17. Assuming that the activity coefficient of Zn in liquid Zn–Cd alloys at 400 °C can be represented by the Margules solution model as,

$$\ln \gamma_{Zn} = x_{Cd}^2 (1 - 0.25 x_{Cd})$$

(a) Write down the difference between the chemical potential of a homogeneous Zn − Cd solution and a mixture of pure Zn and Cd as a function of composition x_{Cd} at 400 °C.

(b) What are the chemical potentials of Zn and Cd in a homogeneous Zn–Cd alloy of composition $x_{Cd} = 0.5$ at 400 °C relative to pure Zn and Cd, respectively?

(c) What is the chemical potential change when 0.25 mol of pure Zn is mixed with 0.75 mol of pure Cd to form 1.0 mol of a homogeneous Zn–Cd alloy at 400 °C?

(d) What is the chemical potential change when 0.5 mol of a homogeneous alloy with composition $x_{Cd} = 0.25$ is mixed with 0.5 mol of a homogeneous alloy with composition $x_{Cd} = 0.75$ to form 1.0 mol of homogeneous alloy at 400 °C?

(e) What is the chemical potential change when 1.0 mol of pure Cd is added to a large quantity of homogeneous Zn–Cd alloy with composition of $x_{Cd} = 0.5$ at 400 °C?

18. In dilute liquid solutions of Sn and Cd, let us assume that the majority component, Cd, obeys Raoult's law and the minority component Sn obeys Henry's law with the Henrian coefficient, k_{Sn}, varying with temperature as (Note: this is neither an ideal solution nor a regular solution)

$$RT \ln k_{Sn} = -7000 + 13.1T$$

(a) Calculate the activity of Sn in the solution with $x_{Sn} = 0.01$ at 1000 K.

(b)   Calculate the chemical potential change of Sn when 1 mol of pure Sn is added to a large quantity of Sn–Cd liquid solution with $x_{Sn} = 0.01$ at 1000 K.

(c)   Calculate the chemical potential change when 0.01 mol of pure Sn is mixed with 0.99 mol of pure Cd to form 1 mol of solution at 1000 K.

(d)   Calculate the entropy change when 0.01 mol of pure Sn is mixed with 0.99 mol of pure Cd to form 1 mol of solution at 1000 K.

(e).   Calculate the heat absorbed or released when 0.01 mol of pure Sn is mixed with 0.99 mol of pure Cd to form 1 mol of solution at 1000 K.

19.   For the Flory–Huggins model of polymer solution,

$$\Delta G/[(N_1 + xN_2)RT] = \chi \varphi_1 \varphi_2 + \varphi_1 \ln \varphi_1 + \frac{\varphi_2}{x} \ln \varphi_2$$

(a)   Plot $\Delta G/[(N_1 + xN_2)RT]$ as a function of φ_2 for $\chi = 1$ and $x = 1$.

(b)   Plot $\Delta G/[(N_1 + xN_2)RT]$ as a function of φ_2 for $\chi = 1$ and $x = 4$.

Chapter 13
Chemical Phase Equilibria and Phase Diagrams

Chemical phase equilibrium is concerned with the equilibrium states with different chemical compositions at a given temperature and pressure, and phase diagrams are graphical representations of thermodynamic equilibrium states at different temperature, pressure, and chemical composition. Most of the experimental phase diagrams are obtained using the ambient pressure of 1 bar, so the discussions will be focused on the determination of chemical equilibrium and construction of phase diagrams showing the equilibrium states at different temperatures and chemical compositions at 1 bar. If one obtains the chemical potentials of all the phases as functions of composition at different temperatures, one can construct a phase diagram by determining the most stable chemical state which has the lowest chemical potential among all phases or combination of phases, at a given temperature and overall composition of a material system. At equilibrium, according to Gibbs, "The potential for each of the component substances must have a constant value in all parts of the given mass of which that substance is an actual component and have a value not less than this in all parts of which it is a possible component".

13.1 Equilibrium States from Chemical Potentials Versus Composition

In order to construct a temperature-composition phase diagram, one must first determine the equilibrium states as functions of composition at a given temperature. We employ examples of model binary systems to illustrate graphically how to obtain equilibrium states from the chemical potential curves as a function of composition at a given temperature.

Example 1 Single-phase solid solution.

Figure 13.1 (top) shows the chemical potentials (molar Gibb free energies), μ^s and μ^l, of binary solid (s) and liquid (l) solutions at temperature T_o as a function of

© Springer Nature Singapore Pte Ltd. 2022
L.-Q. Chen, *Thermodynamic Equilibrium and Stability of Materials*,
https://doi.org/10.1007/978-981-13-8691-6_13

Fig. 13.1 Schematic chemical potential curves showing a stable solid solution over the entire composition range

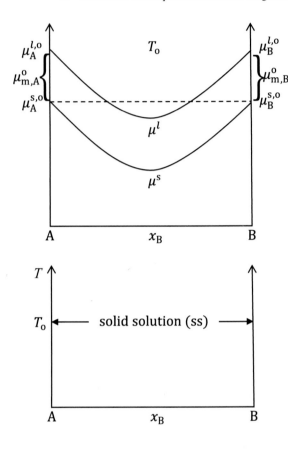

composition (x_B) with pure A and pure B in their stable states at T_o as the standard states; $\mu_A^{s,o}$ and $\mu_B^{s,o}$ are the chemical potentials of pure solid A and B at temperature T_o. Therefore, the chemical potential for the mixture of pure A and B (the horizontal dash line in Fig. 13.1) is given by

$$\mu^o(x_B) = x_A\mu_A^{s,o} + x_B\mu_B^{s,o}$$

On Fig. 13.1, $\mu_A^{l,o}$ and $\mu_B^{l,o}$ are the chemical potentials of pure liquid A and B at temperature T_o, and the chemical potential changes of melting of pure solid A and B are given by

$$\Delta\mu_{m,A}^o = \mu_A^{l,o} - \mu_A^{s,o}$$

and

$$\Delta\mu_{m,B}^o = \mu_B^{l,o} - \mu_B^{s,o}$$

Figure 13.1 shows that the chemical potential μ^s of the solid solution (ss) is lower than the chemical potential μ^l of the liquid solution at all compositions at T_o.

Therefore, throughout the whole composition range at T_o, the homogeneous solid solution (ss) phase is the stable equilibrium state, and the liquid phase is not stable.

In addition, the chemical potential curve of the solid has a positive curvature,

$$\left(\frac{\partial^2 \mu^s}{\partial x_B^2}\right)_{T,p} > 0$$

which implies that the homogeneous solid solution (ss) state (bottom figure of Fig. 13.1) is intrinsically stable through the whole compositional range, and any two-phase mixture has a higher chemical potential than a homogeneous solution.

Example 2   Miscibility gap in the solid state.

Figure 13.2 (top) shows the chemical potentials, μ^s and μ^l, of binary solid (s) and liquid (l) solutions at temperature T_o as a function of composition x_B. The chemical potential of a mixture of pure A and pure B at T_o is given by the dash line. The main difference between this and the previous example is in the chemical potential of solid solution α as a function of composition x_B. In this example, the chemical potential curve of solid solution α shows a double-well shape with both positive and negative

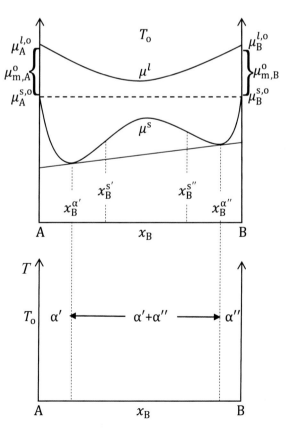

Fig. 13.2   Schematic chemical potential curves showing phase separation in a solid

curvature regions as a function of composition:

$$\left(\frac{\partial^2 \mu^s}{\partial x_B^2}\right)_{T,p} > 0 \text{ for } 0 < x_B < x_B^{s'} \text{ and } x_B^{s''} < x_B < 1$$

$$\left(\frac{\partial^2 \mu^s}{\partial x_B^2}\right)_{T,p} < 0 \text{ for } x_B^{s'} < x_B < x_B^{s''}$$

where composition $x_B^{s'}$ and $x_B^{s''}$ are spinodal compositions at which the chemical potential curve of a solid solution has a zero curvature, i.e., the second derivative of the chemical potential with respect to composition is zero. In the figure, α' and α'' represent B-lean and B-rich solid solutions, respectively. A solution within a region that the chemical potential as a function of composition shows a negative curvature is thermodynamically unstable with respect to compositional fluctuations leading to phase separation into a two-phase mixture. In this case, the equilibrium compositions of the two phases are $x_B^{\alpha'}$ and $x_B^{\alpha''}$. Within the composition ranges $x_B^{\alpha'} < x_B < x_B^{s'}$ and $x_B^{s''} < x_B < x_B^{\alpha''}$, a homogeneous solid solution is metastable. In the metastable regions, the chemical potential has a positive curvature, and hence the solid solution is stable against phase separation into two phases with very small composition differences, but it is unstable with respect to phase separation to two phases with very large composition differences, e.g., two phases with composition $x_B^{\alpha'}$ and $x_B^{\alpha''}$. Therefore, in this example, the equilibrium states exhibit a solid two-phase mixture $(\alpha' + \alpha'')$ for the intermediate compositions between $x_B^{\alpha'}$ and $x_B^{\alpha''}$ (see bottom schematic drawing of Fig. 13.2).

Example 3 Solid–liquid two-phase equilibrium.

In this example, the temperature T_o of interest is higher than the melting temperature of pure B but lower than the melting temperature of pure A. Therefore, at this temperature, the standard state for A is pure solid A, and the standard state for B is pure liquid B.

The schematic chemical potential curves of liquid and solid solutions as functions of composition are shown in the top part of Fig. 13.3. In Fig. 13.3, $\Delta \mu_{m,A}^o$ is the chemical potential change for the melting of pure solid A, and $\Delta \mu_{s,B}^o$ is the chemical potential change for the solidification of pure liquid B. As shown in the bottom part of Fig. 13.3, at T_o, the stable equilibrium state in the composition range $0 < x_B < x_B^s$ is solid solution (ss), while the stable equilibrium state in the composition range $x_B^l < x_B < 1$ is liquid solution (ls). In the intermediate composition range $x_B^s < x_B < x_B^l$, the stable equilibrium state is a two-phase (ss + ls) mixture with equilibrium compositions x_B^s and x_B^l.

The chemical potential of any composition in a heterogeneous system is uniform, not just chemical potentials of A and B, if A and B are free to transport across the interfaces separating the homogeneous regions. The chemical potential values of A and B and all combinations of A and B are given by the common tangent line at equilibrium (Fig. 13.4). For example, a small amount of solution with composition x_B^s is dissolved into the solid, and the chemical potential of this solution is $\mu^s(x_B^s)$. Now if we dissolve a small amount of solution with the same composition x_B^s into the liquid, the chemical potential of this solution is also $\mu^s(x_B^s)$. If a small amount

Fig. 13.3 Schematic chemical potential curves showing equilibrium between a liquid and a solid

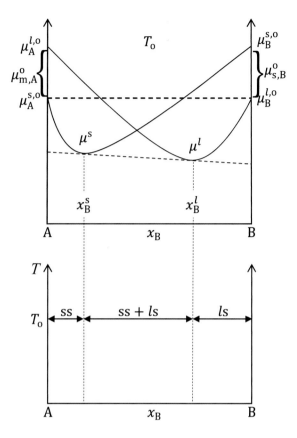

Fig. 13.4 Illustration of uniform chemical potentials in a two-phase system

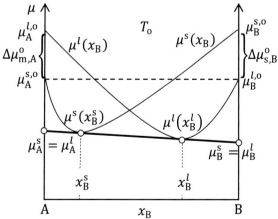

of solution with composition x_B^l is dissolved into the liquid, its chemical potential is given by $\mu^l(x_B^l)$. If a small amount of solution with composition x_B^l is dissolved into the solid, its chemical potential is also given by $\mu^l(x_B^l)$. Actually, for a small amount of solution of any composition x_B dissolved in either the solid or liquid, its chemical potential is given by a point on the tangent line.

Example 4 Solid–solid two-phase equilibrium.

Figure 13.5 shows an example that involves three phases: a liquid phase (l) and two solid phases (α and β). The temperature T_0 is below the melting temperatures of both pure A and pure B. Therefore, at T_0 the standard states for components A and B in two solid solution phases α and β are pure solid A and B, while the standard states for components A and B in liquid are pure liquid A and B. By comparing the chemical potentials of each possible state, it can easily be determined that the equilibrium state between $x_B = 0$ and x_B^α is a single solid solution phase α, the equilibrium state between $x_B = x_B^\alpha$ and x_B^β is a two-phase mixture of α and β phases with their equilibrium compositions given by x_B^α and x_B^β determined from the common-tangent construction, and the equilibrium state for compositions between x_B^β and $x_B = 1$ is a single solid solution phase β.

Fig. 13.5 Schematic chemical potential curves showing equilibrium between two solids with limited mutual solubility

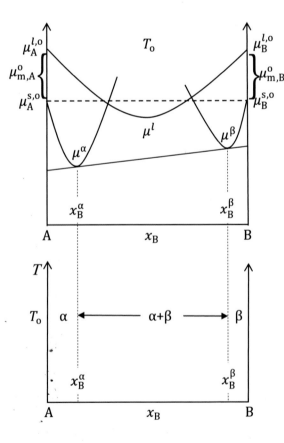

Example 5 Equilibrium between a pure solid and a liquid solution.

Figure 13.6 shows an example in which there is essentially zero solubility of B atoms in solid A. The temperature T_o is below the melting temperature of pure solid A but above the melting temperature of pure solid B. Based on the chemical potential curves in the top half of Fig. 13.6, the solid phase β is not stable throughout the composition range. The equilibrium states are shown in the bottom half of Fig. 13.6. For the composition range of $0 < x_B < x_B^l$, the equilibrium state is a two-phase mixture of pure solid A (the solubility of B in A is very small) and the liquid solution (ls), while for $x_B^l < x_B < 1$ the equilibrium state is a single-phase liquid solution (ls).

Example 6 Eutectic equilibrium.

Figure 13.7 shows an example in which there are three phases, two solid solution phases α and β and a liquid phase (l), at equilibrium at a given temperature, called the eutectic temperature (T_E), and a particular composition called the eutectic composition (x_B^E). This is the lowest temperature at which a liquid can exist in this hypothetical binary system. The eutectic temperature T_E is below the melting temperatures of both pure solid A and B. The chemical potentials for the three phases are

Fig. 13.6 Schematic chemical potential curves showing equilibrium between a pure solid and a liquid solution

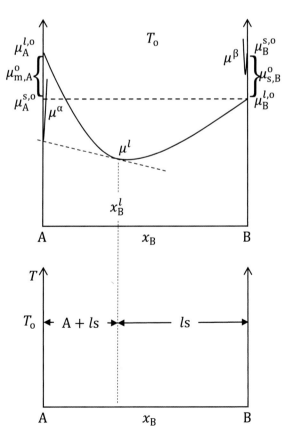

Fig. 13.7 Schematic
chemical potential curves
showing eutectic equilibrium

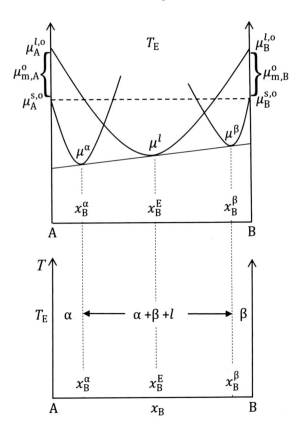

schematically shown in the top half of Fig. 13.7. The common tangent line to the
chemical potential curves of all three phases determines the equilibrium compositions of the three phases. The equilibrium states at this temperature are shown in the
bottom half of Fig. 13.7. For $0 < x_B < x_B^\alpha$, the equilibrium state is a single-phase α
solid solution, while for $x_B^\beta < x_B < 1$, the equilibrium state is a single-phase β solid
solution. For the composition range $x_B^\alpha \le x_B \le x_B^\beta$, three phases, the solid solution
α, the solid solution β, and the liquid solution l, can coexist at their respective equilibrium compositions x_B^α for the α solid solution phase, x_B^β for the β solid solution
phase, and the eutectic composition x_B^E for the liquid phase.

Example 7 Equilibrium between a compound and a solid solution.

Figure 13.8 shows an example that involves the formation of a compound γ at an
intermediate composition x_B^γ in addition to two solid solution phases α and β with
limited solubility as well as a liquid solution phase l. Based on the chemical potential
curves of all four phases in Fig. 13.8, the liquid phase l is not stable throughout the
entire composition range. The equilibrium compositions can be easily obtained by
the common-tangent construction guaranteeing that the chemical potentials of each
component (A and B) are uniform in the system. The stable states at this particular

Fig. 13.8 Schematic chemical potential curves showing equilibrium between a compound and a solution

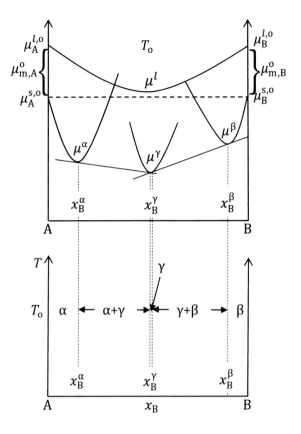

temperature are illustrated in the bottom half of Fig. 13.8. For $0 < x_B < x_B^\alpha$, the single-phase α is the stable state, and the single-phase β is stable in the composition range of $x_B^\beta < x_B < 1$. For $x_B^\alpha < x_B < x_B^\gamma$, the equilibrium state is a two-phase mixture $\alpha + \gamma$, and for $x_B^\gamma < x_B < x_B^\beta$, the equilibrium state is a two-phase mixture of $\gamma + \beta$. At composition x_B^γ, the equilibrium state is the single-phase γ. Since the solubility of A and B in γ is very small, we usually call γ a line compound.

13.2   Examples of Model Phase Diagrams

Once we determine all the equilibrium states for all the temperatures of interest, a plot of the equilibrium states as a function of temperature and composition is called "temperature-composition" phase diagram. Without specifying the pressure for a phase diagram, we assume it is 1 bar.

Example 1 Eutectic diagram with no mutual solubility in two solids.

Figure 13.9 shows a simple eutectic binary phase diagram without mutual solu-

Fig. 13.9 Schematic eutectic phase diagram with no mutual solubility in solid

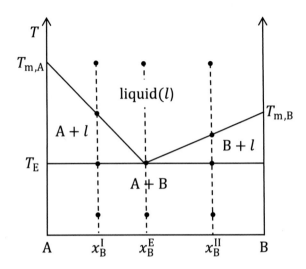

bility in the solid state. The high-temperature phase is a liquid solution (l). At low temperatures, the equilibrium state is a mixture of pure A and B, i.e., there is no mutual solubility between A and B. At intermediate temperatures, depending on the composition, the equilibrium states can be a mixture of pure A and liquid solution, a two-phase mixture of pure B and liquid solution, or a single-phase liquid solution. To illustrate how to utilize a phase diagram, we trace the equilibrium states of a system at several different compositions as the system is heated up from low temperature or cooled down from a high temperature.

On heating of a mixture of pure A and pure B, the evolution sequences of equilibrium states at three representative compositions are

$A + B \rightarrow A + l\left(x_B^E\right) \rightarrow A + l\left(x_B^I\right) \rightarrow l\left(x_B^I\right)$ at overall composition x_B^I,
$A + B \rightarrow A + B + l\left(x_B^E\right) \rightarrow l\left(x_B^E\right)$ at overall composition x_B^E,
$A + B \rightarrow B + l\left(x_B^E\right) \rightarrow B + l\left(x_B^{II}\right) \rightarrow l\left(x_B^{II}\right)$ at overall composition x_B^{II}.

For all compositions, the composition of the liquid that first appears is at the eutectic composition x_B^E at the eutectic temperature T_E. Upon cooling from a high-temperature liquid state, the sequences of appearance for the equilibrium states are simply reversed.

Example 2 Completely soluble liquid and solid phases.

Figure 13.10 shows another example of a simple binary phase diagram. In this example, both the liquid and solid states exhibit homogeneous solutions in which A and B are completely mutually soluble.

Upon heating of a $A - B$ solid solution from low temperature, the evolution sequence of equilibrium states at x_B^I is

$$ss\left(x_B^I\right) \rightarrow ss\left(x_B^I\right) + ls\left(x_B^l\right) \rightarrow ss\left(x_B^s\right) + ls\left(x_B^l\right) \rightarrow ls\left(x_B^I\right)$$

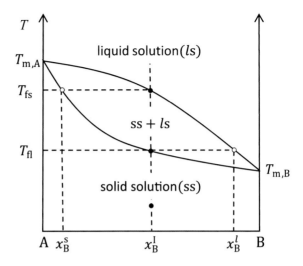

Fig. 13.10 Schematic phase diagram showing complete solubility in both liquid and solid

On cooling of a liquid solution of A and B from high temperature, the evolution sequence of equilibrium states at x_B^l is

$$ls\left(x_B^l\right) \rightarrow ls\left(x_B^l\right) + ss\left(x_B^s\right) \rightarrow ls\left(x_B^l\right) + ss\left(x_B^s\right) \rightarrow ss\left(x_B^l\right)$$

It should be noted that at very low temperatures close to 0 K, a homogeneous solid solution as an equilibrium state is thermodynamically impossible since the third law of thermodynamics dictates that the entropy of an equilibrium state is zero at 0 K. The equilibrium states at 0 K can only be either pure elemental solids or stoichiometric compounds.

Example 3 Eutectic diagram with limited mutual solubility between two solids.

The phase diagram shown in Fig. 13.11 is similar to that in Fig. 13.9. Both of them have a eutectic point. The main difference is the fact that here in this example there is limited mutual solubility between solid A and B while they are completely soluble in the liquid state. To distinguish a solid solution from a pure solid, we call the solid solution as the α phase if the majority of atoms (solvent) in the phase are A atoms, and the minority atoms (solute) are B atoms while the solid solution is called the β phase if the majority atoms (solvent) in the phase are B atoms, and the minority atoms (solute) are A atoms. For the three representative compositions, x_B^I, x_B^E, and x_B^{II}, the equilibrium states and their stable equilibrium compositions during heating are illustrated below.

At x_B^I:

$$\alpha + \beta \rightarrow \alpha\left(x_B^I\right) + \beta\left(x_B^{\beta\alpha}\right) \rightarrow \alpha\left(x_B^I\right) \rightarrow \alpha\left(x_B^I\right)$$
$$+ l\left(x_B^{l\alpha}\right) \rightarrow \alpha\left(x_B^{\alpha l}\right) + l\left(x_B^I\right) \rightarrow l\left(x_B^I\right),$$

Fig. 13.11 Schematic eutectic phase diagram with limited mutual solubility in solid

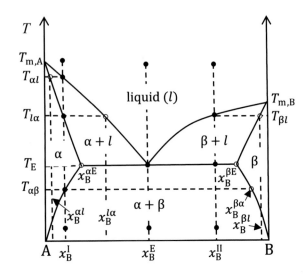

at x_B^E:

$$\alpha + \beta \rightarrow \alpha\left(x_B^{\alpha E}\right) + \beta\left(x_B^{\beta E}\right) \rightarrow l\left(x_B^E\right),$$

and at x_B^{II}:

$$\alpha + \beta \rightarrow \alpha\left(x_B^{\alpha E}\right) + \beta\left(x_B^{\beta E}\right) \rightarrow l\left(x_B^E\right)$$
$$+ \beta\left(x_B^{\beta E}\right) \rightarrow l\left(x_B^{II}\right) + \beta\left(x_B^{\beta l}\right) \rightarrow l\left(x_B^{II}\right)$$

For cooling, the sequences of equilibrium states are reversed to those for heating.

Example 4 Peritectic phase diagram.

Figure 13.12 shows an example of a so-called peritectic phase diagram. At the peritectic temperature T_p, there is a peritectic point at which solid phase α at composition $x_B^{\alpha P}$ and liquid phase at x_B^{lP} react to form the β phase solid at the peritectic composition x_B^P during cooling,

$$\alpha\left(x_B^{\alpha P}\right) + l\left(x_B^{lP}\right) \rightarrow \beta\left(x_B^P\right)$$

or during heating,

$$\beta\left(x_B^P\right) \rightarrow \alpha\left(x_B^{\alpha P}\right) + l\left(x_B^{lP}\right)$$

The peritectic temperature is between the melting temperature of pure solid A and that of pure solid B. For the representative compositions, x_B^I, x_B^P, and x_B^{II}, the respective sequences of equilibrium states during heating are

Fig. 13.12 Schematic peritectic phase diagram with limited mutual solubility in solid

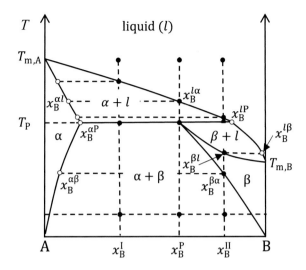

$$\alpha + \beta \rightarrow \alpha\left(x_B^{\alpha P}\right) + \beta\left(x_B^P\right) \rightarrow \alpha\left(x_B^{\alpha P}\right)$$
$$+ l\left(x_B^{lP}\right) \rightarrow \alpha\left(x_B^{\alpha l}\right) + l\left(x_B^l\right) \rightarrow l\left(x_B^l\right),$$
$$\alpha + \beta \rightarrow \beta\left(x_B^P\right) \rightarrow \alpha\left(x_B^{\alpha P}\right)$$
$$+ l\left(x_B^{lP}\right) \rightarrow \alpha\left(x_B^{\alpha l}\right) + l\left(x_B^{lP}\right) \rightarrow l\left(x_B^{l\alpha}\right),$$
$$\alpha + \beta \rightarrow \beta\left(x_B^{II}\right) \rightarrow \beta\left(x_B^{II}\right) + l\left(x_B^{l\beta}\right) \rightarrow \beta\left(x_B^P\right)$$
$$+ l\left(x_B^{lP}\right) \rightarrow \alpha\left(x_B^{\alpha P}\right) + l\left(x_B^{lP}\right) \rightarrow \alpha\left(x_B^{\alpha l}\right) + l\left(x_B^{II}\right) \rightarrow l\left(x_B^{II}\right)$$

Example 5 Intermediate compound.

Figure 13.13 shows an example of a phase diagram that involves the formation of an intermediate compound γ in which both A and B have limited solubility. If A and B have zero solubility or near zero solubility, we call γ a line compound since it will appear in the phase diagram as a vertical line. This phase diagram can be roughly viewed as a combination of two eutectic phase diagrams: one with two solid phases α and γ and the other with γ and β. Therefore, the sequences of equilibrium states during heating and cooling are essentially the same as Example 3 of this section. For example, for the representative compositions x_B^l, the equilibrium states and their equilibrium compositions during heating are

$$\alpha + \gamma \rightarrow \alpha\left(x_B^{\alpha\gamma}\right) + \gamma\left(x_B^l\right) \rightarrow \gamma\left(x_B^l\right) \rightarrow \gamma\left(x_B^l\right)$$
$$+ l\left(x_B^{l\gamma}\right) \rightarrow \gamma\left(x_B^{\gamma l}\right) + l\left(x_B^l\right) \rightarrow l\left(x_B^l\right).$$

Example 6 Ternary phase diagram.

Here we discuss an example of a ternary phase diagram. The representation of chemical phase equilibria in ternary systems is significantly more challenging than

Fig. 13.13 Schematic
eutectic phase diagram with
limited mutual solubility in
solid and formation of a
compound

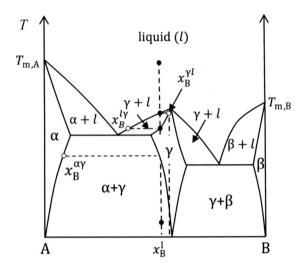

binary systems. There are five intensive variables or potentials: three chemical potentials for the three components, plus temperature and pressure. With the Gibbs–Duhem relation, we have four independent variables. If we fix pressure, we are still left with three independent variables. A ternary phase diagram is typically represented by two compositional variables and temperature. With the knowledge of compositions of two components, we automatically know the composition of the third component. The most often used compositional representation is a triangle or called Gibbs triangle (Fig. 13.14). For example, the three corners of the triangle, A, B, and C represent the three compositions $(x_A, x_B, x_C) = (1.0, 0.0, 0.0)$, $(0.0, 1.0, 0.0)$, and $(0.0, 0.0,$

Fig. 13.14 Composition
representation of a ternary
system

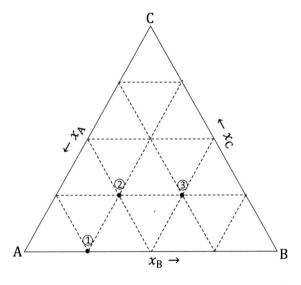

1.0), respectively. The dash lines parallel to the side line BC represent compositions with constant x_A with the composition of A along the BC line being zero and the composition of A at the corner A of the triangle being 1.0. Similarly, the dash lines parallel to AC side line represent compositions with constant x_B, and those parallel to AB represent the compositions with constant x_C. For example, point 1 in Fig. 13.14 represents composition (0.75, 0.25, 0.0), point 2 represents composition (0.5, 0.25, 0.25), and point 3 represents (0.25, 0.5, 0.25).

For a ternary phase diagram, the vertical axis represents temperature. Figure 13.15 is a schematic ternary phase diagram with limited mutual solubilities among the three components in the solid state and complete mutual solubilities in the liquid state. Each of the three binaries exhibits a eutectic point labeled as T_E^{AB}, T_E^{AC}, and T_E^{BC}, and there is one eutectic point for the ternary system labeled as T_E, the lowest temperature, at which a liquid can exist at equilibrium. In Fig. 13.15, $T_{m,A}$, $T_{m,B}$, and $T_{m,C}$ represent the melting temperatures of pure solid A, pure solid B, and pure solid C, respectively. α, β, and γ represent the three solid solution phases with limited solubility, and l represents the liquid phase.

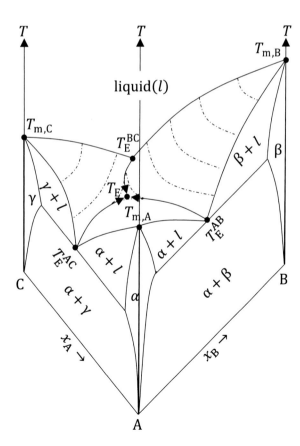

Fig. 13.15 Schematic ternary eutectic phase diagram with limited mutual solubility in solid

13.3 Number of Degrees of Freedom—Gibbs Phase Rule

The Gibbs phase rule for determining the number of degrees of freedom (NDF) is given by Eq. (11.38),

$$NDF = n - \psi + 2,$$

where n is the number of components, ψ is the number of phases, and 2 represents the two variables temperature T and pressure p. However, the majority of temperature-composition phase diagrams are determined under a fixed pressure of 1 bar, so the Gibbs phase rule becomes

$$NDF = n - \psi + 1$$

As examples, let us determine the NDF for the eight points labeled in Fig. 13.16.

At 1: $NDF = n - \psi + 1 = 2 - 1 + 1 = 2$
At 2: $NDF = n - \psi + 1 = 2 - 2 + 1 = 1$
At 3: $NDF = n - \psi + 1 = 2 - 2 + 1 = 1$
At 4: $NDF = n - \psi + 1 = 2 - 2 + 1 = 1$
At 5: $NDF = n - \psi + 1 = 2 - 1 + 1 = 2$
At 6: $NDF = n - \psi + 1 = 2 - 3 + 1 = 0$
At 7: $NDF = n - \psi + 1 = 2 - 3 + 1 = 0$
At 8: $NDF = n - \psi + 1 = 2 - 1 + 1 = 2$

Fig. 13.16 Illustration of degree of freedom calculation at different locations of a eutectic phase diagram with an intermediate compound

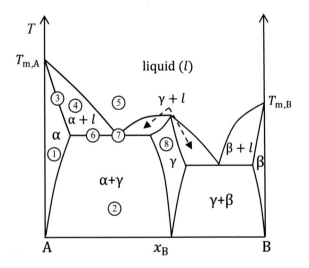

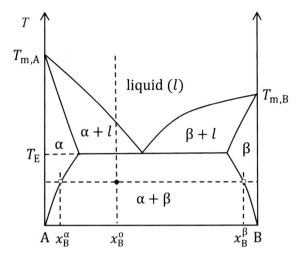

Fig. 13.17 Illustration of applying lever rule to a two-phase state

13.4 Phase Fractions—Lever Rule

To determine the phase fractions, φ^α and φ^β, in an equilibrium two-phase $(\alpha + \beta)$ mixture, we employ the lever rule given by Eq. (12.80),

$$\varphi^\alpha = \frac{x_B^\beta - x_B^o}{x_B^\beta - x_B^\alpha} \text{ and } \varphi^\beta = 1 - \varphi^\alpha$$

where x_B^α and x_B^β are the equilibrium compositions of B in α and β phases, and x_B^o is the overall composition (see Fig. 13.17).

13.5 Estimates of Activity Coefficients in Binary Systems

The activities and activity coefficients in a two-phase mixture with limited mutual solubility can be estimated using the fact that the chemical potentials and thus the activities are uniform in an equilibrium two-phase mixture and the fact that the activity and activity coefficients for the majority component may be approximated by the Raoult's law, and those of the minority component can be approximated by the Henry's law.

Let us assume that the equilibrium compositions for the two phases are $\left(x_A^\alpha, x_B^\alpha\right)$ and $\left(x_A^\beta, x_B^\beta\right)$ (see Fig. 13.17). Since the two phases are at equilibrium, the chemical potentials and thus the activities of each species must be uniform in the system,

$$\mu_A^\alpha = \mu_A^\beta \text{ and } \mu_B^\alpha = \mu_B^\beta$$

$$a_A^\alpha = a_A^\beta \text{ and } a_B^\alpha = a_B^\beta$$

Assuming that the majority component in the β phase is B, and the majority component in the α phase is A, and invoking the Raoult's law for the majority component and the Henry's law for the minority component, we have

$$a_A^\alpha = x_A^\alpha \text{ and } a_B^\beta = x_B^\beta$$

Therefore,

$$a_A^\beta = \gamma_A^\beta x_A^\beta = a_A^\alpha = x_A^\alpha \text{ and } a_B^\alpha = \gamma_B^\alpha x_B^\alpha = a_B^\beta = x_B^\beta$$

The corresponding activity coefficients are

$$\gamma_A^\alpha = 1, \gamma_B^\beta = 1$$

and

$$\gamma_A^\beta = \frac{x_A^\alpha}{x_A^\beta}, \quad \gamma_B^\alpha = \frac{x_B^\beta}{x_B^\alpha} \tag{13.1}$$

13.6 Calculations of Simple Phase Diagrams

Example 1 Melting temperature suppression.

To determine the effect of adding a second component on the melting temperature of a solid, we consider the equilibrium between a pure solid and a liquid with small amount of B, i.e.,

$$A(s, x_A = 1, x_B = 0) = A(l, x_A, x_B)$$

where x_A and x_B are the mole fractions of A and B. At equilibrium, the chemical potential of A atoms in the pure solid A is the same as that of A atoms in the liquid solution, and we assume that the chemical potential of B atoms in solid A is greater than that of pure solid B, so we have

$$\mu_A^s(x_A = 1, x_B = 0) = \mu_A^l(x_A, x_B)$$

We choose the pure solid A as the standard state and approximate the liquid solution as an ideal solution, and we have

$$\mu_A^s(x_A = 1, x_B = 0) = \mu_A^o \tag{13.2}$$

Fig. 13.18 Illustration of
melting temperature
suppression

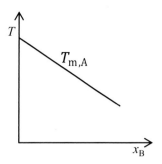

$$\mu'_A(x_A, x_B) = \mu^o_A + \Delta\mu_m + RT\ln x_A \qquad (13.3)$$

where $\Delta\mu_m$ is the chemical potential change for the melting of pure solid A.
Assuming that the heat and entropy of melting are independent of temperature,
or the heat capacities of solid and liquid are assumed to be the same, $\Delta\mu_m$ can be
approximated as

$$\Delta\mu_m \approx \Delta h_m - \Delta s_m T$$

or

$$\Delta\mu_m \approx \frac{\Delta h_m(T_{m,A} - T)}{T_{m,A}} \qquad (13.4)$$

From Eqs. (13.2) and (13.3), we have

$$-\Delta\mu_m = RT\ln x_A$$

Using the approximation (13.4) and the approximation of $\ln(x_A) \cong -x_B$, we get

$$-\frac{\Delta h_m(T_{m,A} - T)}{T_{m,A}} \cong -RT x_B \qquad (13.5)$$

Solving Eq. (13.5) for T

$$T = \frac{\Delta h_m T_{m,A}}{RT_{m,A} x_B + \Delta h_m} = \frac{T_{m,A}}{\frac{RT_{m,A}}{\Delta h_m} x_B + 1} \approx T_{m,A}\left(1 - \frac{RT_{m,A}}{\Delta h_m} x_B\right) \qquad (13.6)$$

which represents the variation of melting temperature with the amount of addition
of B atoms into solid A represented by the mole fraction of B (Fig. 13.18)

Example 2 A binary phase diagram with complete mutual solubility in both solid
and liquid.

To calculate the phase diagram with complete mutual solubility in both solid and liquid, it is sufficient to determine the solidus line and the liquidus line, i.e., the equilibrium compositions of solid and liquid in the solid–liquid two-phase mixture as a function of temperature. For simplicity, let us assume that A and B form ideal solutions in both solid and liquid.

We recall that the equilibrium condition between a solid solution and a liquid solution is

$$\mu_A^s = \mu_A^l, \, \mu_B^s = \mu_B^l$$

If these two conditions are satisfied, the chemical potential of the binary system is minimized.

Since the two-phase field illustrated in Fig. 13.19 is below the melting temperature of A and above the melting temperature of B, we choose pure solid A as standard state for A and choose pure liquid B as standard state for B. Therefore, the chemical potentials of A in solid and liquid are given by

$$\mu_A^s = \mu_A^o + RT \ln x_A^s \tag{13.7}$$

$$\mu_A^l = \mu_A^o + \Delta\mu_{m,A} + RT \ln x_A^l \tag{13.8}$$

where $\Delta\mu_{m,A}$ is chemical potential change for melting of A. The corresponding chemical potentials of B in solid and liquid are

$$\mu_B^l = \mu_B^o + RT \ln x_B^l \tag{13.9}$$

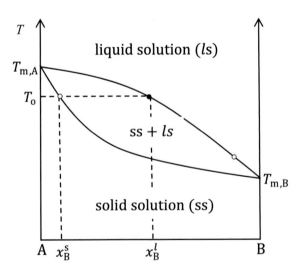

Fig. 13.19 Illustration of calculation of a phase diagram with complete mutual solubility in both liquid and solid

$$\mu_B^s = \mu_B^o - \Delta\mu_{m,B} + RT\ln x_B^s \qquad (13.10)$$

$\Delta\mu_{m,B}$ is the chemical potential change for melting of B.

Using Eqs. (13.7)–(13.10) and the equilibrium condition:

$$\mu_A^s = \mu_A^l, \ \mu_B^s = \mu_B^l,$$

we have

$$RT\ln\frac{x_A^l}{x_A^s} = -\Delta\mu_{m,A}, \quad RT\ln\frac{x_B^l}{x_B^s} = -\Delta\mu_{m,B} \qquad (13.11)$$

The equilibrium compositions as functions of temperature for the solid and liquid can be obtained by solving the above two coupled equations. Let us rewrite Eq. (13.11)

$$\frac{x_A^l}{x_A^s} = \frac{1 - x_B^l}{1 - x_B^s} = e^{-\frac{\Delta\mu_{m,A}}{RT}}, \quad \frac{x_B^l}{x_B^s} = e^{-\frac{\Delta\mu_{m,B}}{RT}} \qquad (13.12)$$

Solving above equations, we have

$$x_B^s = \frac{(1 - e^{-\frac{\Delta\mu_{m,A}}{RT}})}{e^{-\frac{\Delta\mu_{m,B}}{RT}} - e^{-\frac{\Delta\mu_{m,A}}{RT}}}, \quad x_B^l = \frac{\left(1 - e^{-\frac{\Delta\mu_{m,A}}{RT}}\right)}{1 - e^{-\frac{(\Delta\mu_{m,A} - \Delta\mu_{m,B})}{RT}}} \qquad (13.13)$$

which describe the solidus and liquidus lines in the phase diagram (Fig. 13.19).

13.7 Exercises

1. Consider a hypothetical A − B binary phase diagram below. The normal melting temperature of pure A is 1000 K, and the normal melting temperature of pure B is 1200 K. Solid A and solid B have negligible solubility in each other, while liquids A and B form homogeneous solutions through the entire composition range. The phase diagram is a simple eutectic diagram with the eutectic composition at $x_B = 0.4$ and eutectic temperature at 800 K.

   (a) Schematically draw the temperature-composition phase diagram (assuming 1 bar) and label the key temperatures and compositions.
   (b) Schematically draw the chemical potential curves of solid and liquid phases at 1400, 1100, 900, and 700 K (Although the curves are schematic, please pay attention to the key details).
   (c) What is the number of degrees of freedom at the eutectic point?
   (d) What are the volume fractions of pure A and pure B below the eutectic temperature if the overall composition of A in the binary system is 0.4?

(e)   Find the equation describing the activity coefficient of A in the liquid along the liquidus line as a function of composition and temperature.

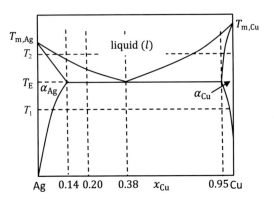

2.   The Ag−Cu binary phase diagram is schematically shown on the right.

(a)   Draw schematic chemical potential curves for α_{Ag}, α_{Cu} and liquid phases at T_1, T_E, and T_2.

(b)   What is the lowest temperature that a liquid can exist at equilibrium in the Ag−Cu binary system?

(c)   What is the highest temperature that a solid can exist at equilibrium in the Ag−Cu binary system for any composition?

(d)   What is the maximum solubility of Cu in Ag at any temperature?

(e)   Why does the solubility of Cu in Ag increase with temperature?

(f)   For the overall composition of 0.20 mole fraction of Cu (see the diagram), what is the number of degrees of freedom for the system at T_1 at equilibrium?

(g)   Estimate the equilibrium compositions of phases at T_1?

(h)   With your estimates from (g), what are the approximate activities of Cu and Ag at T_1?

(i)   What are the approximate activity coefficients of Cu and Ag in α_{Ag} and α_{Cu} at T_1?

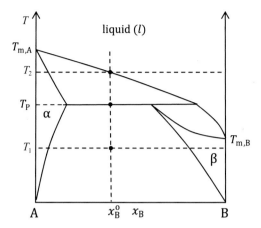

3. (a) Label all of the two-phase coexistence regions and three-phase coex-
         istence points in the diagram. Note that the single-phase regions have
         already been labeled.
    (b) How many degrees of freedom are associated with the two-phase
         coexistence regions and three-phase coexistence points?
    (c) When a liquid solution with overall composition $x_B^o = 0.4$ is slowly cooled
         down from $T_{m,A}$ to T_1, write down the reactions that take place at T_2 and
         T_p during cooling.

Chapter 14
Chemical Reaction Equilibria

Chapter 11 is concerned with the equilibrium states of a material at fixed chemical composition at different temperature and pressure and displaying the equilibrium states in pressure–temperature phase diagrams, while Chap. 13 is on the equilibrium states with respect to chemical compositions of a material at different temperatures and presenting them on temperature-composition phase diagrams. In this chapter, we mainly discuss the equilibrium states among different chemical species at a given temperature and pressure, i.e., chemical reaction equilibria, and for a certain family of reactions, we can also display the reaction equilibria on diagrams, for example, the oxidation reactions on an Ellingham diagram.

14.1 Reaction Equilibrium

We discussed the calculation of driving forces for chemical reactions in Sect. 8.8 by considering a general chemical reaction,

$$\nu_A A + \nu_B B \rightleftharpoons \nu_C C + \nu_D D \tag{14.1}$$

where ν_A, ν_B, ν_C, and ν_D are stoichiometric reaction coefficients. It is reminded here again that in order to compare the thermodynamic stability of the reactants and products, mass has to be balanced between reactants and products for a given chemical reaction, i.e., we have to use the same amounts of different atomic species in the initial state, the reactants, and the final state, the products. This is very similar to determining the equilibrium states for a multicomponent system discussed in Chap. 13 where the equilibrium states are determined by comparing the chemical potentials of different possible thermodynamic states at the same overall chemical composition. Therefore, we could simply compare the chemical potential of the reactants and the chemical potential of the products with reactants and products at the same composition of chemical elements to determine the relative stability of the reactants

© Springer Nature Singapore Pte Ltd. 2022
L.-Q. Chen, *Thermodynamic Equilibrium and Stability of Materials*,
https://doi.org/10.1007/978-981-13-8691-6_14

and products, and thus the driving force and direction for the reaction. As long as the mass is balanced for a chemical reaction, one can either use the chemical potential difference or the Gibbs free energy difference between the reactants and products to compare the relative stability of the reactants and products. In this chapter, we will follow the existing literature using the Gibbs free energies of the reactants and products rather than the chemical potentials although we could very well just use chemical potentials without using the Gibbs free energies to describe the thermodynamics of chemical reactions.

If we assume the chemical potentials of A, B, C, and D are μ_A, μ_B, μ_C, and μ_D, respectively, the fundamental equation of thermodynamics in terms of Gibbs free energy for such a system with four species is given by

$$G = N_A\mu_A + N_B\mu_B + N_C\mu_C + N_D\mu_D \tag{14.2}$$

where N_A, N_B, N_C, and N_D are the number of moles of A, B, C and D in the system. The differential form for the Gibbs free energy at constant temperature and pressure is

$$dG = \mu_A dN_A + \mu_B dN_B + \mu_C dN_C + \mu_D dN_D \tag{14.3}$$

It should be noted that due to the chemical reaction equilibrium, N_A, N_B, N_C, and N_D are not all independent variables. Let us define a reaction extent ξ to represent the degree of reaction with 0 representing no reaction, or the initial state, and 1 the completed reaction, or the final state, and $d\xi$ is the change in the extent of reaction at a given moment of time for the reaction. Therefore, if the initial state consists of $N_A = \nu_A$ mol of A, $N_B = \nu_B$ mol of B, and 0 mol of C, and 0 mol of D, the Gibbs free energy of the reacting system at the reaction extent ξ is given by

$$G(\xi) = (1 - \xi)\nu_A\mu_A + (1 - \xi)\nu_B\mu_B + \xi\nu_C\mu_C + \xi\nu_D\mu_D$$

which can be written as

$$G(\xi) = \nu_A\mu_A + \nu_B\mu_B + \xi(\nu_C\mu_C + \nu_D\mu_D - \nu_A\mu_A - \nu_B\mu_B)$$

The Gibbs free energy at $\xi = 0$ representing the initial state is

$$G(\xi = 0) = \nu_A\mu_A + \nu_B\mu_B \tag{14.4}$$

The Gibbs free energy of the final completely reacted state at $\xi = 1$ is

$$G(\xi = 1) = \nu_C\mu_C + \nu_D\mu_D \tag{14.5}$$

If we define

$$\Delta G = G(\xi = 1) - G(\xi = 0) \tag{14.6}$$

we have

$$G(\xi) = G(\xi = 0) + \xi \Delta G \tag{14.7}$$

Based on the reaction stoichiometric coefficients ν_A, ν_B, ν_C, and ν_D, the changes in the number of moles for each reacting and product species at the reaction extent ξ are given by

$$dN_A = -\nu_A d\xi, \, dN_B = -\nu_B d\xi, \, dN_C = +\nu_C d\xi, \, dN_D = +\nu_D d\xi \tag{14.8}$$

Using the results from Eq. (14.8) in Eq. (14.3), we get

$$dG = (\nu_C \mu_C + \nu_D \mu_D - \nu_A \mu_A - \nu_B \mu_B) d\xi = \Delta G d\xi \tag{14.9}$$

Therefore, we can use either Eq. (14.7) or Eq. (14.9) to derive the variation of the Gibbs free energy of the reacting system with respect to the reaction extent ξ,

$$\left(\frac{\partial G}{\partial \xi}\right)_{T,p} = \Delta G \tag{14.10}$$

The driving force for the chemical reaction is given by

$$D = -\left(\frac{\partial G}{\partial \xi}\right)_{T,p} = -\Delta G \tag{14.11}$$

We write the chemical potentials μ_i in terms of activities a_i,

$$\mu_i = \mu_i^o + RT \ln a_i \tag{14.12}$$

Now we can express the Gibbs free energy difference between the final state and the initial state for the reaction in terms of chemical potentials at the standard state and the activities of the reacting and product species,

$$\Delta G = \left(\nu_C \mu_C^o + \nu_D \mu_D^o\right) - \left(\nu_A \mu_A^o + \nu_B \mu_B^o\right) + RT \ln \frac{a_C^{\nu_C} a_D^{\nu_D}}{a_A^{\nu_A} a_B^{\nu_B}} \tag{14.13}$$

Let us define

$$\Delta G^o = \left(\nu_C \mu_C^o + \nu_D \mu_D^o\right) - \left(\nu_A \mu_A^o + \nu_B \mu_B^o\right) \tag{14.14}$$

and

$$K = \frac{a_C^{\nu_C} a_D^{\nu_D}}{a_A^{\nu_A} a_B^{\nu_B}} \tag{14.15}$$

we have

$$\Delta G = \Delta G° + RT\ln K \tag{14.16}$$

$\Delta G°$ is the free energy change of a chemical reaction when all reactants and products are in their standard states, and K is called the reaction equilibrium constant.

The details for calculating $\Delta G°$ as a function of temperature at 1 bar are outlined in Chap. 8. It is related to the standard enthalpy and entropy changes, $\Delta H°$ and $\Delta S°$, of a reaction,

$$\Delta G° = \Delta H° - T\Delta S° \tag{14.17}$$

$\Delta H°$ and $\Delta S°$ as functions of temperature can be obtained through

$$\Delta H° = \Delta H°_{298\,K} + \int_{298\,K}^{T} \Delta C_p dT \tag{14.18}$$

$$\Delta S° = \Delta S°_{298\,K} + \int_{298\,K}^{T} \frac{\Delta C_p}{T} dT \tag{14.19}$$

where ΔC_p is the difference between the total heat capacity of the products and the total heat capacity of the reactants, i.e.,

$$\Delta C_p = (\nu_C c_C + \nu_D c_D) - (\nu_A c_A + \nu_B c_B) \tag{14.20}$$

where c_A, c_B, c_C, and c_D are the molar heat capacities of A, B, C, and D, respectively. If ΔC_p is constant independent of temperature, we have

$$\Delta H° = \Delta H°_{298\,K} + \Delta C_p(T - 298) \tag{14.21}$$

$$\Delta S° = \Delta S°_{298\,K} + \Delta C_p \ln\frac{T}{298} \tag{14.22}$$

Therefore, for constant ΔC_p, we have

$$\Delta G° = \Delta H°_{298\,K} - 298\Delta C_p + T\left(\Delta C_p - \Delta S°_{298\,K} + \Delta C_p \ln 298\right) - \Delta C_p T\ln T \tag{14.23}$$

The above equation for $\Delta G°$ can be written as a general expression for the standard free energy change of a chemical reaction:

$$\Delta G° = a + bT + cT\ln T \tag{14.24}$$

where a, b, and c are constants which are typically listed in thermodynamic tables.

At equilibrium, the driving force $D = -\Delta G = 0$, and the Gibbs free energy G is minimized, and thus

$$D = -\left(\frac{\partial G}{\partial \xi}\right)_{T,p} = (\nu_A \mu_A + \nu_B \mu_B - \nu_C \mu_C - \nu_D \mu_D) = -\Delta G = 0 \quad (14.25)$$

Therefore, the direction of a reaction is determined by

$$D = -\left(\frac{\partial G}{\partial \xi}\right)_{T,p} = \nu_A \mu_A + \nu_B \mu_B - \nu_C \mu_C - \nu_D \mu_D = -\Delta G \quad (14.26)$$

$$\Delta G = \Delta G^\circ + RT \ln K \begin{bmatrix} = 0 \text{ equilibrium} \\ < 0 \text{ forward} \\ > 0 \text{ backward} \end{bmatrix} \quad (14.27)$$

14.2 Graphical Representation of Oxidation Reactions—Ellingham Diagram

Consider an oxidation reaction,

$$2M(s) + O_2(g) = 2MO(s)$$

where M is assumed to be a metal with valance of $+2$ in the oxide. The standard free energy change of the above oxidation reaction can be written as

$$\Delta G^\circ = \Delta H^\circ - \Delta S^\circ T$$

where the standard enthalpy and entropy changes are given by the enthalpies and entropies of the reaction reactants and products

$$\Delta H^\circ = 2H^\circ_{MO} - 2H^\circ_M - H^\circ_{O_2}$$

$$\Delta S^\circ = 2S^\circ_{MO} - 2S^\circ_M - S^\circ_{O_2}$$

The ΔG° as a function of temperature is schematically shown in Fig. 14.1. It is noted that the intercept of ΔG° with the vertical axis gives the standard enthalpy change of the reaction. Since the entropy of one mole of gas is usually much larger than one mole of solid, we have

$$\Delta S^\circ = 2S^\circ_{MO} - 2S^\circ_M - S^\circ_{O_2} \approx -S^\circ_{O_2}$$

Fig. 14.1 Schematic plot of
standard free energy of
oxidation reaction

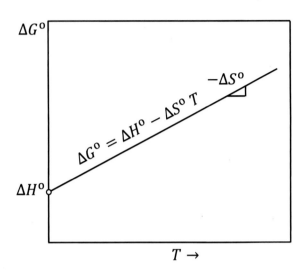

Therefore,

$$\Delta G^\circ = \Delta H^\circ + S^\circ_{O_2} T$$

If ΔH° and $S^\circ_{O_2}$ are assumed to be constant independent of temperature, ΔG° as a function of temperature is linear.

For an oxidation reaction at equilibrium

$$\Delta G = 0 = \Delta G^\circ + RT \ln \frac{a^2_{MO}}{a^2_M a_{O_2}}$$

If we assume metal M and oxide MO are pure solids, the values of their activity are 1 as we choose pure solids as standard states,

$$\Delta G^\circ + RT \ln \frac{1}{p_{O_2}} = 0$$

and hence

$$\Delta G^\circ = RT \ln p_{O_2} \tag{14.28}$$

This equilibrium condition can be graphically represented on a diagram, called the Ellingham diagram. The vertical axis in an Ellingham diagram represents the values of either $\Delta G^\circ(T)$ or $RT \ln p_{O_2}$, and the horizontal axis is temperature T, i.e., we plot both $\Delta G^\circ(T)$ and $RT \ln p_{O_2}$ as a function of temperature for different oxidation reactions on the same diagram.

To construct an Ellingham diagram, we first plot $\Delta G^\circ(T)$ for the oxidation reactions of interest on a diagram (see Fig. 14.2 for illustration). Since the enthalpy

Fig. 14.2 Schematic plots of standard free energy of oxidation reaction in an Ellingham diagram

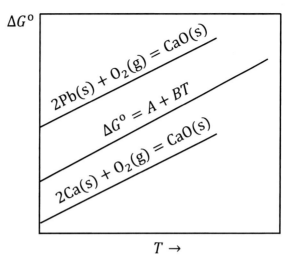

change for an oxidation reaction is negative, i.e., heat is released during an oxidation reaction, the value of $\Delta G°(T)$ is negative at 0 K and at low temperatures. Therefore, the top horizontal line of Fig. 14.2 represents the zero value for $\Delta G°(T)$.

14.2.1 O_2 Scale

We can establish a p_{O_2} scale on the diagram to represent the variation of partial pressure of oxygen with temperature as illustrated in Fig. 14.3. We imagine a set of lines representing $RT \ln p_{O_2}$ at a regular interval of p_{O_2}. However, rather than

Fig. 14.3 Schematic illustration of the oxygen scale in an Ellingham diagram

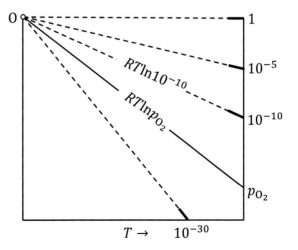

Fig. 14.4 Schematic
illustration of using an
Ellingham diagram to solve
the equilibrium for an
oxidation reaction

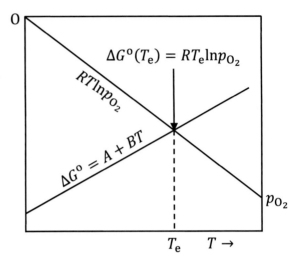

drawing the actual lines of $RT \ln p_{O_2}$, we label ticks on the p_{O_2} scale with each tick representing a particular value of p_{O_2}, and the line of p_{O_2} for a particular value p_{O_2} can be obtained by linking a given tick and the origin labeled as "O" (illustrated Fig. 14.3 as dash lines).

Obviously, there are two variables in the equilibrium Eq. (14.28): the partial pressure of oxygen p_{O_2} and temperature T. Therefore, for a given partial pressure of oxygen p_{O_2}, the intersection between the two lines representing $\Delta G^\circ(T) = A + BT$ and $RT \ln p_{O_2}$ yields the equilibrium temperature T_e; for a given temperature T, the intersection between $\Delta G^\circ(T)$ and the vertical line representing temperature T determines the equilibrium partial pressure of oxygen, p_{O_2}; see Fig. 14.4. Therefore, without doing the calculations using Eq. (14.28), one could estimate the equilibrium temperature between the mixture of metal and oxygen gas and the metal oxide under a given partial pressure of oxygen or find the equilibrium partial pressure of oxygen at a given temperature. Above this equilibrium partial pressure of oxygen at the temperature of interest, the metal will be oxidized, and below it the oxide will be reduced.

Several observations can be made for an Ellingham diagram. All the $\Delta G^\circ(T)$ lines are drawn for the oxidation reaction involving the consumption of one mole of oxygen, and hence most of the $\Delta G^\circ(T)$ lines are approximately parallel to each other since their slopes are approximately equal to the molar entropy of oxygen gas. It is also noted that almost all the lines have the positive slope since their slope is approximately the entropy of one mole of oxygen. However, if a reaction involves other gases, e.g., $2C + O_2 = 2CO$, the slope is negative since one mole of gas reacts with two moles of solid produces two moles of gases, and hence the entropy change is positive, and the slope for $\Delta G^\circ(T)$ is negative in this case.

It should also be noted that when a metal melts, the ΔS becomes more negative, and the slope is larger. When an oxide melts, the ΔS becomes more positive,

Fig. 14.5 Schematic
illustration of slope changes
of standard free energy of
oxidation as a function of
temperature due to melting
and boiling in an Ellingham
diagram

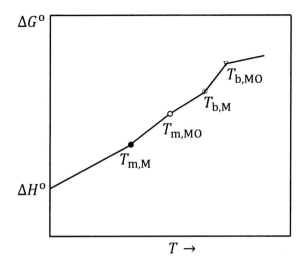

and hence the slope is smaller. These slope changes due to melting or boiling are
schematically illustrated in Fig. 14.5.

The Ellingham diagram can also be used to estimate the driving forces for an
oxidation reaction with one oxide to oxidizing a metal and at the same time reducing
itself. For example,

$$2Pb(s) + O_2(g) = 2PbO(s) \quad \Delta G^o_{PbO} \tag{14.29}$$

$$2Ca(s) + O_2(g) = 2CaO(s) \quad \Delta G^o_{CaO} \tag{14.30}$$

Subtraction of Eq. (14.29) from Eq. (14.30), we have

$$2Ca(s) + 2PbO(s) = 2CaO(s) + 2Pb(s) \quad \Delta G^o = \Delta G^o_{CaO} - \Delta G^o_{PbO}$$

i.e., the driving force for such a reaction can be estimated from the Ellingham diagram
(see the illustration in Fig. 14.6).

14.2.2   H₂/H₂O Scale

The partial pressure of oxygen in a system can be controlled using a mixture of
hydrogen gas H_2 and water vapor H_2O. Therefore, in addition to directly using the
oxygen gas partial pressure scale on an Ellingham diagram, one can also construct
a H_2/H_2O scale. To understand the construction of the H_2/H_2O scale, let us look at
the reaction,

Fig. 14.6 Schematic illustration to obtain the standard free energy of reaction by combining two oxidation reactions on an Ellingham diagram

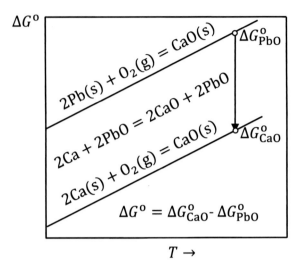

$$2H_2(g) + O_2(g) = 2H_2O(g), \quad \Delta G^\circ(T)$$

where $\Delta G^\circ(T)$ is the free energy change for the above reaction when all the gases are at their standard states, i.e., pure gases at 1 bar. If the gases are not at their standard states, then the equilibrium condition for the above reaction at temperature T is

$$\Delta G(T) = \Delta G^\circ(T) + RT \ln K = 0$$

or

$$\Delta G^\circ(T) = 2RT \ln \frac{p_{H_2}}{p_{H_2O}} + RT \ln p_{O_2}$$

We can rewrite the above equation by moving the first term of the right-hand side to the left-hand side,

$$\left(\Delta G^\circ\right)'(T) = \Delta G^\circ(T) - 2RT \ln \frac{p_{H_2}}{p_{H_2O}} = RT \ln p_{O_2} \qquad (14.31)$$

For a given H_2/H_2O ratio, $(\Delta G^\circ)'$ is only a function of temperature. If $\Delta G^\circ(T)$ is approximately linear, $(\Delta G^\circ)'$ will also be approximately a linear function of temperature, which is rotated by a slope of $2R \ln\left(p_{H_2}/p_{H_2O}\right)$ from $\Delta G^\circ(T)$. A H_2/H_2O scale can be constructed to represent the $(\Delta G^\circ)'$ lines at regular intervals of p_{H_2}/p_{H_2O} as shown in Fig. 14.7.

The above equation has three variables, temperature T, H_2/H_2O ratio, and partial pressure of oxygen p_{O_2}. Therefore, if we fix any two of the three variables, we can determine the third. On an Ellingham diagram, the magnitude of the third can be estimated if the first two of the variables are specified by looking at the intersections

Fig. 14.7 Schematic illustration of H_2/H_2O scale on an Ellingham diagram

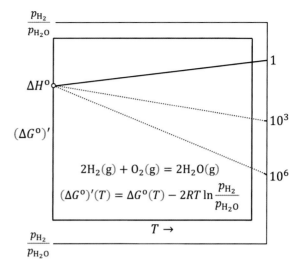

$$2H_2(g) + O_2(g) = 2H_2O(g)$$

$$(\Delta G^\circ)'(T) = \Delta G^\circ(T) - 2RT \ln \frac{p_{H_2}}{p_{H_2O}}$$

of three lines, $(\Delta G^\circ)'(T)$, $RT \ln p_{O_2}$, and $T = T_e$ as illustrated in Fig. 14.8. For example, if we specify p_{H_2}/p_{H_2O} and temperature, we can draw a line representing $(\Delta G^\circ)'(T)$ for a particular p_{H_2}/p_{H_2O} ratio and another line, a vertical line representing temperature $T = T_e$. We then draw a line connecting the "O" point representing the origin of the line $RT \ln p_{O_2}$ and the intersection between the $(\Delta G^\circ)'(T)$ line and $T = T_e$ line; we can obtain the equilibrium partial pressure of oxygen, p_{O_2} from the p_{O_2} scale on the diagram by extrapolating the line to the p_{O_2} scale. On the other hand, if we specify p_{O_2} and T, we can estimate p_{H_2}/p_{H_2O} by connecting the point "H" representing the origin for the line $(\Delta G^\circ)'(T)$ and the intersection

Fig. 14.8 Schematic illustration of using an Ellingham diagram to determine the thermodynamic equilibrium of oxidation using a mixture of H_2/H_2O

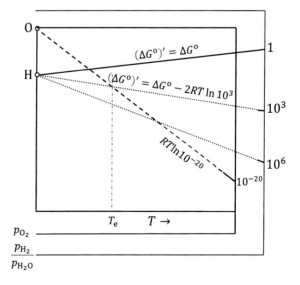

between the line $RT \ln p_{O_2}$ and the vertical line $T = T_e$ and extrapolate the line to the p_{H_2}/p_{H_2O} scale to determine the equilibrium p_{H_2}/p_{H_2O}.

14.2.3 CO/CO₂ Scale

Similar to the H_2/H_2O scale, one can construct a CO/CO_2 scale by considering the reaction

$$2CO(g) + O_2(g) = 2CO_2(g), \quad \Delta G^o(T)$$

where $\Delta G^o(T)$ is the standard free energy of the above reaction when all the gas species are at their standard states, and hence $p_{CO}/p_{CO_2} = 1$. If the gases are not at their standard states, the equilibrium condition is given by

$$\Delta G^o(T) = 2RT \ln \frac{p_{CO}}{p_{CO_2}} + RT \ln p_{O_2}$$

Now we define

$$(G^o)'(T) = G^o(T) - 2RT \ln \frac{p_{CO}}{p_{CO_2}} = RT \ln p_{O_2}$$

Following exactly the same discussion as the H_2/H_2O scale, the construction of the CO/CO_2 scale is schematically illustrated in Figs. 14.9 and 14.10.

Summary on the applications of Ellingham diagrams for estimating the equilibrium conditions for oxidation reactions:

Fig. 14.9 Schematic illustration of CO/CO₂ scale on an Ellingham diagram

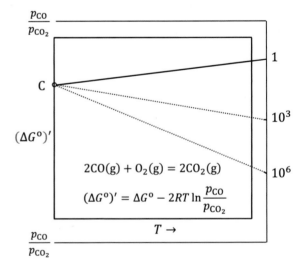

Fig. 14.10 Schematic illustration of using an Ellingham diagram to determine the thermodynamic equilibrium of oxidation using a mixture of CO/CO$_2$

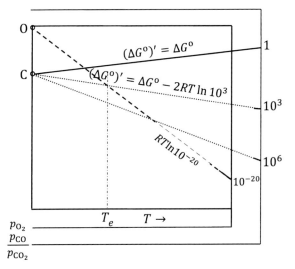

- The equilibrium partial pressure of O$_2$ at a given temperature T.
- The equilibrium temperature at a given partial pressure of O$_2$.
- The free energy change for a reaction which is a combination of two oxidation reactions at a given temperature T.
- The appropriate ratios, H$_2$/H$_2$O or CO/CO$_2$ that will provide a given partial pressure of O$_2$ at a given temperature T.
- The equilibrium temperature for a given ratio, H$_2$/H$_2$O or CO/CO$_2$, and partial pressure of O$_2$.
- The equilibrium partial pressure of O$_2$ for a given ratio, H$_2$/H$_2$O or CO/CO$_2$, and a given temperature T.

14.3 Example

Consider the reaction of H$_2$ gas reacts with O$_2$ gas to form H$_2$O water vapor. The standard enthalpy of formation for H$_2$O water vapor at 298 K and 1 bar is -241.8 kJ/mol. The standard entropies of H$_2$ gas, O$_2$ gas, and H$_2$O water vapor at 298 K and 1 bar are 131.0, 205.0, 188.7 J/(K mol), respectively. Please determine the following:

(a) The heat of reaction at 298 K and 1 bar.
(b) The standard entropy change for the reaction at 298 K and 1 bar.
(c) The standard free energy change for the reaction at 298 K and 1 bar.
(d) The standard free energy for the reaction as a function of temperature at 1 bar by assuming that the heat of reaction and the entropy of reaction are independent of temperature.
(e) The equilibrium partial pressure of oxygen for a fixed ratio of $p_{H_2}/p_{H_2O} = 30$ at 2000 K.

Solution

The reaction equation:

$$H_2(g) + \frac{1}{2}O_2(g) = H_2O(g)$$

(a)   The heat of reaction at 298 K and 1 bar is

$$Q = \Delta H° = -241,800 \text{ J/mol}$$

(b)   The standard entropy of reaction at 298 K and 1 bar is

$$\Delta S° = S_{H_2O}° - S_{H_2}° - \frac{1}{2}S_{O_2}° = 188.7 - 131 - \frac{1}{2} \times 205 = -44.8 \text{ J/(K mol)}$$

(c)   The standard free energy change for the reaction at 298 K and 1 bar is

$$\Delta G° = \Delta H° - 298 \times \Delta S° = -228,450 \text{ J/mol}$$

(d)   Assuming the heat of reaction and the entropy of reaction are independent of temperature, the standard free energy for the reaction as a function of temperature at 1 bar is given by

$$\Delta G° = -241,800 + 44.8T$$

(e)   If we fix the ratio of $p_{H_2}/p_{H_2O} = 30$ at 2000 K, the equilibrium partial pressure of oxygen can be obtained from the following expression,

$$\Delta G = \Delta G° + RT \ln \frac{p_{H_2O}}{p_{H_2} p_{O_2}^{1/2}} = 0$$

Solving the above equation for the partial pressure of oxygen,

$$p_{O_2} = \left(\frac{p_{H_2O}}{p_{H_2}}\right)^2 \exp\left(\frac{2\Delta G°}{RT}\right)$$

Therefore,

$$p_{O_2} = \left(\frac{1}{30}\right)^2 \exp\left(2 \times \frac{-241,800 + 44.8 \times 2000}{8.315 \times 2000}\right) = 1.25 \times 10^{-11} \text{ bar}$$

14.4 Exercises

1. The Gibbs free energy difference $(G^p - G^r)$ between products and reactants for an oxidation reaction of a metal typically becomes increasingly more negative as temperature increases: true or false?

2. For an oxidation reaction of a solid metal to form a solid oxide, the entropy change for the reaction is always positive: true or false?

3. Thermodynamically, at a given temperature, an oxide is more stable with respect to metal and oxygen gas at higher partial pressure of oxygen than lower partial pressure of oxygen: true or false?

4. Thermodynamically, at a given partial pressure of oxygen, an oxide is more stable with respect to metal and oxygen gas at higher temperature than low temperature: true or false?

5. For the reaction, $4Ag(s) + O_2(g) = 2Ag_2O$ (s), please determine whether increasing temperature will favor $4Ag(s) + O_2(g)$ or $2Ag_2O$.

6. Is the chemical potential of O_2 in the air (i) higher; or (ii) equal to, or (iii) lower than the chemical potential of O_2 in equilibrium with Ni/NiO at room temperature?

$$Ni(s) + 1/2O_2(g) = NiO(s), \Delta G^o = -244,550 + 98.5T$$

7. One way to control the partial pressure of oxygen is to use a gas mixture of H_2 and H_2O. Increasing the ratio of partial pressure of H_2O to partial pressure of H_2, i.e., p_{H_2O}/p_{H_2}, increases the partial pressure of oxygen in the gas: true or false?

8. One way to control the partial pressure of oxygen is to use a gas mixture of CO and CO_2. Increasing the ratio of partial pressure of CO to partial pressure of CO_2, i.e., p_{CO}/p_{CO_2}, increases the partial pressure of oxygen in the gas: true or false?

9. Perform a literature search to find the standard Gibbs free energies of reactions for the oxidation reactions of the following chemical species: Ba, C, Ca, CO, Cr, Cu, H_2, Mg, Ni, Si, and plot them on the same diagram of standard Gibbs free energy versus temperature.

10. Add an oxygen scale to the diagram obtained above in exercise 9.

11. Add a H_2/H_2O scale to the diagram obtained above in exercise 9.

12. Add a CO/CO_2 scale to the diagram obtained above in exercise 9.

13. Ni is used as an electrode material in multilayer ceramic capacitors. During fabrication, the dielectric material, $BaTiO_3$, and internal Ni electrodes are

usually cofired around 1300 °C. In order to avoid Ni being oxidized during cofiring, one has to control the partial pressure of oxygen.
Given:

$$2Ni(s) + O_2(g) \rightarrow 2NiO(s), \quad \Delta G^\circ = -480.00 + 0.18889T \ (kJ)$$
$$2CO(g) + O_2(g) \rightarrow 2CO_2(g), \quad \Delta G^\circ = -565.98 + 0.17289T \ (kJ)$$

(a) Determine the maximum partial pressure of oxygen that is allowed in order to avoid oxidation of Ni at 1300 °C.
(b) If one employs CO/CO_2 mixture to control partial pressure of oxygen, determine the minimum CO/CO_2 ratio to avoid oxidation of Ni at 1300 °C.

14. Given

Reactions	ΔG° (J)	Temperature range
$2Pb(s) + O_2(g) \rightarrow 2PbO(s)$	$-438,820 + 201.71T$	$T < 601$ K
$2Pb(l) + O_2(g) \rightarrow 2PbO(s)$	$-447,380 + 215.57T$	601 K $< T < 1161$ K
$2Pb(l) + O_2(g) \rightarrow 2PbO(l)$	$-413,060 + 201.81T$	1161 K $< T < 1808$ K
$2Pb(l) + O_2(g) \rightarrow 2PbO(g)$	$132,020 + 131.39T$	1808 K $< T < 2017$ K

Please do the following:

(a) Determine the heat of oxidation and then determine whether heat is released or absorbed during the oxidation reaction.
(b) Estimate the entropy of one mole of O_2.
(c) At ambient atmosphere (oxygen pressure in the ambient is roughly 0.21 bar), calculate the Gibbs free energy change for the reaction.
(d) Determine the equilibrium temperature for the reaction if the partial pressure of oxygen is 10^{-12} bar.
(e) Determine the equilibrium partial pressure of oxygen above $Pb(l)/PbO(s)$ at 500 K.
(f) Determine the Gibbs free energy change ΔG for the above oxidation reaction of Pb at 400 °C and a partial pressure of O_2 at 10^{-12} bar.
(g) Determine the equilibrium temperature between $Pb(l)$ and $PbO(s)$ at 10^{-12} bar.
(h) Determine the heat of melting of PbO.
(i) If there is approximately no change in the volume of PbO upon melting ($\Delta V_m = 0$), please estimate the pressure of the triple point of PbO between $PbO(s) - PbO(l) - PbO(g)$. You may assume there are no differences among the heat capacities of solid, liquid, and vapor PbO.

Chapter 15
Energy Conversions and Electrochemistry

All energy conversion devices operate by utilizing differences in thermodynamic potentials such as differences in temperature, pressure, chemical potential, electric potential, or gravitational potential. For example, a thermal engine converts thermal energy to other forms of energy such as electric or mechanical energy. It requires differences in the thermal potential, i.e., differences in temperature, to convert thermal energy to other forms of energy. One can drive an electron system out of equilibrium using solar energy, i.e., the photon energy, as photons entering a crystal can excite electrons from the valence band to the conduction band or from lower energy states to higher energy states. The excess electrons and holes arising from the photon excitation will result in the difference in chemical potentials or Fermi-level between the electrons and holes and thus chemical driving force which can be converted to electric energy, the photovoltaic effect. A Li-ion battery operates on the chemical potential difference between Li in the two electrodes; the chemical energy stored in a battery can be converted into electric energy during battery discharge, or vice versa, the electrical energy can be converted to chemical energy during battery charging. A fuel cell converts chemical energy directly into electric energy through the continuous supply of fuels such as hydrogen gas as well as oxygen gas. Electric energy is directly converted to heat when an electric current flows through a material under an electric potential difference.

15.1 Maximum Work Theorem

For all processes leading from a specified initial state to a specified final state of a system, the maximum work theorem states that the work delivered by a system is maximized for a reversible process. For reversible processes, a system and its surrounding only exchange different forms of energy including thermal energy, but there are no dissipative processes which involve the conversion of useful energy to thermal energy through the creation of entropy. The maximum work theorem sets

© Springer Nature Singapore Pte Ltd. 2022
L.-Q. Chen, *Thermodynamic Equilibrium and Stability of Materials*,
https://doi.org/10.1007/978-981-13-8691-6_15

the thermodynamic limit on the maximum work that can be delivered in a process by a system because it is impossible to produce more work than the work produced by a reversible process. It can be used to compute the maximum (or theoretical) thermodynamic efficiency of devices such as engines, refrigerators, heat pumps, and other devices by assuming all the processes involved in a given device are reversible.

15.2 Theoretical Efficiencies of Thermal Devices

Using the first and second laws of thermodynamics and the maximum work theorem, one can calculate the maximum or theoretical thermodynamic efficiency of an engine, a machine, or any device involving the conversion between thermal energy and other forms of energy.

Let us assume that a thermal engine operates between a high-temperature thermal energy source, i.e., a high thermal potential at T_h, and a lower temperature thermal energy sink, i.e., a low thermal potential at T_1. The thermal engine absorbs Q_h amount of thermal energy or heat from the high-temperature thermal energy source, performs W amount of work, and releases Q_1 amount of heat to the low-temperature thermal energy sink. The engine efficiency is defined as the amount of work W that is performed by the system divided by the Q_h amount of heat withdrawn from the high-temperature thermal energy source,

$$\eta = \frac{W}{Q_h} \tag{15.1}$$

Applying the first law of thermodynamics for energy conversation, $W = Q_h - Q_1$, one has

$$\eta = \frac{Q_h - Q_1}{Q_h} \tag{15.2}$$

From the maximum work theorem, the engine efficiency is maximum when all the processes taking place in the engine are reversible, i.e.,

$$\Delta S^{tot} = -\frac{Q_h}{T_h} + \frac{Q_1}{T_1} = 0 \tag{15.3}$$

Therefore, the maximum or ideal or theoretical engine efficiency is

$$\eta = \frac{Q_h - Q_1}{Q_h} = \frac{T_h - T_1}{T_h} \tag{15.4}$$

For the maximum efficiency of a thermal engine, the entropy extracted from the high-temperature thermal energy source should be equal to the entropy delivered to the lower temperature thermal energy sink, i.e., there is only entropy exchange

Fig. 15.1
Temperature–entropy
diagram to illustrate a
reversible or Carnot cycle

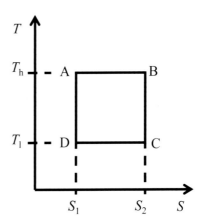

between the thermal energy reservoirs and the engine, but there is no entropy produced. For any real thermal engines, they do produce entropy, and the amount of entropy delivered to the lower temperature thermal energy sink is greater than the amount of entropy extracted from the high-temperature thermal energy source.

The entropy–temperature diagram (Fig. 15.1) is particularly useful and simple to illustrate a perfect thermodynamic engine. The entire cycle of a perfect thermodynamic engine consists of two isothermal (constant temperature) processes and two adiabatic (constant entropy) processes. According to the diagram, the theoretical efficiency of the engine is given by

$$\eta = \frac{\text{area ABCD}}{\text{area AB} S_2 S_1} = \frac{T_h - T_l}{T_h}$$

For example, for a steam engine operating between room temperature 300 and 373 K, the maximum efficiency is given by

$$\eta = \frac{T_h - T_l}{T_h} = \frac{373 - 300}{373} \approx 19.6\%$$

Based on Eq. (15.4), to achieve maximum efficiency, one wants the temperature T_h of the high-temperature heat source to be as high as possible. This is the very reason that we need high-temperature materials for engines that can retain useful mechanical properties at high temperatures for a long period of time. This is largely the driving force for the research and development of superalloys and high-temperature structural materials for engine applications.

The engine efficiency also increases as the temperature of the low-temperature thermal energy sink decreases. However, it should be noted that it is impossible to achieve $T_l = 0$ K. As a result, a certain amount of thermal energy, $T_l \Delta S$, from the high-temperature thermal energy source has to be transferred to the low-temperature sink because the temperature of the low-temperature sink T_l cannot be 0 K. This means that the amount of energy $T_1 \Delta S$ from the high-temperature thermal energy source is not available to do useful work. Therefore, the maximum efficiency of a thermal engine cannot be 100%. This is equivalent to the Kelvin statement for the second law of thermodynamics: It is impossible to perform a process whose sole result is the conversion of heat into an equivalent amount of work.

Using the combination of first and second laws of thermodynamics, one can also measure the efficiency of devices that move heat from a low temperature to a high temperature, e.g., refrigerators. Analogously to engine efficiency, one can obtain the maximum coefficient of performance of a refrigerator by

$$\eta = \frac{Q_1}{W} = \frac{T_1}{T_h - T_1} \tag{15.5}$$

where Q_1 is the amount of thermal energy removed from inside the refrigerator, W is the amount of electric work spent during the process, T_1 is the temperature inside the refrigerator, and T_h is the ambient temperature outside the refrigerator. From Eq. (15.5), it can be seen that the performance of a refrigerator goes to zero as T_1 approaches zero, which is why, even using a perfect device that operates through reversible processes, we cannot create a low-temperature sink at $T_1 = 0$ K.

We can also imagine a heat pump operating to heat a house during winter, which moves heat from a low-temperature thermal energy source, i.e., outside a house at a lower temperature, to a high-temperature thermal energy sink, i.e., inside the house. For the heat pump, the maximum coefficient of performance for a heat pump is

$$\eta = \frac{Q_h}{W} = \frac{T_h}{T_h - T_1} \tag{15.6}$$

where Q_h is the amount of heat added to the house, W is the amount of electric work required for the process, T_h is the temperature inside a house, and T_1 is the outside temperature.

15.3 Voltage Derived from Light-Excited Electron–Hole Pairs: Photovoltaic Effect

The most abundant energy source is the energy irradiated from the sun, i.e., the solar energy. The photovoltaic effect is defined as the creation of an electric voltage and electric current in a solid under the exposure to light. The operation of a photovoltaic cell involves the adsorption of light to generate charge carriers, separation of carriers, and then collection of carriers at electrodes. It involves the conversion of solar energy to chemical energy and then chemical energy to electric energy.

Typical materials for solar cells are semiconductors. Under sunlight, the electrons in the valence band of a semiconductor absorb light energy and are excited to the conduction band. The process of light absorbed in a semiconductor to create an electron–hole pair can be expressed as a reaction process,

$$\text{Null} \xrightarrow{\text{photon}} e^- + h^+ \tag{15.7}$$

During this process, the solar energy is converted to chemical energy stored in the electron system. In order for the conversion to take place, the energy of a photon must be higher than the band gap of the semiconductor. The excited electrons establish equilibrium with the lattice by transferring their kinetic energy to the phonons. Some of the electrons and holes will recombine and annihilate, resulting in light emission. The amounts of electrons and holes in a semiconductor under light are determined by the rates of electron–hole pair creation and recombination. Therefore, this is not a thermodynamic equilibrium system since the concentrations of both electrons and holes are higher than their equilibrium values. It is precisely the nonequilibrium nature of the electron system that can be utilized to convert the chemical energy to electric energy. Here we simply discuss how to determine the available amount of thermodynamic chemical energy.

At thermodynamic equilibrium for the electron system,

$$0 = \mu_e + \mu_h = E_{fe} - E_{fh}$$

where μ_e and μ_h are chemical potentials of electrons and holes, and E_{fe} and E_{fh} are the Fermi levels of electrons and holes. E_{fe} and E_{fh} are equal at equilibrium. For an intrinsic semiconductor, the Fermi level is at or near the middle of the band gap, whereas the Fermi level in an n-type of semiconductor is close to the bottom of the conduction band, and the Fermi level in an p-type of semiconductor is close to the top of the valence band.

Under light, excess electrons and holes are generated above their equilibrium values, E_{fe} and E_{fh} are no longer equal, and they are usually called quasi-Fermi levels. The higher the electron concentration, the closer the E_{fe} moves toward the conduction band edge, E_c, and the higher the hole concentration, the closer the E_{fh} moves toward the valence band edge E_v. The difference between the Fermi level of electrons and that of holes, $E_{fe} - E_{fh}$, is the magnitude of chemical energy stored per electron–hole pair that is available to be converted to electric energy if one could separate the flow of electrons and holes and collect them at two different electrodes. Therefore, the maximum voltage $\Delta\varphi$ that can be derived from this chemical energy $E_{fe} - E_{fh}$ is

$$\Delta\phi = \frac{E_{fe} - E_{fh}}{e} \tag{15.8}$$

where e is the amount of elemental charge of an electron. Assuming Boltzmann distributions for the electrons and holes at electron energy levels, we have

$$\Delta\phi = \frac{E_{fe} - E_{fh}}{e} = \frac{E_g}{e} + \frac{k_B T}{e} \ln \frac{n_e n_h}{N_c N_v} \tag{15.9}$$

where E_g is the band gap for the semiconductor, k_B is the Boltzmann constant, and N_c and N_v are the electron density of states at the conduction band edge and at the valence band edge, respectively. Equation (15.9) can be rewritten as

$$\Delta\phi = \frac{k_B T}{e} \ln\left[\left(\frac{n_e}{n_e^o}\right)\left(\frac{n_h}{n_h^o}\right) \right] \tag{15.10}$$

where n_e^o and n_h^o are the equilibrium concentrations of electrons and holes at a given temperature in the dark. For an n-type semiconductor, $n_e^o \gg n_h^o$. Under light, $n_e \approx n_e^o$ and $n_h \gg n_h^o$. Therefore, for a n-type semiconductor,

$$\Delta\phi \approx \frac{k_B T}{e} \ln\left(\frac{n_h}{n_h^o}\right) \tag{15.11}$$

Similarly, for a p-type semiconductor, $n_h^o \gg n_e^o$. Under light, $n_e \gg n_e^o$ and $n_h \approx n_h^o$, and hence, for a p-type semiconductor,

$$\Delta\phi \approx \frac{k_B T}{e} \ln\left(\frac{n_e}{n_e^o}\right) \tag{15.12}$$

15.4 Electrochemical Reactions and Energy Conversion

For conversion of chemical energy to electric energy, there must be a chemical potential difference in order to convert available chemical energy to other forms of energy. Let us consider a general chemical reaction,

$$\nu_{A_1} A_1 + \nu_{A_2} A_2 + \cdots = \nu_{B_1} B_1 + \nu_{B_2} B_2 + \cdots \tag{15.13}$$

where A_1, A_2, ... are reacting species, and B_1, B_2, ... are product species. Since the amounts of different chemical elements contained in the reacting species are exactly the same as the amounts of the corresponding chemical elements contained in the product species, one can also compare the Gibbs free energy, which, in this case, can be considered as the total chemical potential of the products and that of the reactants to determine the magnitude of the chemical driving force for the reaction,

$$D = -\Delta G = \left(\nu_{A_1}\mu_{\nu_{A_1}} + \nu_{A_2}\mu_{\nu_{A_2}} + \cdots\right) - \left(\nu_{B_1}\mu_{\nu_{B_1}} + \nu_{B_2}\mu_{\nu_{B_2}} + \cdots\right) \tag{15.14}$$

When a chemical reaction reaches equilibrium, the chemical energy (the Gibbs free energy or more precisely the total chemical potential) of the reactants is the same as the chemical energy (or total chemical potential) of the products (Fig. 15.2), i.e., the chemical driving force D for the chemical reaction is consumed

$$D = -\Delta G = 0$$

Fig. 15.2 Chemical reaction at equilibrium

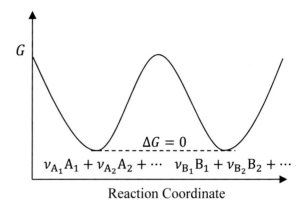

Therefore, when a reaction is at equilibrium, there is no more chemical energy available to be utilized to do any useful work.

For chemical reactions that are not at equilibrium, there are chemical driving forces (Fig. 15.3) ($D = -\Delta G > 0$) that can potentially be converted to other forms of energy. In a normal chemical reaction, the amount of chemical energy $-\Delta G$ is simply converted to thermal energy in the form of heat through the production of entropy. However, rather than converting to heat, the chemical energy difference between the reactants and products can be converted to other forms of useful energy if the chemical reaction is controlled in such a way that it would permit such a conversion.

Below we mainly discuss the conversion of chemical energy to electric energy and vice versa. To connect chemical energy with electric work, let us look at what is going on in a chemical reaction. Chemical reactions, by definition, involve charge transfers due to the difference in the chemical potential of electrons in different elements or molecules, i.e., electrons move from those elements or molecules with higher chemical potential of electrons to those with lower chemical potential of electrons.

Fig. 15.3 Chemical reaction not at equilibrium

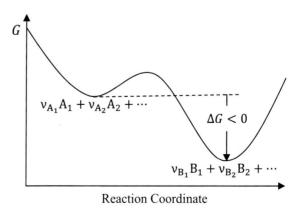

As a result, the electric potential distribution around the elements or molecules will also readjust until the electrochemical potential of the valence electrons is equalized. The total electrochemical potential of the reaction products should be lower than that of the reactants, which is the Gibbs free energy change for the reaction. For example, for the well-known reaction of hydrogen gas H_2 with oxygen gas O_2 to produce water, H_2O,

$$H_2 + \frac{1}{2}O_2 = H_2^+O^{2-} \tag{15.15}$$

This reaction involves electron transfer from H to O because the chemical potential of electrons in the hydrogen molecules is higher than that in the oxygen molecules. As a result of electron transfer from hydrogen to oxygen, the hydrogen atoms have a net positive charge, and oxygen atoms have a net negative charge, which inhibits further electron transfer. The electron transfer stops when the electrochemical potential of electrons is equalized within a H_2O. The chemical potential μ_{H_2O} of reaction product H_2O is lower than the total chemical potential $\mu_{H_2} + (1/2)\mu_{O_2}$ of the reactants $H_2+1/2O_2$, and the difference $\mu_{H_2}+(1/2)\mu_{O_2}-\mu_{H_2O}$ is the driving force $D = -\Delta G$ for the reaction.

Instead of direct electron transfer between atoms or molecules, one could alternatively consider the above reaction in two steps to accomplish the electron transfer process. One is the reduction reaction, a reaction consuming electrons,

$$\frac{1}{2}O_2 + 2e^- = O^{2-} \tag{15.16}$$

and the other is an oxidation reaction, a reaction producing electrons,

$$H_2 + O^{2-} = H_2O + 2e^- \tag{15.17}$$

The overall reaction is

$$H_2 + \frac{1}{2}O_2 = H_2O \tag{15.18}$$

The above analysis implies that if we physically separate the reaction (15.18) into the above two separate reactions (15.17) and (15.16), the reaction requires both electronic and ionic transfers between the two reactions to accomplish the overall reaction. Therefore, if one can design an apparatus to separate the ionic transport and electronic transfer during a chemical reaction, one could draw electricity out of a chemical reaction from the electron transfer process. This is the principle of an electrochemical cell, in this particular example, a fuel cell. A fuel cell is a device that converts the chemical energy of a fuel, e.g., H_2, into electricity through a chemical reaction with oxygen. Continuous operation of fuel cells requires a constant supply of fuel and oxygen. A schematic fuel cell is illustrated in Fig. 15.4. The fuel is H_2. It contains an electrolyte that only conducts ions, the oxygen ions in this case, but not

Fig. 15.4 Illustration of a H_2 fuel cell

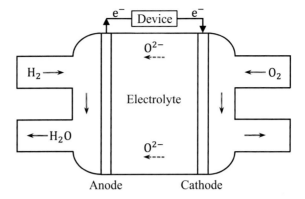

electrons, and two electrodes, cathode and anode. The two separate reactions, called electrode reactions, take place at the interfaces between the electrolyte and the two electrodes. The electrode at which the reduction reaction takes place is the cathode, and the electrode at which the oxidation reaction takes place is the anode.

Another example is the following reaction,

$$Li_xC + CoO_2 \leftrightarrow C(graphite) + Li_xCoO_2 \tag{15.19}$$

The driving force for the reaction is

$$D = -\Delta G = \mu_{Li_xC} + \mu_{CoO_2} - \mu_C - \mu_{Li_xCoO_2} \tag{15.20}$$

Li atoms have lower chemical potential in CoO_2 than in graphite, and hence there is a finite driving force for the reaction. If we can separate the above reaction into two separate reactions, one is the reduction reaction,

$$CoO_2 + xLi^+ + xe^- = Li_xCoO_2, \tag{15.21}$$

and the other is the oxidation reaction,

$$Li_xC = C + xLi^+ + xe^-, \tag{15.22}$$

we can convert chemical energy to electric energy or vice versa. This is the thermodynamic principle of Li-ion batteries. In this case, CoO_2 at which reduction reaction takes place is the cathode, and graphite at which the oxidation reaction takes place is the anode. A schematic drawing of the components in a lithium-ion battery is shown in Fig. 15.5.

Li-ion batteries are rechargeable batteries which can restore the original composition of the electrodes by connecting the battery to an external electric power source (Fig. 15.5). During discharging, the chemical energy is converted to electric energy,

Fig. 15.5 Illustration of a
Li-ion battery

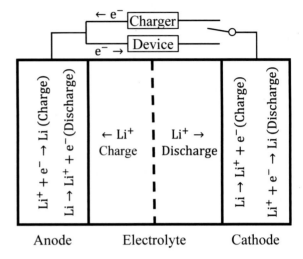

and during charging the electric energy is converted to chemical energy. When a
battery is fully discharged, all the lithium atoms are in the cathode.

15.4.1 Electrode Reactions and Electrode Potentials

It is reminded that electric potential is the amount of electric energy per unit of
electric charge with the SI unit of Volt (V) while chemical potential is the chemical
energy (Gibbs free energy in Joule) per mole or per particle of matter such as atoms
and electrons. The total potential for a charge species such as electrons and ions is the
electrochemical potential. For example, the electrochemical potential $\tilde{\mu}_e$ of electrons
is given by

$$\tilde{\mu}_e = \mu_e - \mathcal{F}\phi \tag{15.23}$$

where $\mathcal{F}$ is the Faraday constant 96,480 C/mol, μ_e is the electron chemical potential
(in unit of $G = $ J/mol), and ϕ is the electric potential.

For the fuel cell example in the last section, the reaction at the cathode is

$$\frac{1}{2}O_2 + 2e^- \rightarrow O^{2-} \tag{15.24}$$

Since the reaction involves electrons and ions, at equilibrium the total electrochemical potential of the reactants is equal to that of products, i.e.,

$$\frac{1}{2}\mu_{O_2}^c + 2\tilde{\mu}_e^c = \tilde{\mu}_{O^{2-}}^c \tag{15.25}$$

where $\tilde{\mu}_e^c$ and $\tilde{\mu}_{O^{2-}}^c$ are the electrochemical potentials of electrons and O^{2-} at the cathode, respectively. In terms of chemical and electric potentials in the electrolyte solution and electrode, we have

$$\frac{1}{2}\mu_{O_2}^c + 2\mu_e^c - 2\mathcal{F}\phi_{electrode}^c = \mu_{O^{2-}}^c - 2\mathcal{F}\phi_{electrolyte}^c \tag{15.26}$$

where $\phi_{electrolyte}^c$ is the electric potential on the electrolyte side of the electrode/electrolyte interface at the cathode, and $\phi_{electrode}^c$ is the electric potential at the electrode side of the cathode/electrolyte interface. Rearranging the above equation, we have

$$2\mathcal{F}\left(\phi_{electrode}^c - \phi_{electrolyte}^c\right) = -\left(\mu_{O^{2-}}^c - \frac{1}{2}\mu_{O_2}^c - 2\mu_e^c\right) \tag{15.27}$$

Therefore, the electric potential drop from electrode to electrolyte across the electrode/electrolyte interface at the cathode at equilibrium, i.e., the cathode electrode potential E^c, is given by

$$E^c = \left(\phi_{electrode}^c - \phi_{electrolyte}^c\right) = -\frac{\left(\mu_{O^{2-}}^c - \frac{1}{2}\mu_{O_2}^c - 2\mu_e^c\right)}{2\mathcal{F}} = -\frac{\Delta G^c}{2\mathcal{F}} \tag{15.28}$$

where ΔG^c is the Gibbs free energy (or total chemical potential) change for the cathode reaction (Eq. 15.24).

The reaction at the anode of the fuel cell is

$$H_2 + O^{2-} \rightarrow H_2O + 2e^- \tag{15.29}$$

At equilibrium for the anode reaction,

$$\mu_{H_2}^a + \mu_{O^{2-}}^a - 2\mathcal{F}\phi_{electrolyte}^a = \mu_{H_2O}^a + 2\mu_e^a - 2\mathcal{F}\phi_{electrode}^a \tag{15.30}$$

where $\phi_{electrolyte}^a$ is the electric potential at the electrolyte side of the electrode/electrolyte interface at the anode, and $\phi_{electrode}^a$ is the electric potential on the electrode side of the anode/electrolyte interface. Rearranging the above equation, the voltage drop from electrolyte to electrode across the electrode/electrolyte interface at the anode is

$$E^a = \left(\phi_{electrolyte}^a - \phi_{electrode}^a\right) = -\frac{\left(\mu_{H_2O}^a + 2\mu_e^a - \mu_{H_2}^a - \mu_{O^{2-}}^a\right)}{2\mathcal{F}} = -\frac{\Delta G^a}{2\mathcal{F}} \tag{15.31}$$

where ΔG^a is the Gibbs free energy (or total chemical potential) change for the anode reaction.

For a Li-ion battery cell with CoO_2 as cathode and graphite as anode, the overall reaction is

$$Li_xC + CoO_2 = C + Li_xCoO_2, \tag{15.32}$$

and the two electrode reactions are

$$CoO_2 + xLi^+ + xe^- = Li_xCoO_2 \text{ at cathode} \tag{15.33}$$

$$Li_xC = C + xLi^+ + xe^- \text{ at anode} \tag{15.34}$$

At equilibrium, the voltage drop from electrode to electrolyte at the electrolyte/cathode interface is

$$E^c = \phi^c_{electrode} - \phi^c_{electrolyte} = \frac{\mu_{CoO_2} + x\mu^c_{Li^+} + x\mu^c_e - \mu_{Li_xCoO_2}}{x\mathcal{F}} = -\frac{\Delta G^c}{x\mathcal{F}} \tag{15.35}$$

The voltage drop from electrolyte to electrode at the anode is

$$E^a = \phi^a_{electrolyte} - \phi^a_{electrode} = \frac{\mu_{Li_xC} - \mu_C - x\mu^a_{Li^+} - x\mu^a_e}{x\mathcal{F}} = -\frac{\Delta G^a}{x\mathcal{F}} \tag{15.36}$$

15.4.2 Standard Hydrogen Electrode (SHE) and Standard Electrode Potential

To define and quantify the voltage drop across an electrode/electrolyte interface, the standard hydrogen electrode is usually chosen as the reference electrode. The standard hydrogen electrode is a platinum electrode with an ideal electrode/electrolyte interface where the activity of H^+ is 1, and the fugacity or pressure of hydrogen gas is at 1 bar. Its potential jump across the electrolyte–electrode interface serves as a reference for all other electrode reactions, i.e., the standard hydrogen electrode potential is set to be zero volt at equilibrium, and the potentials of any other electrodes are measured relative to the standard hydrogen electrode at the same temperature.

For the hydrogen electrode reaction

$$2H^+ + 2e^- (Pt) \rightarrow H_2 \tag{15.37}$$

At equilibrium,

$$2\tilde{\mu}_{H^+} + 2\tilde{\mu}^{Pt}_{e^-} = \mu_{H_2} \tag{15.38}$$

where $\tilde{\mu}_{H^+}$ and $\tilde{\mu}_{e^-}$ are the electrochemical potentials of H^+ and electrons. The electric potential drop across the electrode/electrolyte interface at the electrode is given by

$$E_{H_2/H^+} = \frac{\mu_{H_2} - 2\mu_{H^+} - 2\mu_{e^-}^{Pt}}{2\mathcal{F}} = -\frac{\mu_{H_2}^o - 2\mu_{H^+}^o - 2\mu_{e^-}^{Pt}}{2\mathcal{F}} - \frac{RT}{2F} \ln \frac{p_{H_2}}{a_{H^+}} \quad (15.39)$$

If we define

$$E_{SHE}^o = -\frac{\mu_{H_2}^o - 2\mu_{H^+}^o - 2\mu_{e^-}^{Pt}}{2\mathcal{F}} \quad (15.40)$$

We have

$$E_{H_2/H^+} = E_{SHE}^o - \frac{RT}{2\mathcal{F}} \ln \frac{p_{H_2}}{a_{H^+}} \quad (15.41)$$

Now we set the standard potential $E_{SHE}^o = 0$, the electrode potential of the hydrogen electrode at nonstandard conditions is given by

$$E_{H_2/H^+} = \frac{RT}{2\mathcal{F}} \ln \frac{a_{H^+}}{p_{H_2}} \quad (15.42)$$

The standard electrode potential of Li relative to the standard hydrogen electrode is given by

$$Li^+ + e^- \rightarrow Li(s), \quad E^o = -3.04 \text{ V} \quad (15.43)$$

A negative value for the standard potential for Li indicates that it is more difficult to reduce Li^+ to Li metal than H^+ to H_2, or it is easier to oxidize Li than H_2.

The standard electrode potential of oxygen relative to standard hydrogen electrode potential is

$$O_2 + 4H^+ + 4e^- \rightarrow 2H_2O, \quad E^o = 1.23 \text{ V} \quad (15.44)$$

15.4.3 Cell Reactions and Cell Voltage

An electrochemical cell consists of a cathode, an anode, an electrolyte, a separator separating the cathode and anode, and metallic contacts to an external device. A schematic example of a cell using a short-hand notation is given by

$$M2(I)|Pt|H_2, H^+ \| M1^{z+}|M1|M2(II) \quad (15.45)$$

where M1 is a metallic electrode, $M1^{z+}$ is the corresponding ion of valence z, and M2(I) and M2(II) are the metallic connects of the same metal M2 to an external device. A vertical line represents a phase junction, and double vertical lines represent a separator in the electrolyte.

Any two materials in contact create a contact electric potential to achieve uniform electrochemical potential of electrons if the electrons are allowed to move from one material to another due to their chemical potential differences in the two materials, while the ion cores are fixed.

At the junction M1|M2(II), at equilibrium

$$e^-(M2) \leftrightarrow e^-(M1) \tag{15.46}$$

$$\mu_e^{M2(II)} - \mathcal{F}\phi^{M2(II)} = \mu_e^{M1} - \mathcal{F}\phi^{M1} \tag{15.47}$$

where $\mu_e^{M2(II)}$ and μ_e^{M1} are chemical potentials of electrons in metal M2 and electrode M1, respectively, and $\phi^{M2(II)}$ and ϕ^{M1} are electric potentials in M2 and M1, respectively. Hence the potential drop at the junction M1|M2(II) at equilibrium is

$$\phi^{M2(II)} - \phi^{M1} = \frac{\mu_e^{M2(II)} - \mu_e^{M1}}{\mathcal{F}} = \frac{\mu_e^{M2(II)}}{\mathcal{F}} - \frac{\mu_e^{M1}}{\mathcal{F}} \tag{15.48}$$

For the electrode reaction at $M1^{z+}$|M1 electrolyte/electrode interface at equilibrium,

$$M1^{z+} + ze^-(M1) = M1 \tag{15.49}$$

$$\mu_{M1^{z+}} + z\mathcal{F}\phi^{electrolyte} + z\mu_e^{M1} - z\mathcal{F}\phi^{M1} = \mu_{M1} \tag{15.50}$$

where $\mu_{M1^{z+}}$ and μ_{M1} are chemical potentials of $M1^{z+}$ and M1, respectively, and $\phi^{electrolyte}$ is the electric potential in the electrolyte. The electrode potential for $M1^{z+}$|M1 is then

$$\phi^{M1} - \phi^{electrolyte} = \frac{\mu_{M1^{z+}} + z\mu_e^{M1} - \mu_{M1}}{z\mathcal{F}} = \frac{\mu_{M1^{z+}} - \mu_{M1}}{z\mathcal{F}} + \frac{\mu_e^{M1}}{\mathcal{F}} \tag{15.51}$$

For the Pt electrode at equilibrium,

$$2H^+ + 2e^-(Pt) \leftrightarrow H_2 \tag{15.52}$$

$$2\mu_{H^+} + 2\mathcal{F}\phi^{Electrolyte} + 2\mu_{e^-}^{Pt} - 2\mathcal{F}\phi^{Pt} = \mu_{H_2} \tag{15.53}$$

where μ_{H^+} and μ_{H_2} are chemical potentials of H^+ and H_2, respectively, and $\mu_{e^-}^{Pt}$ is the chemical potential of electrons in Pt. Therefore, the Pt electrode potential is

$$\phi^{\text{Electrolyte}} - \phi^{\text{Pt}} = \frac{\mu_{H_2} - 2\mu_{H^+}}{2\mathcal{F}} - \frac{\mu_{e^-}^{\text{Pt}}}{\mathcal{F}} \tag{15.54}$$

At the M2(I)|Pt junction at equilibrium,

$$e^-(\text{Pt}) \leftrightarrow e^-(\text{M2}) \tag{15.55}$$

$$\mu_e^{\text{Pt}} - \mathcal{F}\phi^{\text{Pt}} = \mu_e^{\text{M2(I)}} - \mathcal{F}\phi^{\text{M2(I)}} \tag{15.56}$$

where ϕ^{Pt} and $\phi^{\text{M2(I)}}$ are electric potentials in Pt and the left metallic contact M2, respectively. Hence, the potential drop at this junction is

$$\phi^{\text{Pt}} - \phi^{\text{M2(I)}} = \frac{\mu_e^{\text{Pt}}}{\mathcal{F}} - \frac{\mu_e^{\text{M2(I)}}}{\mathcal{F}} \tag{15.57}$$

Summing up the Eqs. (15.48), (15.51), (15.54) to (15.57), we have

$$\phi^{\text{M2(II)}} - \phi^{\text{M2(I)}} = \frac{2\mu_{M1^{z+}} + z\mu_{H_2} - 2\mu_{M1} - 2z\mu_{H^+}}{2z\mathcal{F}} \tag{15.58}$$

Therefore, the open-circuit voltage or potential drop for the entire cell is independent of the chemical potentials of electrons in electrodes or in the metallic contacts as long as the two metallic contacts are the same metal M2.

Now consider the overall cell reaction by combining the two electrode reactions,

$$2H^+ + 2e^-(\text{Pt}) \leftrightarrow H_2 \tag{15.59}$$

$$M1^{z+} + ze^-(\text{M1}) = M1 \tag{15.60}$$

The overall reaction is

$$2M1^{z+} + zH_2 = 2M1 + 2zH^+ \tag{15.61}$$

The thermodynamic driving force for the above overall reaction is

$$D = -\Delta G = 2\mu_{M1^{z+}} + z\mu_{H_2} - 2\mu_{M1} - 2z\mu_{H^+} \tag{15.62}$$

Comparing Eqs. (15.58) and (15.62), we have the relation between the thermodynamic driving force and the open-circuit electric voltage of an electrochemical cell,

$$E = \phi^{\text{M2(II)}} - \phi^{\text{M2(I)}} = \frac{D}{2z\mathcal{F}} = -\frac{\Delta G}{2z\mathcal{F}} \tag{15.63}$$

where $2z$ represents the number of moles of electrons involved in the reaction.

For a general chemical reaction

$$\nu_{A_1} A_1 + \nu_{A_2} A_2 = \nu_{B_1} B_1 + \nu_{B_2} B_2 \tag{15.64}$$

which involves the transfer of z moles of electrons. The chemical energy or the Gibbs free energy change for the reaction is

$$\Delta G = \Delta G^\circ + RT \ln \frac{a_{B_1}^{\nu_{B_1}} a_{B_2}^{\nu_{B_2}}}{a_{A_1}^{\nu_{A_1}} a_{A_2}^{\nu_{A_2}}} = \Delta G^\circ + RT \ln K \tag{15.65}$$

Therefore, the voltage variation with activity is given by

$$E = E^\circ - \frac{RT}{zF} \ln K \tag{15.66}$$

which is often referred to the Nernst equation. E° in the above equation is the standard cell voltage when all the species involved in the reaction are in their standard states,

$$E^\circ = -\frac{\Delta G^\circ}{z\mathcal{F}} \tag{15.67}$$

where ΔG° is the standard Gibbs free energy of reaction involving z moles of electrons.

For example,

$$H_2(g) + \frac{1}{2} O_2(g) = H_2O(l) \tag{15.68}$$

$\Delta G^o = -237.2$ kJ at room temperature. Hence,

$$E^\circ = -\frac{\Delta G^\circ}{z\mathcal{F}} = -\frac{-237,200}{2 \times 96,480} = 1.23 \text{ V} \tag{15.69}$$

15.4.4 Standard Cell Voltage from Standard Electrode Potentials

The standard electrode potentials for many electrode reactions with respect to the standard hydrogen electrode are available in the literature. Therefore, the standard voltage of an electrochemical cell can be obtained from the standard electrode potentials. For example,

$$4H^+ + 4e^-(Pt) \leftrightarrow 2H_2, \quad E^\circ = 0 \tag{15.70}$$

$$O_2 + 4H^+ + 4e^- \rightarrow 2H_2O, \quad E^\circ = 1.23 \text{ V} \tag{15.71}$$

Therefore, the standard cell voltage for the overall reaction,

$$2H_2 + O_2 = 2H_2O, \quad E^\circ = 1.23 \text{ V} \tag{15.72}$$

As a matter of fact, as long as the electrode potentials for the two electrodes are measured against the same reference electrode, one can combine the two electrode potentials to obtain the overall cell potential. For example, using metallic Li as the reference electrode, the CoO_2 cathode electrode potential is given by

$$CoO_2 + xLi^+ + xe^- = Li_xCoO_2, \quad E = -3.9 \text{ V} \tag{15.73}$$

The graphite anode electrode potential is

$$C + xLi^+ + xe^- = Li_xC, \quad E = -0.2 \text{ V} \tag{15.74}$$

Therefore, the cell voltage for the overall reaction for the Li-ion battery is

$$Li_xC + CoO_2 = C + Li_xCoO_2, \quad E = -3.7 \text{ V} \tag{15.75}$$

15.4.5 Temperature Dependence of Cell Voltage

The temperature dependence of a cell voltage can be obtained by the temperature dependence of the Gibbs free energy of the overall chemical reaction,

$$\frac{\partial E}{\partial T} = -\frac{1}{z\mathcal{F}}\frac{\partial \Delta G}{\partial T} = \frac{\Delta S}{z\mathcal{F}} \tag{15.76}$$

where ΔS is the entropy change for the overall reaction. If all the reaction species are at their standard states,

$$\frac{\partial E^\circ}{\partial T} = \frac{\Delta S^\circ}{z\mathcal{F}} \tag{15.77}$$

i.e., the temperature coefficient of a cell is proportional to the entropy change. The heat involved in a cell is given by

$$Q = \Delta H = \Delta G + T\Delta S = -z\mathcal{F}\left(E - T\frac{\partial E}{\partial T}\right) \tag{15.78}$$

15.4.6 Voltage Derived from Chemical Composition Difference

A chemical composition difference leads to a chemical potential difference that can be converted to a voltage by designing an electrochemical cell. For example, let us consider two Zn-based alloy electrodes. Either the difference in the activity of Zn in the two Zn-based alloy electrodes or the concentration difference of Zn ions near the cathode and anode leads to a chemical potential difference in Zn or Zn ions and thus a voltage (Fig. 15.6). For this cell, the half-cell reactions are

$$Zn(I) = Zn^{2+}(I) + 2e^-(I)$$
$$Zn^{2+}(II) + 2e^-(II) = Zn(II)$$

The overall reaction is

$$Zn(I) + Zn^{2+}(II) + 2e^-(II) = Zn(II) + Zn^{2+}(I) + 2e^-(I)$$

At equilibrium,

$$\mu_{Zn}^I + \mu_{Zn^{2+}}^{II} + 2\mathcal{F}\phi_{electrolyte}^{II} + 2\mu_e^{II} - 2\mathcal{F}\phi_{electrode}^{II} = \mu_{Zn}^{II} + \mu_{Zn^{2+}}^I$$
$$+ 2\mathcal{F}\phi_{electrolyte}^I + 2\mu_e^I - 2\mathcal{F}\phi_{electrode}^I$$

The contribution of chemical potentials of electrons disappears if external metallic connections of the same metal are considered. Therefore, the cell voltage

$$E = \phi_{electrode}^{II} - \phi_{electrode}^I = \frac{\left(\mu_{Zn}^I - \mu_{Zn}^{II}\right) + \left(\mu_{Zn^{2+}}^{II} - \mu_{Zn^{2+}}^I\right)}{2\mathcal{F}} \qquad (15.79)$$

In terms of activities of Zn and Zn ions,

Fig. 15.6 Illustration of a concentration electrochemical cell

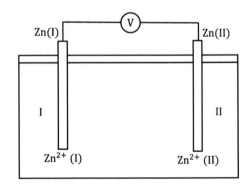

$$E = \frac{RT}{2\mathcal{F}} \ln \frac{a_{Zn}^{I} a_{Zn^{2+}}^{II}}{a_{Zn}^{II} a_{Zn^{2+}}^{I}} \tag{15.80}$$

If both electrodes are pure Zn, we have

$$E = \frac{RT}{2\mathcal{F}} \ln \frac{a_{Zn^{2+}}^{II}}{a_{Zn^{2+}}^{I}} \tag{15.81}$$

Therefore, if one of the two activities, $a_{Zn^{2+}}^{I}$ or $a_{Zn^{2+}}^{II}$, is known, we could use this concentration cell to measure the other activity by measuring the cell voltage.

On the other hand, if the Zn ion concentrations are uniform in the electrolyte, we have

$$E = \frac{RT}{2\mathcal{F}} \ln \frac{a_{Zn}^{I}}{a_{Zn}^{II}} \tag{15.82}$$

If Zn^{II} is pure, and Zn^{I} is an alloy, the activity of Zn in the alloy as a function of voltage is given by

$$a_{Zn}^{I} = e^{\left(\frac{2\mathcal{F}E}{RT}\right)} \tag{15.83}$$

Therefore, one can use a concentration cell to measure the activity of a species in an alloy as a function of composition using an electrochemical cell by measuring the cell voltage.

15.4.7 Voltage Derived from Partial Pressure Difference

For gases, a partial pressure difference leads to a chemical potential difference from which a voltage can be derived. Such a voltage can in turn be used to measure the partial pressure of a gas, e.g., oxygen gas. An electrochemical cell is schematically shown in Fig. 15.7. The solid electrolyte $Y_2O_3-ZrO_2$ conducts oxygen ions. Such a cell typically uses two porous Pt electrodes to allow oxygen gas to permeate through and react at the electrode/electrolyte interfaces. The two half-cell reactions are

$$O_2(II) + 4e^-(II) = 2O^{2-}(II)$$
$$2O^{2-}(I) = O_2(I) + 4e^-(I)$$

The overall reaction is

$$O_2(II) + 4e^-(II) + 2O^{2-}(I) = O_2(I) + 4e^-(I) + 2O^{2-}(II)$$

Fig. 15.7 Illustration of an
O_2 gas sensor

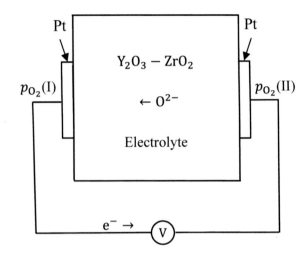

At equilibrium

$$\mu_{O_2}^{II} + 4\mu_e^{II} - 4\mathcal{F}\phi_{electrode}^{II} + 2\mu_{O^{2-}}^{I} - 4\mathcal{F}\phi_{electrolyte}^{I} = \mu_{O_2}^{I} + 4\mu_e^{I}$$
$$- 4\mathcal{F}\phi_{electrode}^{I} + 2\mu_{O^{2-}}^{II} - 4\mathcal{F}\phi_{electrolyte}^{II}$$

The chemical potentials of electrons disappear when external metallic contacts are considered. Assuming $\mu_{O^{2-}}^{I} = \mu_{O^{2-}}^{II}$, and $\phi_{electrolyte}^{I} = \phi_{electrolyte}^{II}$, we have

$$\mu_{O_2}^{II} - 4\mathcal{F}\phi_{electrode}^{II} = \mu_{O_2}^{I} - 4\mathcal{F}\phi_{electrode}^{I}$$

The cell voltage due to the difference in the partial pressure of oxygen at the two electrodes is given by

$$E = \phi_{electrode}^{II} - \phi_{electrode}^{I} = -\frac{\mu_{O_2}^{I} - \mu_{O_2}^{II}}{4\mathcal{F}} = -\frac{RT}{4\mathcal{F}} \ln \frac{p_{O_2}^{I}}{p_{O_2}^{II}} \qquad (15.84)$$

Therefore, if one of the two partial pressures of oxygen is known, the partial pressure of oxygen at the other electrode can be determined by measuring the cell voltage.

15.5 Exercises

1. When a system undergoes an isothermal expansion, the relationship between the amount of work delivered by a reversible process (W_r) and that by an irreversible process (W_{ir}) going from the same initial state to the same final state is

(A) $W_r < W_{ir}$, (B) $W_r = W_{ir}$, (C) $W_r > W_{ir}$, (D) Cannot be determined

2. For an engine to reach the ideal, maximum thermodynamic efficiency, all the processes involved in the engine cycle have to be reversible, true or false?

3. The higher the temperature of the high-temperature heat source, the higher a thermal engine efficiency is, true or false?

4. For an engine operating between a high-temperature heat source at 1500 K and a low-temperature heat sink at 300 K, what is the maximum possible efficiency?

5. For a refrigerator, the inside temperature is controlled at 277 K, and outside is 295 K. What is the maximum amount of heat that can be removed from inside of the refrigerator and released to outside with 1 J of electric work?

6. For a refrigerator, the inside temperature is controlled at 0.1 K, and outside is 295 K. (i) What is the maximum coefficient of performance for the refrigerator? (ii) What is the minimum amount of electric work required to remove 1 J of heat from inside the refrigerator?

7. Consider a house that needs to be maintained at temperature 25 °C at all times. The house is believed to be losing heat at 150,000 kJ/h when the temperature outside is 0 °C. Determine the minimum power needed, in kW, to operate a heat pump to keep the house at 25 °C.

8. Consider using an electrochemical cell to measure the oxygen partial pressure. The reference partial pressure of oxygen corresponds to that in the air. Write an expression for the partial pressure of oxygen to be measured as a function of temperature and the cell voltage measured.

9. An electrochemical cell was built using pure Cu and an equilibrium Ag–Cu alloy with an overall 50 atom% of Cu as electrodes with some fused salt as electrolyte which has uniform Cu^{2+} concentration.

   (a) Write down the two half-cell reactions at the two electrodes.
   (b) Assume that Cu behaves in a Raoultian manner in the Cu-rich solid solution with equilibrium composition of $x_{Cu}^{Ag-Cu} = 0.98$, what is the cell voltage at 600 °C?

10. Is the chemical potential of lithium in pure lithium metal exposed to air at room temperature (i) higher; or (ii) equal to, or (iii) lower than the chemical potential of lithium in Li_2O at room temperature and 1 bar in the air?

11. Is the chemical potential of lithium in pure lithium metal at room temperature and 1 bar (i) higher; or (ii) equal to, or (iii) lower than the chemical potential of lithium in $LiFePO_4$ at room temperature and 1 bar?

12. Is the chemical potential (or Fermi level) of electrons in pure lithium metal in the air (i) higher; or (ii) equal to, or (iii) lower than the chemical potential (Fermi level) of electrons in Li_2O in the air at room temperature and 1 bar?

13. Is the chemical potential of lithium in pure solid lithium metal at room temperature and 1 bar (i) higher; or (ii) equal to, or (iii) lower than the chemical potential of pure solid lithium at 0 °C and 1 bar?

14. If you design an electrochemical cell using pure lithium metal and $LiFePO_4$ as two electrodes, which one would you label as cathode and which one as anode?

15. If you design an electrochemical cell using pure lithium metal and $LiFePO_4$ as two electrodes, write down the two electrode reactions.

16. If the open cell voltage of a Li-ion battery is 3.2 V, what is the chemical potential difference of Li in the two electrodes (the Faraday constant $\mathcal{F}$ is 96,480 C/mol)?

17. Write down an expression for the open-circuit voltage for an oxygen–hydrogen fuel cell as a function of temperature, partial pressure of hydrogen, partial pressure of oxygen, and activity of water.

18. The standard electrode potentials with respect to the standard hydrogen electrode for the following reactions are given by

$$CO_2(g) + 2H^+ + 2e^- \rightarrow CO(g) + H_2O(l), \quad E^\circ = -0.11 \text{ V}$$
$$O_2(g) + 4H^+ + 4e^- \rightarrow 2H_2O(l), \quad E^\circ = 1.23 \text{ V}$$

What is the theoretical standard voltage that one can generate if a fuel cell employs CO as a fuel?

19. The standard electrode potentials with respect to the standard hydrogen electrode for the following two electrode reactions are given below:

$$Li^+ + e^- \rightarrow Li(s), \quad E^\circ = -3.0401 \text{ V}$$
$$Li^+ + C_6(s) + e^- \rightarrow LiC_6(s), \quad E^\circ = -2.84 \text{ V}$$

Determine the standard voltage for the overall reaction.

20. For the two electrode reactions for the Daniel cell, $Zn(s)|Zn^{2+}(aq,1M)||Cu^{2+}(aq,1M)|Cu(s)$

$$Zn^{2+} + 2e^- \rightarrow Zn(s), \quad E^\circ = -0.7618 \text{ V}$$
$$Cu^{2+} + 2e^- \rightarrow Cu(s), \quad E^\circ = +0.337 \text{ V}$$

(a) Determine the standard cell voltage.
(b) Express the cell voltage as a function of activities of Zn^{2+} and Cu^{2+}.

Index

© Springer Nature Singapore Pte Ltd. 2022

453

L.-Q. Chen, *Thermodynamic Equilibrium and Stability of Materials*,
https://doi.org/10.1007/978-981-13-8691-6

Printed in the United States
by Baker & Taylor Publisher Services